Reflections of a Technocrat
Managing Defense, Air, and Space Programs during the Cold War

DR. JOHN L. MCLUCAS

with

KENNETH J. ALNWICK AND LAWRENCE R. BENSON

FOREWORD

by

MELVIN R. LAIRD

Air University Press
Maxwell Air Force Base, Alabama

August 2006

Disclaimer

Opinions, conclusions, and recommendations expressed or implied within are solely those of the author and do not necessarily represent the views of Air University, the United States Air Force, the Department of Defense, or any other US government agency.

All photographs are courtesy of US Government or family photos except as noted.

Air University Press
131 West Shumacher Avenue
Maxwell AFB, AL 36112-6615

Published by Books Express Publishing
Copyright © Books Express, 2011
ISBN 978-1-78039-968-3

Books Express publications are available from all good retail and online booksellers. For publishing proposals and direct ordering please contact us at: info@books-express.com

Contents

Chapter   *Page*

DISCLAIMER . . . .   *ii*

FOREWORD . . . .   *vii*

ABOUT THE COAUTHORS . . . .   *xiii*

ACKNOWLEDGMENTS . . . .   *xv*

INTRODUCTION . . . .   *xix*
    Notes . . . .   *xxv*

1   FROM COUNTRY BOY TO COMPANY
    PRESIDENT   1
      Whose Son Am I?   1
      Attending Davidson College and Tulane
        University   6
      Employing Radar in the Navy   9
      Growing a High-Tech Enterprise   16
      Notes   29

2   CAREER BROADENING IN SCIENCE
    AND TECHNOLOGY   31
      Defense Management under McNamara   31
      Promoting Innovations in Tactical Weapons . .   37
      My Role in the Controversial TFX Program . . .   40
      Progress and Problems with Other Systems . .   43
      An Interlude with NATO in Paris   45
      Closing a Circle at the MITRE Corporation . . .   49
      MITRE Goes to War—Igloo White and
        Other Projects   54
      Accomplishments and Angst   57
      Notes   60

3   MY AIR FORCE YEARS: PEOPLE AND
    POLITICS   63
      Joining the Laird-Packard Team   63
      My Roles as Undersecretary   65

Chapter		Page

Defense Management in the First Nixon
Administration   72
Serving as Secretary   80
Notes   94

4   MODERNIZING THE FORCE: NEW SYSTEMS
FOR FUTURE AIR SUPREMACY   97
Reforming Acquisition Management   97
Sustaining Basic and Applied Research   106
Developing Weapon Systems   110
Notes   147

5   WHAT CAN NOW BE TOLD: THE NATIONAL
RECONNAISSANCE AND AIR FORCE SPACE
PROGRAMS   169
Directing the National Reconnaissance
Office in Its Golden Age   170
Key NRO People and Organizations   176
Selected Reconnaissance Programs   183
Some Thoughts on Secrecy and
Bureaucracy   188
The Air Force in Space   191
The NRO and Air Force in Retrospect   204
Notes   207

6   FACING OTHER ISSUES—OVERSEAS AND
AT HOME   213
Trying for Peace with Honor in Vietnam   213
Turning toward Europe and the Middle East   229
The Air Force and American Society   234
My Farewell to Arms   250
Notes   253

7   MANAGING CIVIL AVIATION AND COMMERCIAL
SPACE PROGRAMS   257
Heading the Federal Aviation Administration . . .   257
Connecting the Global Village at Comsat
Corporation   270
Chairing the Arthur C. Clarke Foundation . . .   278
Launching the International Space University . . .   280
Notes   282

Chapter   *Page*

8   PROMOTING SPACE, SCIENCE, AND
      TECHNOLOGY   285
         Defending the Free Exchange of Scientific
            Information   286
         Advancing International Cooperation in Space . . .   289
         Overseeing QuesTech, Inc.   292
         Fostering Space Commerce   293
         Mission to Planet Earth   298
         Looking Farther into the Space Age   302
         Final Years   305
         Notes   307

Appendix

A   Selected Organizational Affiliations   315

B   Letter on Declassifying the Existence of the
         National Reconnaissance Program   317

      ABBREVIATIONS.   323

      BIBLIOGRAPHY.   331

      INDEX   353

Photographs follow page 151.

THIS PAGE INTENTIONALLY LEFT BLANK

Foreword

In the final decade of the twentieth century, our nation focused attention on the noble and courageous men and women who, 50 years earlier, had participated in World War II. Thanks to a best-selling book by newscaster Tom Brokaw, the Americans who came of age in the 1930s and early 1940s became known as "the greatest generation." Many of those who fought or supported the war effort humbly disavowed such a superlative. No one, however, can deny the achievements of our citizens who grew up in the hard times of the Great Depression. These young people helped win the largest conflict in world history and went on to make further contributions to our country during the long Cold War that followed.

Despite all of its death and destruction, World War II accelerated the growth of scientific knowledge and the march of technology. As one consequence of this, our military services—in partnership with the nation's universities—trained many young Americans, especially those with a background in science and engineering, to operate and maintain the new technologies so important to the war effort. Others contributed as civilian scientists and engineers. After the war, many veterans who had been exposed to these new technologies took advantage of the G. I. Bill to seek advanced degrees in science, engineering, mathematics, and related disciplines. When the Cold War set off a prolonged arms race and space competition with the Soviet Union, this well-educated cadre of the greatest generation was ready to provide the technical and managerial expertise needed to meet the Soviet challenge. Combining patriotism with a desire to be on the cutting edge of technology, these "technocrats" played key roles in the defense industry, university and federal research centers, the military services, and other government agencies.

Dr. John L. McLucas was one of the finest examples in this group of influential public servants. After a poignant childhood in the rural South, John took advantage of educational opportunities by earning a master's degree, served as a radar officer in the Navy, earned a PhD in physics, became president and chief executive officer of an innovative technology company, managed research programs in the Pentagon, served as the top scientist in the North Atlantic Treaty Organization (NATO), and headed an eminent electronic systems engineering corporation. These were some of the achievements listed on John's résumé in early 1969 when I, as the newly appointed secretary of defense, was looking for some good people to help me lead our nation's military through a very difficult period.

One of my preconditions for accepting Richard Nixon's request for me to take charge of the Department of Defense was being granted full authority to choose my own team of appointees, both civilian and military. I selected them based on competency and compatibility, not politics. Being a nontechnical person myself, I made sure I hired some top managers with scientific and engineering expertise, starting with my deputy, the esteemed David Packard. To be secretary of the Air Force (the most technically oriented of the services), I selected Bob Seamans. Then a professor at the Massachusetts Institute of Technology, Bob had recently served as deputy administrator of NASA. Dave, Bob, and I wanted someone else astute in technology and experienced in defense matters to work with Bob as undersecretary of the Air Force and with Dave and me as director of the National Reconnaissance Office—a job that was then totally secret. I was extremely pleased when John McLucas—at a considerable sacrifice to himself and his family—agreed to return to Washington and fill this dual position. I believe his willingness to serve the country rather than pursue only private interests was a hallmark of the postwar generation of executives that John exemplified.

With inherited responsibilities for the divisive war in Vietnam, which was proving difficult to end on favorable terms, with Americans in uniform being ostracized by many of their fellow citizens, and with the defense budget declining steadily, the Pentagon of 1969 certainly did not seem like a pleasant place to work. I was determined, however, to help our defense establishment weather this difficult period, restore the public's faith in our military personnel and institutions, and prepare for the continuing Cold War challenges we knew lay ahead. Well aware of the acrimony and infighting that had often tarnished defense policies in the past, and realizing that a dictatorial approach to running the Pentagon quickly becomes counterproductive, I was determined to instill a sense of teamwork among the services and between them and the Office of the

Secretary of Defense (OSD). I intended to take full advantage of the military experience of our top-ranking generals and admirals, whose advice had often gone unsought in previous administrations. I also wanted to restore authority to the service secretaries and their staffs, who, since the 1950s, had lost much of their prestige and influence as a result of centralized decision making at OSD. As one example of my philosophy, Dave Packard and I met regularly, both as a group and individually, with the service secretaries and undersecretaries to develop policy and keep each other informed. I also worked closely with the Joint Chiefs of Staff to make operational decisions and to seek their opinions on defense issues. I called this cooperative relationship with the services "participatory management." Bob Seamans and John McLucas were two of our most capable partners in this endeavor.

Building on some of the planks in Richard Nixon's 1968 campaign platform, we focused on a number of specific goals for defense policy and management. These goals included reducing American forces and casualties in Vietnam as quickly as possible by de-Americanizing the war, phasing out the unfair and unpopular selective service system in favor of an all-volunteer force, reinvigorating NATO's force posture in Europe, improving our intelligence about Soviet military capabilities, reforming the defense procurement process, investing in needed research and development, better using reserve and national guard components as part of what we dubbed "total force," expanding equal opportunity within the Department of Defense, helping promote social progress with a program we called "domestic action," and working in close harmony with Congress on budgetary requirements. As so well documented in chapters 3 through 6 of this book, John McLucas played an important role in our efforts to meet all of these objectives.

In January 1973, just before I left the Pentagon, I had the pleasure of awarding John the Department of Defense Medal for Distinguished Public Service. As I stated in his citation, "Dr. John L. McLucas has attained singular recognition as a leader, administrator, and scientist." The two paragraphs quoted below are from a certificate I signed as secretary of defense, awarding John L. McLucas the Department of Defense Medal for Distinguished Public Service (First Bronze Palm) "for exceptionally dis-

tinguished service from March 1969 to January 1973," and at-
test to John's technical and humanitarian achievements.

> He was deeply involved in all aspects of the statutory responsibility of the
> Secretary of the Air Force. . . . His direction of the Air Force space pro-
> gram has brought significant gains to the United States. His leadership,
> coupled with his impressive knowledge of electronics, has opened new
> horizons and capabilities in weapon delivery accuracy. Dr. McLucas
> stimulated new developments in reconnaissance concepts and opera-
> tions that led to major advancements in this crucial defense area.

> Because of [his] personal interest and leadership in promoting domestic
> action programs, the Air Force has made exemplary progress in attaining
> and exceeding the human relations goals of the Department of Defense.
> His sensitivity to the complexities of equal opportunity issues in the
> armed forces and his perceptive and forthright leadership in this area
> have brought a new awareness to the field of personnel use, and en-
> hanced the effectiveness of manpower management programs.

I was not the only secretary of defense to think highly of
John's talents. Several months after I left the Pentagon, Dr.
James R. Schlesinger, recently named as defense secretary,
asked Dr. McLucas to be his Air Force secretary. As expected,
John moved into this position without missing a stride. Indeed,
he performed so well during the next two and one-half years
that, when Pres. Gerald Ford needed an experienced and
trusted manager to run the Federal Aviation Administration on
short notice, he called upon John McLucas.

Dr. McLucas's contributions to American technology and scientific progress continued until his death in December 2002. As an
executive in the Communications Satellite Corporation, he helped
link together the global community with fast and reliable connections. He served on the corporate boards of advanced technology
companies, as the chairperson of government committees, and as
a top officer in prestigious professional and cultural organizations.
Especially noteworthy was his advocacy of space technology for
monitoring Earth's environment, expanding commercial opportunities, and fostering science education. Until his final days, he sat
on the board of several high-tech companies and was an active advisor to the International Space University, which he was largely
responsible for establishing.

Late in life, John decided that a record of his career might be
of historical value to those interested in defense technology,
national security policy, and air and space developments through

several decades of the twentieth century. This interesting and informative book is the result. Since relatively few of his fellow technocrats have written detailed autobiographies, I believe the John McLucas story will expand the historical record of the institutions with which he was associated while shedding new light on some important chapters of the Cold War era.

House of Representatives (1952–68)
Purple Heart Veteran of World War II

THIS PAGE INTENTIONALLY LEFT BLANK

About the Coauthors

Kenneth J. Alnwick   Lawrence R. Benson

Col Kenneth J. Alnwick, USAF, retired, is president of the Alnwick Design Group, Ltd., which provides war-gaming, simulation, and analysis services to clients in the metropolitan Washington, D.C., area. He was born in 1937 and graduated from the US Air Force Academy in 1960. He received a master of history degree from the University of California at Davis in 1968. He served four tours in Southeast Asia between 1963 and 1975, flying special operations combat missions from South Vietnam, Laos, and Thailand. Other assignments included teaching at the USAF Academy's Department of History and serving on the Air Staff in the Pentagon (Plans and Policy). At Maxwell AFB, Alabama, he served on the Air War College faculty for two years and graduated there in 1982. He then became the founding director of the Air University's College of Aerospace Doctrine, Research and Education (CADRE), winning the McKay Trophy for this contribution. Colonel Alnwick completed his military career as director of the War Gaming and Simulation Center and deputy director of the Strategic Concepts Development Center at the National Defense University in Washington, D.C. After retiring from the Air Force in 1986 and working a brief stint as vice president for operations at EAST, Inc., he became project manager, vice president, and director of gaming and simulation programs for Kapos Associates Inc. (KAI) and Emergent Government Services, providing analysis, gaming, and simulation services to numerous government agencies. He became a private consultant in 2001. In addition to helping write this book, he also advised Thomas Reed, another former secretary of the Air Force, on his Cold War memoir, *At the Abyss: An Insider's History of the Cold War*. Ken coaches Lacrosse at Falls Church High School, where his wife Judy teaches, and he is a trustee of the Air Force Historical Foundation.

Lawrence R. Benson retired in early 2000 as chief of the Air Force Historian's Pentagon office. Born in 1943, he attended the University of Maryland at College Park, where he received a master's degree in 1967 having specialized in American military and diplomatic history. While in graduate school, he worked as a teaching assistant to Prof. Gordon W. Prange, a prominent expert on World War II. After serving in the US Army with a tour in South Vietnam, Mr. Benson became a civilian employee of the Air Force in 1971. Over the next 30 years, he worked in a variety of administrative and historian positions at 10 locations in the United States, Turkey, and Germany, advancing in grade from GS-7 to GS-14. His continued education and professional training included completion of the Air War College seminar program in 1991. He has researched and written numerous official histories, monographs, articles, book reviews, and studies on a range of topics related to military history, international relations, military operations and training, and air and space technology, including a history of Air Force acquisition management. Larry and his wife Carolyn, who retired in 2000 as the legislative liaison specialist in the Air Force's Senate Liaison Office, now live in Albuquerque, New Mexico.

Acknowledgments

One of the most rewarding aspects of helping John McLucas complete *Reflections of a Technocrat* was the opportunity this gave us to meet and work with many of his friends and colleagues. Like so much else in John's life, this book is the result of a cooperative effort in which many contributed. We attempt here to at least mention the names of some of these people, even if we can't give each of them the personalized recognition that John might have wished.

Before either of us became involved in this project, John received assistance from two former government historians. Alfred M. Beck, formerly with the Army and Air Force history centers, helped John prepare a rough draft of what eventually became the sections on John's childhood and Navy service. Then Nick A. Komons, a retired historian of the Federal Aviation Administration (FAA), helped John compose a draft about his time at the FAA before illness forced Nick to withdraw from the project. As for our division of labor as coauthors of the final manuscript: Larry Benson was responsible for the first half of chapter 1 through World War II, chapters 3 through 6 on John's Air Force years, chapter 8 on his postretirement activities, most front and back matter, compiling the photos, and performing the editorial chores. Ken Alnwick was responsible for the introduction, John's career from 1946 through 1968 in chapters 1 and 2, and his service with the FAA and Comsat in chapter 7. Although most of our drafts benefited from John's vigilant proofreading, we are responsible for any remaining errors of fact, interpretation, or omission.

In helping to recall his time at sea during World War II, John was helped by at least one of his former shipmates, Milton Richards, who had been his best radar technician. Chapter 1's account of the founding and growth of Haller, Raymond, and Brown (HRB) benefited greatly from information provided by Edwin Keller, a longtime officer with HRB-Singer and its unofficial historian, and Joseph Amato, an HRB telemetry pioneer who became a senior official at the National Security Agency. Both the Pentagon and NATO portions of chapter 2 benefited from the memory of Brig Gen Edwin "Spec" Powell, US Army, retired, with Col George Munroe, US Air Force, retired, also helping on the NATO period. John's boss and mentor during much of the 1960s, the Honorable Harold Brown, kindly reviewed the first half of chapter 2. Frank Mastrovita, the archivist at MITRE Corporation, provided essential assistance in researching background information for the last part of chapter 2, while Bob Everett, a longtime MITRE executive, shared his extensive memories. F. Robert Naka—former MITRE chief scientist, deputy National Reconnaissance Office (NRO) director under McLucas, and later a chief scientist of the Air Force— contributed to this as well as to chapter 5 on space programs. Cindy Hardy, secretary to MITRE's president, Martin Faga, also provided assistance and hospitality. Ruth Liebowitz, historian of the Air Force's Electronic Systems Center, later reviewed and provided several helpful suggestions on this section.

Relevant portions of chapters 3 through 6 benefited from the review and comments of many of John's former colleagues, most notably three of his distinguished superiors in the Pentagon: the Honorable Melvin Laird, James Schlesinger, and Robert Seamans. Some of the people who worked under John also reviewed drafts and contributed their knowledge to various parts of the story. These included the aforementioned Bob Naka as well as Jack Stempler, Jimmie Hill, and retired Air Force generals Lew Allen Jr., Keith McCartney, W. Y. Smith, and William Usher. Several historians also reviewed chapters and provided helpful comments, including Jacob Neufeld and George Watson of the Air Force History Office, George "Skip" Bradley and Rick Sturdevant of the Air Force Space Command, Kenneth Werrell of the Air University, and R. Cargill Hall of the NRO, who also supplied useful unclassified publications and assisted in coordinating the somewhat inscrutable security review and redaction of chapter 5. John also appreciated the hospitality and the interest in the book extended by the Honorable Peter D. Teets, director of the NRO and undersecretary of the Air Force.

In addition to Nick Komons, the FAA section benefited from the memories of Gene Weithoner, Jeff Cochran, Jack Stempler, and Joseph Laitin. While John was still alive, Joseph Pelton, Edward Martin, and Burton Edelson—former colleagues at Comsat and cohorts in later space endeavors—provided essential help in drafting chapter 7 on Comsat and the Arthur C. Clarke Foundation. To John's sorrow, both Joe Laitin and Burt

Edelson departed this Earth shortly before his own death. In helping Larry Benson write the final chapter on McLucas's later years, comments by Joe Pelton and Ed Martin were again valuable, as were reviews of relevant portions by two visionary space entrepreneurs, David Thompson and Charles Trimble, and a visionary atmospheric scientist, Randolph Ware. All three were valued associates and friends of John. Cargill Hall also provided constructive editorial comments on improving an early draft of the final two chapters.

Many more of John's friends and former colleagues helped him lay the foundation for this book by sharing their memories in interviews and phone conversations during the last decade of his life. Those whose transcripts were available to us are listed in the bibliography. The audiotapes of these, as well as many of John's solo dictations, were skillfully transcribed by Mim Eisenberg, providing us with invaluable information and insight into his activities and thoughts. We are also grateful to David Chenoweth and Matthew Doering of the Air Force and NRO history offices, respectively, for some of the photographs. Although already mentioned in the introduction, we must again emphasize the hospitality and support received from Harriet McLucas and the encouragement given to this project by Pam, John, Susan, and Rod McLucas.

Finally, for meticulously transforming our manuscript into a finished product, we extend our thanks and admiration to the Air University Press and this book's publication team, especially Emily Adams and Darlene Barnes. We believe John McLucas would have been proud of the result.

THIS PAGE INTENTIONALLY LEFT BLANK

Introduction

Reflections of a Technocrat is an autobiography that ends as a biography. John McLucas died on the first of December 2002, at the age of 82, with all but the last chapter remaining to be started. He had been preparing to do a memoir, on and off, for many years, but only in the late 1990s, as declining health caused him to cut back on other commitments, did he devote a large part of his energies to getting the job done. To help complete this project, he engaged me—Ken Alnwick—a retired Air Force pilot and defense analyst, and my associate, Larry Benson, a recently retired Air Force historian. We are both grateful for the opportunity of getting to know John and his gracious wife, Harriet, as well as to research and help write about the many people, institutions, technical achievements, and national security issues with which he was associated. Chief among his numerous affiliations was the US Air Force. He began his civilian career with the Army Air Forces right after World War II and continued to advance the Air Force mission as a reserve officer, defense contractor, government executive, and valued consultant for the rest of the century.

Not long before John died, he and Harriet decided the time had come to move out of their spacious home in Alexandria, Virginia, to a more manageable apartment in The Fairfax, a pleasant retirement community at nearby Fort Belvoir. In anticipation of the move, John decided to donate the bulk of his papers, professional library, and much of his memorabilia to two schools he admired: the Air Force Academy and Embry-Riddle Aeronautical University, with additional papers offered to the Comsat Alumni Association. His files helped shed light on every phase of his career up to and including recent activities as a director at Orbital Sciences Corporation, chairman of the Arthur C. Clarke Foundation, trustee of the Air Force Historical Foundation, and contributing member of several other public service organizations. The process was not easy. Each dusty box released a flood of memories as we went through the agonizing process of deciding what to keep, what to send to the repositories, what to give away, and what to relegate to recycling bins or the county landfill.

Often, one or more of John's children—Pam, Susan, John, or Rod—would come to town and help with the sorting process. On one of these occasions, Rod and I sat down around the table with John and Harriet in their sunny kitchen to talk about progress on the book and John's reasons for wanting to write it. He had started contemplating an autobiography some 25 or 30 years ago. John felt grateful for the educational opportunities and the acts of charity and encouragement that came along at the right time for him to lead an interesting and even influential life.

His career choices led him into a wider variety of scientific, technical, and defense management positions than almost any of his contemporaries. In the field of aviation, for example, John was the only person to have held the positions of both secretary of the Air Force and administrator of the Federal Aviation Administration. As someone who had a great respect for history, he wanted to add to the record of what had transpired during his tenure in these and other important positions and to share his perspectives on what it meant to lead technology-oriented organizations. He was also a staunch advocate for the peaceful and unifying aspects of space endeavors and saw this book as another venue to espouse this cause, which had engaged him for over two decades and been the theme of his previous book, *Space Commerce*.

John McLucas accrued almost 50 years' experience on both sides of the government procurement table. From that perspective, he also hoped to analyze various acquisition strategies, such as the value of prototyping (which he had employed with great success to bring the A-10 and F-16 aircraft to fruition) and a willingness to accept risks when appropriate. As preparations for the book proceeded, he conducted numerous interviews with senior active duty and retired defense officials on the subject of acquisition reform. He was never quite able, however, to coalesce this research into a more comprehensive theory for the defense acquisition process beyond his passionate support for competitive "fly-offs" of major weapon systems and for reinvigorating the roles of the service secretaries and their military chiefs as a counterpoint to what he saw as an unhealthy concentration of power in the Office of the Secretary of Defense.

John once told us that, when he was sworn in at the White House as the FAA administrator, several newspaper accounts referred to him as a *technocrat*. The latest *Random House-Web-*

ster's Dictionary defines this term as "a proponent, adherent, or supporter of technology" or "a technological expert, especially one concerned with management or administration." Both of these descriptions certainly fit John McLucas. Regardless of its formal definition, he especially liked the way the term was used by Nick Komons, a former FAA historian, in a book about the long-running controversy over the need for three pilots in the cockpit of long-haul airliners and John's role in resolving it. Komons contrasted John's approach to this issue with that used in setting the existing policy after World War II.

> Unlike James M. Landis, who presided over the 1947 hearings [on airliner crew size], McLucas came from a technical not a legal background (he held a PhD in Physics). Nor was he a partisan. He was part of that army of technocrats that had first appeared in Washington during World War II and had continued to be prized and recruited in peacetime, particularly [at] the Pentagon. Many of these experts, because they were valued for their technical competence, not their politics, had served under both Democratic and Republican administrations.[1]

In recent decades, the word *technocrat* has frequently been used as a polite synonym for bureaucrat. In this book, however, we apply the term more narrowly to identify those at the higher echelons of government and business who combine managerial competence with scientific and technical knowledge, often credentialed with advanced academic degrees. In this context, McLucas's education, intellectual interests, technical knowledge, and management skills allowed him to rise to positions of responsibility in commercial and not-for-profit companies, a regulated industry, professional associations, and the federal government. In all these positions, he lived by the credo that scientists and engineers must assume responsibility for their decisions and that "engineers need to have a greater understanding of the relationships between user and machine, between the individual and the technology."[2] John— and others like him in industry, government, and academe— used their mastery of the technological milieu to help shape the world in which we now live. On two occasions he left executive positions in the private sector because he thought that being a civilian public servant could be just as important as serving in uniform. As he explained to other young business leaders during his first tour in the Pentagon, "I believe that a man has an obligation to serve his country through government service if he has that combination of background, education, and personal characteristics which will enable him to be effective."[3]

The John McLucas story is also enlightening because, in a small way, it validates the American experience—that quality about us that is so hard for many outsiders to understand. As imperfect as it may be, the United States is still one of the few places on Earth where a boy (or now a girl) of humble origins can make it to the upper echelons of society through education and honorable work. John was a quintessential self-made man, yet one who also took advantage of educational and cultural opportunities to become something of a modern Renaissance man. He loved art and music but proudly wore the mantle of science and engineering. He was also a master at the skill of getting things done, be it in the lab, in the boardroom, or in a government office.

John once reflected on this in a memo I found in one of the dusty boxes in his office. Since he did not live to write the preface to this book, this introspective essay is worth quoting here almost in its entirety.

Sept. 30, 1993

Today Harriet and I went to Colin Powell's retirement. It was a real privilege—a kind of day neither of us will forget. There are many things wrong with the world, with our part of it that we call America. But listening to his speech, we both had lumps in our throats and mist in our eyes. To use the corny expression—Is this a great country, or what?—doesn't cover the half of it. Colin did so many things right today, it would take pages to cover them all—even superficially.

When I think of him as a role model, I think first that all of us are role models; little do we know—there is no way to know—who and where are the people whose lives we've influenced for good or ill. It is not only the young who are consciously looking for models, for mentors, for hope for their futures who are influenced as we pass before them. Our colleagues, our loved ones and many whom we don't even know are affected by what we do. As John Donne said, "No man is an island."

Most of us like to think that we had to overcome obstacles to success in order to become what we became. Most of us realize we could have done a better job, could have achieved more, could have helped more people along the way and could have had more fun if we had not let others' affronts affect us. But listening to Colin's speech makes me feel good about him, about me, and about my dear wife. All three of us have had very different lives but they are all the result of having grown up and lived in an incredible place. Harriet says she deserves little credit for her success, having been spoiled by loving parents. She was an only child and the apple of her parents' eyes; she never caused them any trouble and couldn't have borne the agony of letting them down. She has successfully passed along the love she knew and is one of the most loved people I know. Just about everyone thinks she is great and appreciates being within her orbit of affection. Not the least appreciative is myself. I spend at least a little time each day saying thank you for her love—even when I fail to let her know how good I feel about our being together.

I think much less often about my own "case." I'm usually focusing on some project I want to complete, or start or push ahead. Sometimes I think about the obstacles I've overcome and—once in a while, I'm bitter about them. But all in all, I realize I've been wonderfully blessed and have had my own success story. Without trying to exaggerate the discrimination I've experienced, I've known its hurt at various strategic times in my life. I've been disadvantaged and no doubt all those negative things have caused me to lower my sights too much.

I was the orphan kid from the hills who spoke funny (Appalachian Mountain English). Later I was the kid from the farm going to school in a town of a thousand people where those kids thought of me as having cow manure between my toes (not too far wrong). Later, I was the kid from a small-town school going to college with sons of wealthy planters, too poor to afford fraternity dues and too shy to be pledged by any of them anyway.

Still later I was the kid in graduate school with too little solid science background from my liberal arts college and even later the kid at MIT without adequate background and coming from the South where everyone was a hayseed, an ignoramus or a redneck. I was for years at a disadvantage for having come from the wrong place, failed to have learned enough or somehow hadn't picked up the proper social graces. . . . On attempting to get into graduate school after the war, Harvard and MIT turned me down and I went to Penn State. I worked my way through to a PhD in physics with the help of the GI bill and the lowered tuition of a land grant school "for the industrial classes." I survived, took a job in a local R&D organization, became its president and learned how to succeed by chairing the Chamber of Commerce, helped build the town library and the community swimming pool, chaired the local theater, chaired Penn State's engineering school advisory committee, joined the right professional societies, joined the YPO [Young Presidents Organization] and finally proved to myself that a kid from nowhere could succeed. I went to the Pentagon in an intermediate-level job, graduated to a senior State Department job at NATO Headquarters, and returned to head a think tank founded by MIT. The country kid had finally made it out of [the] mill pond and into the outer world.

Three presidential appointments later, I left the government to head a satellite communications company and somehow have been a senior government advisor, NAE [National Academy of Engineering] member and finally a senior citizen. I've not done too badly, but as Colin Powell also acknowledges, I've had dozens of supporters, mentors, good friends and family who've helped me at every stage. What I am I owe to them all—and to a lot of hard work and optimism.

Colin has succeeded—probably beyond his wildest dreams. He has overcome obstacles Harriet and I know little about. I know what it is like to be a minority, but not the hurt of being a black minority. But I do know he had the essential advantages that could turn around the black experiences of despair in this country. His immigrant parents looked on America as the land of opportunity, which it is. They taught him the disciplines of concentration, determinations and faith in himself. They inspired him to honor his parents, his soul and his country. His wife has shared that faith and together they have built an American success story that caused many in today's audience to choke up, to go home with a lump in the throat and a renewed faith in this great land of ours. "This land was made for you and me." Hallelujah and Amen.[4]

Reflections of a Technocrat is the product of a collaboration between John McLucas and Larry and me as his support team. John's previous essays on various topics, his extensive papers and other materials, and transcripts of interviews and oral histories provided the heart of the source material, although my trips to State College, Pennsylvania (home of what had once been HRB-Singer) and Bedford, Massachusetts (where the MITRE archive resides), Larry's forays to the Pentagon, the Air Force history office, and the University of New Mexico libraries, and our joint explorations of the Internet also yielded essential information. We had a lot of help along the way, starting with Harriet McLucas. We also owe a debt of gratitude to a host of John's friends and colleagues who gave generously of their time and knowledge. If John had been able to write a preface, he would have given them all proper credit. We have tried to recognize these many contributors in the preceding acknowledgments section. Here, however, we would especially like again to thank Mim Eisenberg, John's skilled transcriber and long-distance friend.

John left his mark on the lives of many people and on the nation that he served for over half a century. He became one of the relatively small coterie of public servants who placed an indelible stamp on recent American history. He eschewed the spotlight, but that did not mean his accomplishments went unnoticed. As news of his death spread through a wide circle of friends and associates, his family was flooded with words of sympathy for their loss and praise of John's accomplishments.

Typical of many was a letter of condolence to Harriet McLucas from Secretary of the Air Force James G. Roche a few days after John's final flight: "His lifetime of service to our nation, as Secretary of the Air Force and in a variety of public service and academic positions, protected and defended the highest ideals that form the foundation of our great nation. His timeless contributions to America will not be forgotten. We are very proud to remember him as one of our own."[5] John McLucas's lasting contributions in the fields of science and technology have been posthumously recognized by the Air Force, which has memorialized its annual USAF Basic Research Award in his honor.

For two years, Larry and I knew John as a colleague, friend, employer, and careful editor. For me, living less than three miles away, this grew to be a special bond forged in the almost weekly visits with him and Harriet, in some ways akin to that between me and my father. In his inimitable way of cutting to the heart of things, John once wrote, as a young man at Davidson College, that in the end we were put on this world for two reasons: (1) do some good, and (2) have some fun. Looking back some 60 years later, he explained this was simply a rephrasing of the first question in his *Child's Catechism and the Shorter Catechism.* "Question: What is the chief end of man? Answer: To glorify God and enjoy Him forever." At its core, this book is a reflection of John McLucas's quest to achieve these simple yet powerful goals he had set for his life on Earth.

KENNETH J. ALNWICK
Alexandria, Virginia

Notes

1. Nick A. Komons, *The Third Man: A History of the Airline Aircrew Complement Controversy, 1947–1981* (Washington, D.C.: Department of Transportation, Federal Aviation Administration, 1987), 107.

2. Dr. John L. McLucas, "The Future of Communications and the Role of the Engineer in a Technological Society" (Honorary Engineering Lecture, Pennsylvania State university, State College, Penn., 17 April 1980), 9.

3. John L. McLucas, "When Should a Man Go into Government Service," *YPO Enterprise: The Magazine of the Young Presidents' Organization* 13, no. 3 (April 1963): 4.

4. Handwritten essay by John McLucas, 30 September 1993.

5. James G. Roche, secretary of the Air Force, to Mrs. Harriet McLucas, letter, 4 December 2002.

Chapter 1

From Country Boy to Company President

In the summer of 1970, just before my 50th birthday, I took a short vacation from my job in the Pentagon, where I had been serving as the Air Force's second-ranking civilian for about 15 months. While visiting the Carolinas, land of my birth and childhood, I went to see my father's first cousin, Harold Cousar. Uncle Harold, as I called him, had been like a father to me. He was now 93 years old and had recently moved into a nursing home. When I said I had come to see how he was getting along, he asked who I was. It became obvious that his failing eyesight didn't recognize my face, or his failing memory, my name.

"You say you're John McLucas," he said.

"Yes, John McLucas," I answered.

He thought for a while and then asked, "Whose son are you?"

I said I was his cousin Luther's son and that some time after Luther died, I had lived with him for many years. We talked more, but I'm not sure he ever quite figured out who I was. I left in some dismay, never to see him again. He died later that year.

Whose Son Am I?

That encounter troubled me greatly. (In those days we understood little about Alzheimer's disease and similar disorders that can so tragically afflict the elderly.) Uncle Harold had been the most significant bridge to my past, so his question seemed especially poignant. I had begun thinking about how many of the people who had been important in my early life were now gone, and that I too was a mortal man.

Whose son am I? For the purpose of this book, I will try to answer by reminiscing a little about my early life. Because I grew up in a different environment than most of my professional peers—in a setting that no longer exists—I hope that briefly telling the story of my childhood might help put what came later into a better perspective.

1

Whose son am I? On my father's side, I can trace my ancestry to Mull, an isle off the west coast of Scotland. After Bonnie Prince Charlie, last of the Stuart dynasty, lost the Battle of Culloden to the English in 1746, more and more Scots began crossing the Atlantic to seek a better life. Many of those reaching North Carolina came up the Cape Fear River and settled in and around what is now Scotland County. My McLucas ancestor, Archibald, arrived there from the Mull village of Torassay in the early 1780s, just after the 13 colonies won their independence. His son John eventually moved just across the state line, where the McLucas clan became well established in the vicinity of Clio, South Carolina, near where Interstate 95 runs today. For two centuries, generations of McLucases have been buried in the family graveyard on John's old farm. In one of the graves lies my father, John Luther McLucas. There is a plot nearby reserved for me.

Known by his middle name of Luther, my father led a rather rootless life after leaving the family farm near McColl, South Carolina. By his early forties, he was buying cattle in the mountains of North Carolina and shipping them to South Carolina, hoping to make a few dollars per head. In 1918, just east of Mount Mitchell in the village of Conley Ridge, North Carolina, he met a young woman named Viola Conley, who became my mother. She was only 18 when she married Luther.

So I am Luther's son, and I am Viola's son. He took her back to the family homestead near McColl to start a family: my sister, Jean, born in 1919, and me in 1920. When I was one year old, my parents returned to North Carolina in hopes that the mountain air would improve his rapidly failing health. He died about nine months later, apparently of tuberculosis. My mother soon remarried Nathan Boone, a widower with five children from the nearby hamlet of Booneford, North Carolina.

My father didn't leave me much in any tangible sense, but he did bequeath an extended family. Shortly before I was to start first grade, Uncle Hugh McLucas, one of my father's younger brothers, took Jean and me by train to South Carolina for what my mother thought would be a summer visit. But my father's relatives had decided, or soon would decide, that we deserved better than to be raised as "hillbillies." So they kept us in what they considered part of the civilized world. Meanwhile, my mother continued raising a family of Boones: five stepchildren and eventually seven more of her own.

Was this a kidnapping? I've never been able to decide. In any event, my sister and I grew up knowing little about either of our parents. I don't remember being too upset, but in the late 1980s, my Aunt Willie, Uncle Hugh's widow, told me about an incident she recalled from my childhood. One day, having noticed all the kids playing in the yard except me, she searched all over and finally found me in a closet trying to pack a suitcase. Maybe I really did want to go back to my mother, and more than once, I dreamed about riding on a train and wondering where it was going.

When Hugh brought us down from North Carolina, Jean and I lived as part of his family, which already included three children from Willie's first marriage. From there we attended elementary school in McColl, a few miles away. After a couple of years, Aunt Effie, my father's oldest sister, took Jean and me to live in her rented house in McColl. She worked as the town clerk for Mayor Donnie McLaurin, another of our cousins. Intellectually stimulating, Effie had graduated as valedictorian from Presbyterian College in Clinton, South Carolina. I also have fond memories of Aunt May, a nurse who served in France during World War I with the American Expeditionary Force. She told exciting stories about her time in Paris and flying over France in an airplane. How could I have imagined that some day I too would live in Paris or fly in airplanes all over the world as the leader of our country's military and civil aviation agencies?

Then came the Great Depression. Three more relatives had to move in with Aunt Effie. Hugh lost the old family farm, but when I was entering fourth grade, Jean and I moved back in with him and Aunt Willie on a rented farm. With two new children of their own, it was so crowded that, in desperation, the McLucas clan decided to allocate Jean and me to other relatives. She went to live with another aunt and her husband in Mt. Airy, North Carolina (hometown of actor Andy Griffith). After that, I didn't see much of my sister except during occasional summer gatherings. I moved in with Harold and Donella Cousar, who owned a large farm about 15 miles from McColl, where I had spent much of the previous three summers. Shortly thereafter, I saw my mother one last time. She and Nate came to McColl in an unsuccessful attempt to persuade me to return with them to North Carolina. When my mother died in 1940 giving birth to her 10th child, a stillborn, I had not seen her in the intervening decade. My trip to Conley Ridge for her funeral was my first visit there since my abrupt departure 15 years earlier.

Whose son am I? Most of all, I consider "Uncle" Harold and "Aunt" Donella my foster parents, although the arrangement was purely informal. For 10 years, as I grew into a young man, they were my role models. Harold was rather easygoing, but he had high standards of morality. A college graduate when few family farmers were, he continued to pursue learning all his life and would often read the newspaper and other literature out loud to us, especially about new scientific advances. Donella was of average intellect, conscientious, and honest. She had a kind heart but could be strict when needed. The year I moved in with the Cousars, their son, Harold Jr., went off to Clemson University. They also took in two teenage boys: Charlton Jr., a nephew of Harold; and Eugene, a nephew of Donella. Two years later, Charlton won a full scholarship to the Citadel and went on to a career with the Army Corps of Engineers.

Living in the "Deep South," I grew up in the era of strict segregation between the races. Uncle Harold was a man of his time, having been born only 10 years after the end of slavery. Although believing that Negroes should "stay in their place," he felt they should be treated with Christian charity. He scrupulously tracked his farmhands' accounts and tried to take good care of them and their families as well as our full-time maid. He also frowned on the hateful attitudes and actions of many of our white neighbors.

Living and working on the Cousar farm was an important part of my education. In addition to milking the cows and doing other daily chores, I spent many hours working the fields with our farmhands as we plowed the soil, sowed the seed, "laid by" the corn, picked the cotton, gathered the tobacco, and stayed up late at night to cure the leaves in big tobacco barns. This was a time of transition to mechanized farming, and Uncle Harold had an old-fashioned tractor with steel wheels. It was often in need of repair, so we often relied on our horses and mules. I feel fortunate to have grown up so close to nature. I can't help but feel sorry for most of today's youth who, despite all their material advantages, have no connection with rural life.

Uncle Harold was interested in technology, and I learned about plumbing and electricity while helping him install these modern conveniences. My toys were also educational. I inherited an erector set from Harold Jr. and eventually acquired enough parts to build a skyscraper with an electric motor to lift an elevator. One year my Christmas present was a chemistry set of the type no longer sold because of safety concerns. Inevitably, I became interested in automobiles and bought an old Model T for all of seven dollars. The body was in bad shape, but I modified it with wooden boards into a primitive pickup truck that I could drive on the farm. Once, Harold allowed me to pick out a calf to raise. In due course he was a grown steer, and I sold him for $150. Having already read every page of the *Book of Knowledge,* I used the proceeds to buy my first set of the *Encyclopedia Britannica.*

As this might indicate, I was academically inclined. I really enjoyed school and still fondly remember many of my teachers. Although McColl Elementary School gave me a solid foundation in the three R's, Harold had me transferred to the more highly regarded Latta Elementary School, about seven miles away. I later attended Latta High School. I took as many academic courses as possible. I especially remember Mr. Doane James, who taught general science, chemistry, and physics. A small but wiry man who had also doubled as our scoutmaster, he helped me capitalize on my chemistry set and instilled an ambition to become some type of scientist. I also liked math. South Carolina held a special state contest for ninth and 10th graders. The first year I received a medal for algebra and the next, for geometry. I was not a bit athletic, but my awards in the academic arena were enough to convey me from obscurity to minor celebrity. I eventually graduated number two in my high school class.

After Harold Jr. graduated from Clemson in 1936, he returned to live on the farm and manage an ice-cream plant in Dillon, South Carolina. Over Aunt Donella's protest, he hired me to work there for the summer after my junior year, and I did so again after my senior year. The plant had a variety of devices and equipment that interested me. I was exposed to some basic laws of physics, such as how compressors and inert gasses could be used for refrigeration. New refrigerated delivery trucks were gradually replacing the older trucks that relied on blocks of dry ice. Even during the Depression, the pace of technological progress was becoming obvious to me.

So, whose son am I? Beyond my extended family, I am in many ways a son of the Carolinas, having been brought up in those states and later attending one of their best small colleges. Although spending most of my later life no farther south than the suburbs of Washington, D.C., I still consider myself a Southerner at heart. To me, that means taking time to smell the roses and share conversations, stories, and everyday activities with friends and family. Looking back, I wish now that I had been better able to follow this lifestyle later in my career, when I became something of a workaholic. Yet, being a Southerner also sometimes meant feeling defensive among my northern friends. I know that their negative impressions of southern whites had much basis in fact, but these impressions were also another form of prejudice. It's been small comfort to find that prejudice applies to people in every region, although the objects of such bigotry may vary with geography. Despite the old South's sad legacy of exploitation and intolerance, life there had redeeming qualities, and I feel fortunate to have been raised when and where I was.

Attending Davidson College and Tulane University

Most of my fellow Latta graduates who were continuing their education went to Clemson. I chose instead to go to Davidson, a small Presbyterian college in a town of the same name on the rolling Piedmont terrain north of Charlotte, North Carolina. Two Davidson graduates whom I respected convinced me that it was the best choice, and I think this has made a difference in my life. Davidson offered a more cosmopolitan student body than Clemson and whetted my appetite for an advanced education. I was able to attend thanks to Aunt May, who had a modest but steady income as a nurse with the Veterans Administration. She gave me $250 a semester to cover basic tuition, leaving me to take care of room, board, books, and other expenses.

Doing this meant working in a variety of part-time jobs. I served meals and washed dishes at a boarding house and for a while cleaned latrines at the local high school. I also taught adult education classes in English and math. Later a friend made me his partner in selling Radio Corporation of America (RCA) products, including records and radios, for which he held the local franchise. One of my most prestigious jobs was in the Office of the Registrar. Among other functions, it kept the minutes of the Student Council, which enforced Davidson's honor system. At the end of the year, when typing up some of the council's minutes from its secretary's longhand notes, I discovered that I had been suspected of having shared answers with another student, who later admitted copying without my knowledge. I would recall my brush with Davidson's honor code many years later when dealing with the strict honor code at the US Air Force Academy.

During my junior year, I became truly serious about my studies and future prospects. I decided physics offered more intellectual challenges than chemistry and was accepted into the physics honor society Sigma Pi Sigma, which I have belonged to ever since.

The summers after my sophomore and junior years allowed me to apply some practical mathematics and geometry while making good use of a recently acquired 1935 Ford. Franklin D. Roosevelt's New Deal agricultural programs required an efficient way to ensure farmers did not exceed quotas of certain crops for which they received subsidy payments. With a recommendation from Uncle Harold, the county hired me at 60 cents an hour to help monitor crop acreage. Airplanes shot overhead photographs of farmland from an altitude of 15,000 feet. On the ground, we studied big prints of these images to identify fields for measuring the extent of various crops. This application of aerial photography was my first experience with remote sensing, a sphere of technology that occupied a good fraction of my later life.

As another small source of income, I had signed on for the Army's Reserve Officer Training Corps (ROTC). At the start of my junior year in 1939, I was in a barbershop when the radio announced that Germany was invading Poland. Suddenly our military indoctrination didn't seem quite so abstract. The war became even more real in the spring of 1940 when France fell to Germany, and British forces barely escaped to their island. Because I was only 20 when I graduated in May 1941, and the Army's minimum age for commissioning second lieutenants was 21, my military commitment could be deferred. Even classmates who received commissions were not immediately called to active duty.

I was fortunate that the head of the physics department at Davidson recommended me to Daniel Elliott, a friend of his who ran the physics program at Tulane University. Professor Elliott gave me a graduate fellowship that covered tuition and paid an unbelievable stipend of $55 a month. Upon arrival in New Orleans, he began trying to make me and three other new graduate assistants into cultured men of the New South. His wife was on the board of the New Orleans Symphony, which engaged us as ushers to escort patrons to their seats during performances. This exposure to high culture introduced me to the joys of classical music. On a more practical level, Dr. Elliott arranged for a shop course to teach us how to use lathes, drill presses, and other tools—all valuable skills for an engineer to be familiar with. Our normal duties had us tutoring undergraduates, setting up experiments, and grading tests and papers. In addition, the Navy had a contract with Tulane to teach new officers the rudiments of certain scientific subjects as part of its educational initiative called "V" programs. As a graduate assistant, I shepherded a roomful of young Navy personnel through the basics of physics.

As with nearly every American of my generation, I remember the exact circumstances on hearing the news that brought us into the war. I was spending Sunday afternoon, 7 December 1941, in the home of a girl named Chris Paris, listening to the New York Symphony on the radio when a breathless announcer told us of the Japanese attack on Pearl Harbor. Despite the subsequent declaration of war against the Axis powers, its effect on my life was hardly immediate. By now the right age for an officer's commission, I was informed that my first duty was to keep instructing naval officers while pursuing a master's degree in physics with a minor in electrical engineering.

So I continued teaching, studying, and working on my thesis, which dealt with the acoustics of streetcar noise on the streets of New Orleans. Among my professors was Dr. Joseph Morris, a jolly man who usually had a cigar in hand or mouth. In addition to theories and principles of electronics, he described a recent invention called radar (radio direction and ranging). Radar was one of the most classified technologies of the time; so secret we said it had to be spelled backwards. Despite this, Dr. Morris introduced us to the burgeoning military fields of radar surveillance and fire control. One day he captured our attention by revealing the purpose of his frequent visits to Washington, where he served on a subcommittee of the National Defense Research Committee (NDRC).[1] Morris was involved in development of the proximity fuse, which used a tiny radar mounted in an artillery shell to detonate a warhead when it came close to a targeted aircraft. His enthusiasm helped convince me to accept a Navy offer to enter its radar officer training program.

Employing Radar in the Navy

In June 1943, with degree in hand, I signed on with the US Navy. Following a visit back on the farm with the Cousars, I was sworn in at Charleston, South Carolina. Along with a hundred other recent civilians, I learned the rudiments of seamanship, navigation, naval protocol, and military discipline during a three-month training stint at the Naval Indoctrination Center, Fort Schuyler, New York. Next, I studied electrical engineering and electronics at Princeton University for another four months. I then learned the specifics of Navy shipboard systems with a semester at the Massachusetts Institute of Technology in Boston (beginning an intermittent professional relationship with MIT). Several of us stayed another month for a specialty course in the latest airborne radars.

One of my classmates had a girlfriend at prestigious Wellesley College who arranged blind dates for me and another friend. We met the young women at an officers' club in downtown Boston. My date was Patricia Knapp of Warren, Pennsylvania, a liberal arts student with tastes considerably different from mine. There was, however, an immediate and grand attraction between us, and her letters would follow me around the Pacific for the following two years.

My next orders were to a ship, aboard which I was expected to put all this knowledge about electronics and radar to good use. This vessel was the USS *Saint George*, just out of the builder's yard at Tacoma, Washington. Getting there took a four-day, cross-country train trip, which introduced me to the magnificent scenery of the northern Rockies and Pacific Northwest. Then I was aboard the Navy's newest seaplane tender in time for her commissioning ceremony on 24 July 1944. I took my place as the *Saint George*'s radar officer among a crew of 1,077 under the command of Capt Robert Armstrong. In my section, I supervised four to five radar technicians, who maintained and calibrated the equipment, and six to eight operators, who sat at the scopes interpreting the blips projected on cathode ray tubes.

Not the most glamorous fighting vessel of World War II, the *Saint George* was one of four *Kenneth Whiting*–class tenders designed to service long-range patrol aircraft. Her hull number, AV-16, indicated an auxiliary ship with an aviation-related function. She was 492 feet long, 69.5 feet in the beam, and 12,000 tons displacement. The ship could perform many of the functions of a small aircraft carrier, except having planes operate from the deck. Normally the big flying boats would set down nearby and taxi alongside for refueling, rearming, repairs, personnel exchanges, and the all-important mail deliveries. On her starboard side toward the stern was a 12-ton crane for lifting seaplanes up onto the aft deck for major overhauls and patches to their watertight hulls.

Early in our shakedown exercises from the port of San Pedro, I got on the good side of Captain Armstrong, who had not sailed with radar before. The coastal area of southern California often experienced dense morning fog. We had mounted a radar display on the bridge, which allowed him to see and avoid other ships despite zero visibility. Everything was routine until 12 October 1944, when we hastily left San Pedro to go to the aid of a downed PB2Y-1 Coronado. Although this four-engine patrol plane (based on the B-24 Liberator heavy bomber) sank before our top speed of 18 knots could bring us to the scene, we continued to Hawaii and the central Pacific. While stopping at Eniwetok, our first sight of a coral atoll, we joined up with our first PBM-3 Mariner seaplanes. Designed and built by the Glenn L. Martin Company, the various PBM-3 models had a wingspan of 118 feet, a length of 80 feet, and a top speed of about 200 miles per hour (mph). Most important, these twin-engine aircraft had a range of about 3,000 miles when on patrol (less if carrying bombs or depth charges), and their tall, boat-shaped fuselage could weather fairly heavy seas.[2]

We eventually reached Saipan, one of the large islands in the Marianas chain and a key base for attacking Japan. Within hours of our arrival, the Japanese staged a spirited air raid that prompted the *Saint George* and some nearby vessels to employ their armament against real targets for the first time. Our ship, full of greenhorns, quickly learned that what went up often came down in unexpected places. Those of us inside the ship thanked our lucky stars we had the cover of steel to catch the errant shell fragments.

One duty of junior officers was censoring outgoing mail according to various rules about classified information. Since there were 1,000 people writing letters and a dozen or so censors, we normally did not know the sailor writing a particular letter, but once I was fortunate to find out what one of my own men—a senior radar technician in his early fifties—really thought of me. In a letter to his wife, he complained that his boss was a young know-it-all ensign who kept trying to push his people around. He couldn't wait to be reassigned somewhere else—anywhere else. This hit me between the eyes. After my initial shock and denial wore off, I took his appraisal to heart. When the ship finally sailed home for decommissioning, he said how much he had enjoyed working for me. Normally in the military, criticism flows only from higher rank to lower. In this case, I'm glad the opposite occurred. I think this feedback helped make me a better manager in the future, and I sometimes sought out the opinions of subordinates before hiring or promoting their supervisors.

Although the atomic bomb would end the war, many historians consider radar and signals intelligence as key technologies in the Allies' victory. I was not privy to the secrets of code breaking, but I felt privileged to help apply the wonders of radar toward winning the war. There were three major radar systems on the *Saint George*, each serving a different purpose. Both of its two five-inch guns had small radars to measure distance to targets. To detect aircraft, it had an air-search radar, called the "SK," which used a yagi dipole antenna. To watch for ships and other objects on the surface was the function of the "SG" radar, fed by an oblong dish antenna about 130 feet above the waterline. We also could send out a radar beacon for planes with a compatible receiver to home in on.

11

We learned a lot about how to use radar during our first several months in the Pacific. Some of the problems encountered turned out to be the result of human foible. After the Saipan raid, for example, some of my radar operators got overly enthusiastic, causing the ship to be called to general quarters based on false alarms. I had to stress the need to carefully assess each radar image before assuming it might be Japanese. We also found that radars did strange things on their own volition. Sometimes the ship's visual lookouts could spot land before we did. At other times, our radar might show islands or other large objects far beyond its normal range. This phenomenon is called anomalous propagation. In the Marianas, it usually resulted from layers of cool air near the surface, which caused radar signals to reflect in unexpected ways. In my later career, we would often encounter complex variations of such ducting phenomena higher in the atmosphere when using much more sophisticated equipment, including long-range, over-the-horizon, and airborne radars. Troubleshooting radar equipment in the vacuum-tube era could also be challenging. Sometimes we operated our search radars continuously for days or weeks on end, and they would get out of adjustment. We then had to find time to stand down and quickly upgrade their performance. I took pride in sometimes being able to solve problems that stumped my technicians, perhaps confirming the axiom that showing technical expertise can help supervisors earn the respect of their subordinates.

I was fascinated by the rapidly advancing progress in radar technology. One day, when I had some time off, I went to visit Saipan's new ground controlled approach (GCA) facility at an airfield onshore. I observed as several fighters made practice approaches under ground control—to me, a beautiful thing to witness. After the war, this system was refined for use at military bases to permit landings under bad weather conditions. Years later I got to work closely with the inventor of GCA, Dr. Luis Alvarez.[3] When we returned to Saipan in March 1945, there was more time to look around at what was happening in my area of professional interest. By that time a huge new microwave early warning (MEW) radar was in place. It was sited on a hill, which put its radar horizon at least 30 miles away. Radars were being built on a grand scale that seemed to portend the future.

During the final months of 1944, the *Saint George* cruised ever closer to the line of active combat in support of Adm Chester Nimitz's Central Pacific campaign. She stopped at Kwajalein in the Marshall Islands in late October, just as Gen Douglas MacArthur's forces were retaking the Philippines. Proceeding west to Eniwetok in early November, the *Saint George* took aboard the last of her full crew complement: a patrol aircraft technical service unit (PATSU). We now could develop our full potential, eventually serving as mother ship for as many as 50 seaplanes. After three months of relatively uneventful assignments, sending out patrols from well-protected anchorages at Peleliu and Ulithi Atoll, we received orders to support what would become the last great battle of the war: the bloody campaign to take Okinawa, largest of Japan's Ryukyu Islands.

This put us in the thick of the deadliest naval action in American history. In the company of four other seaplane tenders, the *Saint George* arrived on 28 March at Kerama Retto, a string of small islands just off the southwestern coast of Okinawa. After days of shelling and bombing by Allied warships and aircraft, American marines and soldiers hit the beaches on Easter Sunday, 1 April 1945. A week later, those of us at sea were engaged in the most concerted action most of us could imagine. When Americans elsewhere celebrated V-E day on 8 May, the bitter battle for Okinawa continued to rage. It was not over until 22 June, after all but 7,400 of the 130,000 defenders had fought to the death.

We on the *Saint George* had three major missions: to use our radars as part of the air control network, to use our repair facilities to fix damaged American ships, and to use our patrol planes to detect any Japanese ships. One of our PBMs helped pinpoint the last suicidal foray of the superbattleship *Yamato*. Because the *Saint George* had well-equipped repair shops and a high boom, my team kept busy removing and repairing antennas from destroyers damaged while on dangerous radar picket duty north of Okinawa. I remember the executive officer of one battered destroyer telling me, "Don't hurry too much with your repairs. We're not that anxious to get back out there."

With Japan's surface navy virtually destroyed, the main threat came from the air, especially the infamous kamikazes. On 29 April during one of the frequent air raids, our ship's gunners shot down their first enemy plane as it headed for another ship. Whenever a Japanese aircraft flew in at low level, gunners on all

the surrounding vessels in the anchorage began shooting at it—
sort of like the proverbial circular firing squad. Inevitably, this
kind of firepower caused what has become known as friendly-
fire casualties.

Our turn to suffer a strike from the sky came soon enough.
On 6 May we were riding at anchor as usual when a Mitsubishi
12M3 fighter began approaching from the southwest at about
2,000 feet. It was an impressive attack; every gun around was
shooting at the plane, but it could not be stopped. As caught
on film, the last heavy shell from the *Saint George* apparently
struck just in front of the oncoming fighter but could not deflect
its momentum. On the aft deck sat the hulk of a Mariner being
stripped for parts. The enemy plane sliced through it and plowed
into the base of the big crane, the toughest spot the doomed pilot
could have chosen. Stunned silence followed the impact. Two
men working on the PBM had unfortunately sought cover under
the crane, where they died instantly. The enemy plane's engine
carried through the deck plating and into some crew spaces,
killing a pilot who was writing a letter to his wife. Three others
were badly injured, but the ship survived without crippling damage. Soon there was a barge with a large crane tied up alongside
the *Saint George* to substitute for its inoperable crane.

Samuel Eliot Morison, the official naval historian of the war,
calculated that Allied ships near Okinawa faced 1,465 kamikaze and 1,351 other aircraft attacks from 6 April through 22
June. The US Navy counted 34 of its ships sunk, 368 damaged,
more than 4,900 sailors killed or missing, and more than 4,800
others wounded (many horribly burned). It was by far the cost-
liest battle in American naval history, and an experience I will
never forget.[4]

After the battle, the *Saint George* returned to Guam for sorely
needed repairs. It was here that we heard about the atomic
bombs dropped on Hiroshima and Nagasaki, leading to the
Japanese surrender. I can still remember my feelings of awe
and wonder upon hearing the news. Perhaps it was my knowledge of physics that impressed me more than most of my shipmates with the revolutionary significance of this magnificent
but terrible invention.

Under the command of a new skipper, whose often inebriated
behavior made us regret the departure of Captain Armstrong,

we returned to Okinawa on 21 August 1945, standing by for further orders. There we rode out a powerful typhoon on 16 and 17 September. The storm was hard enough on ships, but the seaplanes really suffered. Right after this reminder of nature's power, the *Saint George* was among the first contingent of ships assigned to occupation duty in Japan. On 20 September we sailed to the Japanese port of Wakayama Wan. From there our aircraft patrolled Japan's Inland Sea and ran passenger, mail, and courier services among Tokyo, other cities, and Okinawa. In November we moved farther west to the port of Sasebo. Without much real work to perform, some of us went sightseeing, driving Jeeps around the countryside and using our PBMs to view what remained of Japanese cities. I especially remember flying at low level over Nagasaki and being stunned by the sight of almost utter destruction.

Continued flight operations occasionally caused peacetime casualties, including a close call for me. An admiral from another group wanted to get in some flight time over the scenic Inland Sea. I was operations officer by that time and booked myself to go along on his mission. By chance, the skipper ordered me to stay behind to brief another visiting brass hat. The PBM flew out over the Inland Sea, never to be seen again. Such are the twists of fate.

After months doing little of apparent value, many of us grew increasingly impatient to return home and get on with our lives. I even wrote to my congressman, a young L. Mendell Rivers, to ask his help in being demobilized. Probably without his intervention, the *Saint George* finally arrived at San Diego on 25 March 1946.[5] As with most everyone who has spent time in harm's way aboard a warship, I have a lasting affection for the *Saint George*. In the company of this rather humble vessel and her crew, I can rejoice, as Winston Churchill once wrote, that in my youth I had been shot at without result. I can also thank the Navy for giving me a valuable education in both the theoretical and practical aspects of electronics and radar. Like other fortunate members of the World War II generation, my wartime training and experience, combined with the educational benefits of the GI Bill, helped lay the foundation for a rewarding career.

Upon my return from Japan, the United States seemed caught up in a surge of optimism. Most Americans were eager to enjoy the fruits of peace and not yet fully aware of the new role on the world stage in which history had cast their nation. Much of the great armed host that contributed so much to defeating the Axis powers had already melted away, and I was eager to join the massive demobilization. For a well-educated young officer returning to the civilian world, life held more than a little promise.

My orders had me reporting to Charleston to "process out" of the Navy where I had entered it. I boarded a train headed east and before long was back on the Cousars' farm. Patricia Knapp soon joined me, with Harold and Donella making the young Yankee woman feel very welcome. Then Pat and I headed north to Warren, Pennsylvania, for me to meet her family. Six weeks later, I was back in Warren for our wedding. After a honeymoon at Niagara Falls and a resort in Quebec, we arrived at my newly rented apartment in Boston, where I had recently started my first full-time civilian job. Three months later, Pat was pregnant with our first child, whom we would name Pamela. The postwar "baby boom" was under way. Becoming a mother interrupted Pat's pursuit of a career for the next 35 years.

Growing a High-Tech Enterprise

My transition from military service to civilian life was less abrupt than for many veterans. I had accepted an offer to become a junior engineer, designing parts for new radars at the Cambridge Field Station of the Watson Laboratories. Although operated by the Army Air Forces, the lab (located in New Jersey) was still under contract to the Army Signal Corps. Its field station consisted of a few hundred people housed in an old warehouse a couple of blocks from the MIT campus. For me, this job marked the beginning of more than 50 years' association with the Air Force. Although the work was interesting enough, it soon convinced me that I needed a stronger theoretical background for thriving in a field with such potential. Yet in later years, I would look back on my 15 months at the Cambridge Field Station—designing circuits, building breadboards, and testing them—as one of the best learning experiences of my life.

In the spring of 1947, I began applying to various graduate schools, including Pennsylvania State University in State College, a charming town nestled in the shadow of Mount Nittany in what the locals call Happy Valley. Pat's hometown was not too far to the northwest. The university offered me an assistantship that, with the GI Bill stipend thrown in, amounted to $3,600 a year. Little did I know that by going to Penn State, I would also be choosing our home for the next 15 years. After some unsuccessful attempts to find a place to live, I bought a lot on the edge of town and put up a three-bedroom prefabricated frame unit. It took all my savings of $3,000 and a $10,000 mortgage obtained with some help from my father-in-law (a well-respected factory owner). Although greatly appreciating the assistantship, I was wishing we had a bit more to live on.

One day in early 1948, after making it through the first semester on a tight budget, Dean George Haller of the Department of Physics and Chemistry asked if I might like a part-time job working for something called "H, R and B." Scrutinizing my blank expression, he explained that these initials stood for Haller, Raymond, and Brown, a small partnership he had started that was doing consulting work for the Air Force. When he doubled his initial offer of one dollar an hour, I eagerly accepted. Starting in February I began splitting my time between course work, teaching physics to undergraduates in exchange for free tuition, and working 16 to 20 hours per week at Haller, Raymond, and Brown (HRB). My first boss there was Dick Raymond—the "R" of HRB. He was managing a $5,000 contract for Wright Field on improving the use of chaff (aluminum strips) as a radar countermeasure.* A year later Wright Field renewed the contract for $15,000. We must have done something right.

George Haller, a local native whose father managed Penn State's farm, had headed the Aircraft Radio Section of the Signal Corps lab at Wright Field during the war and, among other things, developed an aerial trailing wire antenna. The government retained rights to this invention for military use but let him keep the rights for commercial development. Although this

---

*Wright and Patterson Fields near Dayton, Ohio, were combined into Wright-Patterson Air Force Base (AFB) in 1948. Long the center of aviation research, development, testing, and logistics for the Army Air Forces, it also became the site of Air Materiel Command headquarters. For the sake of simplicity, the various labs and other organizations on the west side of the base with which HRB did business are referred to generically in this chapter as "Wright Field," which was the name most of the people at the company kept using.

didn't make him rich, it did encourage him to seek extra money through consulting. During the war, Haller also built one of the earliest airborne electronic intelligence (ELINT) systems, code-named Ferret. This name ultimately came into use for all airborne ELINT systems—and even satellites in later years. ELINT and other intelligence-related technologies became the backbone of HRB's work for the US government. Because of the high-level security classifications involved, much of this work was awarded under sole-source contracts.

Dick Raymond first met George Haller while working on electronic countermeasures (ECM) at Harvard University's innovative Radio Research Laboratory. In 1943 they went to North Africa together and began to put some of their ELINT and airborne jamming developments into use. After the war they converged on State College, where Dick became a professor of physics and, ultimately, my dissertation supervisor. Walter Brown was also a friend and collaborator of George Haller at Wright Field, where Brown served as the assistant chief of the Special Projects Laboratory. At war's end, he too came to Penn State for graduate work. In 1946 these three veterans formed HRB. Using their extensive contacts at Wright Field, they landed some contracts to work on electronic mapping devices for the Air Force. This work started HRB on its road to success. The company began operating from Brown's house trailer, then expanded into Raymond's garage. In early 1948, shortly after I appeared on the scene, the company converted from a partnership to a corporation.

To further supplement my income and continue my military status (I was still a Naval Reserve officer), I signed up with a new Air National Guard (ANG) outfit in town: the 112th Air Control and Warning (AC&W) Squadron. Walt Brown, a reserve major, was the founder and commander of the unit. It was more than coincidence that the 112th was assigned to the 153d AC&W Group in Harrisburg, which was created and commanded by Col George Haller. As a radar officer, I participated in regular drills, encampments, and exercises along with several of my HRB colleagues and other members of the Penn State academic community. From time to time, HRB had access to the unit's radars to help run system tests. I served with the 112th for two years, first as a lieutenant and later as a captain.

When the Korean War buildup began, the unit was called to active duty to eventually serve as a source of replacements for other radar units scattered along the eastern seaboard and overseas. At the time the 112th was federalized, HRB was so deeply involved in classified work for the Air Force that Haller, Brown, and McLucas were given what amounted to deferments. Looking back, I have always felt a bit uncomfortable with that decision, but it seemed to be the most logical choice in terms of overall national defense. In any event, that marked the less-than-heroic end of my uniformed service. Yet on several subsequent occasions, the Pennsylvania ANG has seen fit to honor me as one of its distinguished alumni. Over the years I've been fond of pointing out to some audiences that I had the honor of wearing the uniforms of three services: the Army's during my ROTC days at Davidson, the Navy's during World War II, and the Air Force's while in the Guard.

By the summer of 1950, I had completed my course work and produced a dissertation titled "The Effects of Noise on Electronic Circuits" (knowledge that would come in handy for pulling television signals into mountain valleys). When awarded my PhD in August, I was actively exploring the job market. Meanwhile, Haller, Raymond, and Brown were coming under pressure from the Penn State administration to choose between running their business and remaining as full-time faculty. Much of HRB's day-to-day operations were already being supervised by Raymond Miles as chief engineer and Reginald Eggleton as general manager. Desiring to continue their affiliations with Penn State, the triumvirate decided to make full-time job offers to three more of their employees: Robert Higdon to be HRB's new president, Phil Freed to be vice president for finance, and John McLucas to be vice president for technical development. All three of us accepted these offers, and I called off my job search. I was now 30 years old and saw HRB as my best opportunity for advancing in some promising new technologies. The salary was $6,000 a year, which matched other offers I had received thus far from Corning Glass Company and Goodyear Aircraft. By now we had lived in State College for three years, and it was nice not having to move.

With the extra push we were able to contribute as full-time managers, the new corporation grew nicely. The three founders continued to provide technical direction in chosen areas, but its

19

management was now largely in the hands of Higdon, Freed, and McLucas.[6] Everyone then was on a first-name basis. Almost all the professional employees had current or previous ties to Penn State, and all the nonprofessional employees were from the surrounding area. We were able to keep labor costs down because, for anyone seeking technology-related employment in the State College area, HRB was the only game in town. It was financed almost entirely by local stockholders and banks. As the company expanded, we took over and remodeled various storefronts and apartments in town. Ultimately, we built an industrial park on the outskirts of State College to consolidate our expanding business. It is now called Science Park, and what used to be HRB is still there, operating as a division of Raytheon.

Our company stressed flexibility. Its informal motto was "HRB will do anything, so long as the description of work is not too specific." Many employees, with our blessing, also used the work they were doing at HRB as a springboard to move on or found companies of their own. My favorite example is William J. Perry, who would much later become a highly respected secretary of defense. In 1951 he was fresh out of Stanford University with a master's degree and had come to Penn State to seek a doctorate in math. I hired young Bill for $2.50 an hour, starting him off on his distinguished career. He told me later that he modeled his own successful company, Electronic Systems Laboratory (ESL), on HRB.

As one of its early forays into the private sector, HRB decided in 1948 to profit from the exploding field of commercial television. The company applied for a Federal Communications Commission (FCC) license to operate a TV station, but newspaper interests in Johnstown, Pennsylvania, outbid us. Undeterred, HRB became a licensed dealer for Dumont Television Corporation, an early manufacturer of television sets. Reception was so poor in the valleys of central Pennsylvania that a few of us began applying our technical knowledge to improve the situation. We stacked four fairly cheap but effective Japanese-designed directional antennas on top of one another for more gain, put amplifiers up on the towers to improve the signal-to-noise ratio, and did everything else we could to get a picture of half-decent quality to our customers. This has been credited as the first cable powering system in the United States. We eventually built a large main antenna up on Mount Nittany and established a subsidiary, State College Television (SCTV), to string cables into town on specially erected poles. This centralized network became one of the first municipal cable television systems in the country.

Although always looking for commercial opportunities, HRB's core business remained focused on some of the most highly classified areas of national defense. We were one of the first companies to move into the post–World War II signals intelligence (SIGINT) business. Through its connections with the Air Force, Army, and later with the National Security Agency (NSA), HRB worked on numerous close-hold projects. In addition to SIGINT, two devices in particular occupied us: Reconofax and Rafax. The operating principles derived from these projects would lead to other applications in the future.

Reconofax was the name given to a scanning apparatus that lay at the heart of an aerial reconnaissance system first developed by Dick Raymond. Toward the end of the war, photoreconnaissance pioneer George Goddard at Wright Field developed a special low-altitude camera that moved the film past a focal-plane slit at a rate coordinated exactly with the speed of the aircraft. HRB's contract with Goddard's office required us to build an electronic camera to do the same job as his mechanical cameras. Our application employed a sensitive photomultiplier-detection tube and a rotating mirror with additional equipment to transmit a video signal to the ground and print out the picture before the airplane landed, in what would later be called "near real time."

HRB then refined new versions of the system and made it sensitive enough to take pictures by moonlight. Next we wanted to get images in total darkness. When I began working on this problem, I realized that the basic concept was sound, but we needed to use film sensitive to light beyond the visual range, specifically, infrared (IR). I rigged up a quick demonstration using part of the spectrum called near-infrared and shot a daylight picture. This was just for show, because near-infrared didn't work at night, but it did the trick by giving me something to persuade customers of the potential of an IR scanning system. Even though HRB was rapidly expanding into such areas, Dick Raymond became attracted to bigger challenges. In 1953 he joined the RAND Corporation in Santa Monica, California, and visited us only once in the next year to see our latest Reconofax camera. His interest was sparked by a study RAND was do-

21

ing for an Air Force space system—probably the pioneering WS 117-L reconnaissance satellite program—that might also have a use for this technology.

A physicist at Wright Field thought I was on to something big and gave us a contract for $60,000 to build a demonstration camera, using true infrared. That camera eventually captured one of the first infrared photos of Manhattan Island, taken at 11:00 p.m. on 9 January 1958, from 4,000 feet. I still keep a copy of that photo on a wall in my home office. It brought HRB several million dollars in business. Over time, HRB developed scanners for various airplanes, including the U-2. During the Vietnam War, steadily improved models of Reconofax were used on some old RB-26s and later on jet-powered RB-57E Canberras. These proved to be the most effective reconnaissance aircraft for detecting skillfully camouflaged enemy facilities.[7]

At the same time Dick Raymond invented Reconofax, Walt Brown was developing Rafax, a device that made possible the transmission of radar data. Later, when the Air Force began building a highly complex air defense network, the Semi-Automatic Ground Environment (SAGE), a significant fraction of its data had to flow over existing commercial telephone longlines. Because pure radar signals were far too complex to transmit over conventional wires, Brown's device had the potential to help get around this problem. The heart of Rafax was a mechanical scanner that took a reading off the cathode-ray tube at a stepped-down rate. This reduced the image to a signal that could be transmitted over the limited bandwidth of a telephone line and recorded on a facsimile display for printing. Raymond's early results looked terrible, so I worked with two or three of our best engineers to come up with various improvements to the original Rafax, eliminating the fax printers in the process. We put everything on magnetic tape, a relatively new idea at the time. I also patented two or three other related devices that made the whole process much more feasible.

In the early 1950s, Brig Gen Hal Watson, founder of the Air Technical Intelligence Center (ATIC), tasked Haller and Raymond to explore technologies for picking up data on Russian missile activities. Drawing in part on the work HRB was doing with signal propagation and reception with our experimental television work, we sent a small team with antennas and other equipment to the mountain village of Cloudcroft, New Mexico, to see what kind of telemetry we could capture from the nearby test centers at White Sands Missile Range and Holloman AFB. Air Force officials were shocked by what we learned and immediately stamped a sensitive communications intelligence (COMINT) classification on our report—thus opening the door to a new business area for us. We next put up a more extensive array of antennas on Haller's farm near State College and, with the help of some experimental equipment from the Signal Corps, were able to monitor telemetry from as far away as Cape Canaveral, Florida. After successful tests in Utah, where our team set up in a motel under the guise of telephone employees, we were ready to go after bigger game.

One could say without too much exaggeration that HRB helped pioneer the remote sensing of Soviet missile activity. This spawned what eventually became a major segment of the defense industry, devoted to the acquisition, interpretation, and analysis of electronic signals from missile tests and launch sites around the globe. As a key player, HRB soon had people working at almost every government site involved with missile development, and we often switched personnel among jobs to achieve synergy and expand their knowledge base. We even had a contract to produce a Russian-English standardized dictionary of technical terms. In view of rising concerns about a looming "missile gap," the demand for information became almost insatiable, and each breakthrough opened the door to a whole new set of questions.

In the mid-1950s, teams composed of Air Force Security Service cryptologists, Army Signal Corps communicators, and HRB engineers and technicians were deployed overseas. They went to highly secure US listening posts, such as those in Turkey and other places on the periphery of the Soviet Union. We were in position to help monitor the first Soviet intercontinental ballistic missile (ICBM) launch in August 1957. Throughout this very sensitive endeavor, HRB enjoyed a close and mutually supportive relationship with its government sponsors. Our employees were highly motivated and, as a bonus to the government, were able to facilitate the exchange of information among various agencies as we rotated our people in and out of different contracts and job sites. I could argue that HRB was instrumental in the pooling of knowledge among compartmented government intelligence bureaucracies seeking an accurate picture of the Soviet missile threat.

Over the next decade, HRB would build and service many specially designed receivers and antennas—both ground based and airborne—for collecting signals. In the 1950s and 1960s, almost every US collection site had one or more pieces of HRB-developed equipment. At the height of the Cold War, we had people stationed at sites throughout Europe, Asia, and North Africa. By the time I left HRB in 1962, our small experiment at Cloudcroft had grown into an operation employing over 200 people at dozens of locations around the world.

Despite the lure of such challenging work for the intelligence community and other government clients, some of us envisioned our television business as a profitable and stable underpinning for HRB's future growth. When unable to convince the board of the wisdom of this approach Haller, Brown, and I bought the TV department from HRB in 1953 and set it up as a separate business entity called Community Engineering Corporation (CECO). I became president of CECO, but Walt Brown was the outfit's technical brains. Haller's main interest was as an investor. Our business plan depended on VHF amplifiers powered by reduced-voltage 60 Hertz AC power fed through coaxial TV cables. Community Engineering's technicians kept building and testing new products as others went into production. Much of the equipment was installed by SCTV as the operating arm of the company.

In the summer of 1955, tragedy struck both CECO and HRB when Walt Brown drowned in the surf off Redbank, New Jersey. Despite his sometimes mercurial oversight, we sorely missed him. We also lost George Haller that same summer when he went to work for General Electric in Syracuse as vice president for military systems. Before leaving, George had brought in a Navy retiree, Gus Detzer, as vice president for administration, and put Mahlon Robb, our banker, onto the board of directors. Haller and Raymond continued to be the largest HRB stockholders, along with Brown's widow.

Inevitably, the business side of managing the expanding company began to distract me from the joys of science and engineering. At the same time, running CECO in my spare time was getting to be a bit overwhelming. I hired a chief engineer to worry about the technical side of CECO so I could pay more attention to a recurring management headache: meeting the payroll. Over the next five years, HRB and CECO continued to grow, each in its own way. CECO's stock was still closely held, but we gradually took on more employees including several professionals.

In my first eight years with HRB, it expanded into a thriving enterprise with 220 full-time employees, having shown a profit every year. Not all of its corporate evolution was smooth or logical, though its strategies seemed smart at the time. To exploit the opportunities in new fields with more customers, HRB set up a manufacturing facility in State College and a lab in Solona Beach, near San Diego. It also opened offices in Washington, D.C., and Tucson, Arizona. The company became known among influential government officials as an innovator that could respond quickly with quality products at reasonable prices. How had we done it? One way was by keeping abreast of customer needs through both formal and informal contacts. As my responsibilities grew, I considered maintaining and expanding these contacts as one of my primary functions. There is also an old saying in business that "the best way to get a contract is to have a contract," and we had a head start in several technical areas. Of course, none of this could have come about if we had not been able to deliver on time and within budget. Most of the credit for this certainly must go to the superb qualifications and work ethic of the workers we were able to attract and retain. HRB was among the first companies of its type to embrace employee stock ownership (in 1953) and resist laying people off when contracts were slow in coming.

Despite HRB's successes, not all was going well. By 1956 we were having trouble raising enough working capital to sustain growth without giving up the control that went with stock ownership. Along came an offer from the West Coast. Elwood "Pete" Quesada, a retired and highly respected USAF general who was a friend of Pres. Dwight D. Eisenhower, came to visit HRB. He was associated with Topp Industries, an aggressive young Los Angeles company supplying military equipment (principally aircraft instruments) to the Air Force. Quesada apparently liked what he saw and got Topp's management interested in making a deal with HRB. This provoked some controversy within HRB's workforce, many of whom were reluctant for us to lose our independence. I recognized the pros and cons of selling the company, but I was deeply concerned about the drawbacks of our existing situation. To make a long story short, HRB became part of Topp Industries, with my colleagues and me staying on as managers of its new subsidiary. Most of us traded our shares in HRB for Topp stock, while Haller, Raymond, and Brown's widow converted their holdings to cash. Unfortunately, the merger did not provide as much access to capital markets as we had hoped. More disconcerting to us in management, the Topp people thought they should actually run HRB—something we didn't think they had the smarts to do. This was a prescription for bad feelings on both sides.

HRB's growing pains could not all be blamed on Topp. Internally, some fundamental differences in corporate philosophy had been emerging within the company. Besides being strapped for cash, HRB had begun to experience increasing tension between the engineering staff and the administrators. By the time of the Topp deal, the "green-eyeshade" types in the front office and on the board had adopted a policy that we engineers should conform to irksome rules governing a more prosaic production-oriented workplace. The front office executives didn't seem to comprehend that it was HRB's scientists and engineers, not them and the support staff, who landed the contracts and generated the ideas that kept the company going. I also had to defend the notion that HRB, as a local enterprise, owed something to the surrounding community. State College's chamber of commerce had offered me its presidency, but I turned this down rather than oblige myself to justify why this might be a positive circumstance. The same arguments prevailed over memberships in professional organizations, especially if they required even the most menial duties away from the company.

Despite my disappointment with HRB's changing corporate culture, life in my adopted hometown of State College had become a source of great satisfaction to my growing family and me. Having outgrown our first house, we moved into a larger place closer to the center of town. Later we had the good fortune to renovate a handsome farmhouse with a small pond on a 15-acre plot of land a few miles out of town. There we had an ideal view of Mount Nittany and an adjacent barn for the kids' horses and cats. By the time we acquired this estate, our family had grown to its ultimate size. In addition to Pam in 1947, we added Susan in 1949, John

C. in 1952, and Rod in 1955. Pat and I were active in the Parent-Teacher Association (PTA) and alternated presidencies when our kids moved from one school to another. Our little farm became a playground for our children and their friends, and we would often have dozens of them and their parents over for "swim days" in the summer or ice-skating parties in the winter.

Back at work, Topp Industries continued to be a disappointment. Its headquarters contained fewer seasoned managers than first thought, and we had to try to create our own solutions rather than wait for help from the West Coast. On the personal level, however, not all of Topp's changes were bad for me. In the fall of 1957, Bob Higdon left to become a vice president at Topp headquarters, I was promoted to be the third president of HRB, and my chief nemesis, Gus Detzer, was out of a job. Taking charge gave me welcome influence on HRB's management philosophy and alleviated many of my recent frustrations.

Before long it became apparent that Topp's corporate headquarters was now wrestling with cash flow problems of its own. Thus, in February 1958, we kissed Topp good-bye, and HRB became a subsidiary of Singer Sewing Machine of New York. Singer was cash rich and eager to diversify into defense technology. The move proved mutually beneficial. Singer provided the financial stability that would secure HRB's growth over the next 30 years. With competent management from above, I was able to devote more time to community activities. I finally had a chance to head the chamber of commerce, chair the library board, serve on the local planning commission, run the little theater group, and take part in many other civic activities, to include serving on or chairing advisory councils at Penn State's colleges of engineering and business administration. I also became a partner in a small real estate company.

In addition to these local activities, I joined a civic-minded national business group called the Young Presidents' Organization (YPO). The YPO had been founded in 1950 to serve as a forum where relatively youthful business leaders could network and share ideas with their peers.[8] My affiliation with YPO encouraged me to act more "presidential." I joined Toastmasters to help overcome my innate shyness and began accepting speaking engagements. Although never a great orator, this training was to prove valuable in my subsequent career. Perhaps one of

27

the most important lessons my YPO associates taught me was to have more confidence in my own instincts. Scientists and engineers can never have enough data, but the YPO philosophy was that it was better to make a decision with imperfect data than to make no decision at all. YPO also provided special opportunities for overseas travel. I took two YPO-sponsored trips: the first to the Soviet Union in 1959 and the second to South America in 1961. Both were real "eye-openers" for me.

A year before the Russian trip, potential legal problems had surfaced with implications for my cable television business. CECO, now operating in 104 municipalities in Pennsylvania, Ohio, and West Virginia, had expanded its manufacture of amplifiers and related devices. Without my knowledge, one of the HRB labs bought a needed piece of equipment from CECO. Advised of this by one of my erstwhile colleagues, Singer's management warned me that my status at CECO posed a conflict of interest. I quickly had to dump my stock and find a replacement to run CECO. I immediately thought of Jim Palmer, a talented project manager I had hired in 1954. I sold him all my stock for only $7,000—a price that was roughly its book value, since CECO (including SCTV) was still a struggling venture despite its expansion. Jim eventually sold SCTV to a company called TCI for $7 million, but he did himself proud by using part of this windfall to endow the Penn State Arts Center. He continued as president of CECO for the next quarter century, changing its name to C-COR in 1964 to avoid a trade name dispute with another company. C-COR continued to prosper, and I eventually rejoined the company in 1980 as a director.[9]

Even without CECO, my position at HRB-Singer and community activities kept my plate full enough. Because few in Singer's management could review our highly classified work, we were able to retain a large degree of autonomy and much of our State College–based company culture. We were providing a growing variety of intelligence services, some from a new facility in Reston, Virginia, including efforts to apply computers to such activities. Meanwhile, infrared imagery finally appeared poised to become an important tool for tactical reconnaissance, and HRB was one of the companies most active in developing such equipment. The company had annual sales of $10.5 million and 860 full-time employees. As 1962 began, I was reasonably proud of my accomplishments as president of HRB-Singer and very happy with our life in State College.

Notes

1. Under the chairmanship of Vannevar Bush, the NDRC mobilized many of the best scientific brains in the country to work on various technologies important to the war effort.

2. For the only book about this seaplane, see Richard A. Hoffman, *The Fighting Flying Boat: A History of the Martin PBM Mariner* (Annapolis: Naval Institute Press, 2004).

3. Being developed about the same time as GCA was the Instrument Landing System (ILS), which sent out beams in the sky that a pilot could follow without intervention by ground controllers and thus became the favored installation for civil aviation.

4. To augment memories, we have referred to several histories on the Okinawa invasion and the latter stages of the Pacific war, most notably Samuel Eliot Morison, *History of United States Naval Operations in World War II*, vol. 14, *Victory in the Pacific* (Boston: Little, Brown, and Company, 1962).

5. For helping confirm milestones in the life of the *Saint George*, we are indebted to the Naval Historical Center's *Dictionary of American Naval Fighting Ships*, vol. 6 (Washington, D.C.: Government Printing Office [GPO], 1976).

6. More details about the evolution of HRB can be found in its 50th anniversary publication, *The History of HRB: 1947–1997, 50 Years of Excellence*, ed. Edward R. Keller (State College, Pa.: HRB, April 1997). Keller worked for HRB and its successors from 1948 to 1994.

7. René J. Francillon, *Vietnam: The War in the Air* (New York: Crown Publishers, 1987), 93–94. As an interesting sidelight, coauthor Kenneth J. Alnwick piloted the first combat mission in Vietnam to conduct operational tests of the Reconofax camera in 1963. The large device, cooled by liquid nitrogen, was carried in the bomb bay of his RB-26 to take nighttime low-level images in an attempt to locate Vietcong cooking fires in denied areas.

8. Today the YPO is an international organization with more than 8,000 members in 75 nations. See http://www.ypo.org/.

9. TCI is now a unit of Dielectric Communications, a producer of high-power TV and FM broadcast systems, http://www.tcibr.com, May 2003. C-COR, which went public in 1981, was renamed "C-COR.Net" in 1998 and focuses on various broadband technologies, http://www.c-cor.net, May 2003. The HRB division of Singer later became HRB Systems, a wholly owned subsidiary of defense contractor E-Systems, which was in turn absorbed by the even larger Raytheon Corporation in 1995.

THIS PAGE INTENTIONALLY LEFT BLANK

Chapter 2

Career Broadening in Science and Technology

Although generally pleased with Singer, my long-term status in its HRB subsidiary was becoming less certain. The concept of Singer's management for my future career development was to move up to the headquarters in New York City. Although I understood their reasoning, the prospect of leaving our "happy valley" in Pennsylvania for the concrete canyons of Manhattan had little appeal to either my family or me. In January 1962, however, I suddenly was presented with an unexpected opportunity for an even more drastic career change, one that appealed to my sense of public service—but would still keep me in the field of science and technology.

Under the aura of its "New Frontier" image, the John F. Kennedy (JFK) administration was inspiring many well-educated and fairly youthful professionals to go to Washington. I had felt serious misgivings with the Eisenhower administration's emphasis on massive nuclear retaliation, so Kennedy's flexible response strategy of developing a wider range of military options seemed a more realistic way to meet the challenges the United States was facing around the world. When informed that Dr. Harold Brown, the Pentagon's director of Defense Research and Engineering (DDR&E),[1] was interested in considering me for a new Pentagon office called Tactical Warfare Programs (TWP), I was immediately intrigued.

Defense Management under McNamara

From previous endeavors I knew Dr. Eugene Fubini, who had become Brown's deputy director for research and information systems. I still remember Gene's answer when I asked if I would have enough authority in Washington's infamous bureaucracy to get much done: "John, in this job, you won't be authority-limited; you'll be wisdom-limited." Soon I was in Washington to meet Harold Brown. Besides impressing me as being extremely

31

intelligent (Brown had earned a PhD in physics at Columbia University when he was only 21), he convinced me why running his new office would be a significant enough challenge to take a government job at about half my current pay. So, despite our deep roots in State College, I decided to grab the brass ring and accept his offer. Of course, I first wanted to try to leave HRB-Singer in a rational way, easing its transition to new management, and that took some time. After obtaining a leave of absence from Singer, I began cleaning up affairs in State College and put our beloved farmhouse up for rent. I located a nice hillside house in Arlington, Virginia, a short drive south of the Pentagon, where Pat and the children would join me at end of the school year.

On Monday, 14 May 1962, I reported to my new office on the E-ring of the Pentagon's third floor (Room 3E1040). My desk was already covered with papers, and my new staff introduced me to the office routine. I quickly learned the significance of being at the upper echelons of the Office of the Secretary of Defense (OSD), where Harold Brown was the third-highest-ranking official. Among the items on my desk was an invitation to attend a reception in the White House the following evening. Pat hurried down from State College to accompany me. The reception made it abundantly clear that I wasn't in central Pennsylvania anymore.

During my first few months on the job, I remember feeling totally overwhelmed by all the things I was supposed to know. Since I had an electronics background, and the job dealt more with airplanes and weapons, I had to do a lot of learning in a hurry. Fortunately for me, Harold allowed some time before giving me full responsibility for major decisions. Indeed he urged me to get out of the building to look at operations in the field, visit contractor plants, attend firepower demonstrations, and even fly in the latest jet fighters. My formal title was "Deputy Director (Tactical Warfare Programs)." My immediate counterpart was Fred A. Payne, who headed the Strategic and Defensive Systems office. As Harold put it, "John, you deal with limited war, and Fred deals with unlimited wars." This sat well with me. I felt strategic warfare had become a "never-never land," while any real battles would be fought with tactical systems like the ones I was taking under my wing. They encompassed programs as different as the already controversial Tactical Fighter

32

Experimental (TFX) being pushed by Robert S. McNamara and the mobile medium-range ballistic missile (MMRBM) much desired by the North Atlantic Treaty Organization (NATO). Also on our agenda were smaller missile systems, aircraft capable of vertical and/or short takeoff and landing (V/STOL), a giant new cargo airplane, Army tanks, Navy destroyers, submarines, and various guns, bombs, and vehicles.

My office reflected the new thrust in weapons priorities and oversight under McNamara. Within the military-industrial complex (to use Eisenhower's famous term), there are always more ideas than resources. In view of this, we were charged with evaluating research and development (R&D) programs to make sure they were worthwhile and affordable. We also tried to foster promising areas of research for possible development in the future. In addition to technical issues, the DDR&E got involved in certain areas of management, such as how the military departments were organized for R&D, at what level specific programs should be supervised, and what types of incentives and overhead charges were reasonable for industrial and university contracts. But even though the dollar value of the DDR&E's programs averaged somewhere around $7.5 billion, our jobs were not unlike those of many other government bureaucrats. I was mainly a coordinator and overseer, without line responsibilities to "direct" any of the programs TWP carried in its portfolio.

Occasionally, on my way from one appointment to another, I would encounter Secretary McNamara striding down the hall with his head up and eyes focused straight ahead—apparently oblivious to everything around him, including John McLucas. In fact, I was somewhat disappointed to see so little of McNamara, who remained a rather remote figure to those of us at the middle levels of the OSD. Be that as it may, I think my colleagues and I sincerely believed in making McNamara's disciplined approach to R&D and procurement work. Ever since he arrived on the scene, there has been a certain rhythm to the Pentagon's calendar, driven by the Planning, Programming, and Budgeting System (PPBS). Prior to McNamara, the OSD basically allocated overall budgets to the services to spend as they saw fit. Instituted by Charles J. Hitch, DOD comptroller, the PPBS was at the heart of McNamara's reforms. It combined force structure and cost data to project foreseeable implications of decisions into the future (the "out-years" of the Five-Year Defense Plan or FYDP—a related innovation).[2]

Many of us in the OSD were relatively young, with solid academic credentials and records of success in chosen fields. Unfortunately, like Mr. McNamara, some of my colleagues tended to be a bit brash and imperious in dealing with the existing bureaucracy. In response, McNamara's detractors called his new cadre of Pentagon civilians "the Whiz Kids" (a name inspired by a popular quiz show of the 1940s and previously applied to McNamara's team of young veterans hired to modernize Ford Motor Company after World War II). Especially at first, a number of McNamara's appointees seemed to go out of their way to offend senior military officers, and we all paid a price for this.

Meetings and briefings punctuated work in the Pentagon. In addition to gatherings on specific topics, I normally attended two weekly meetings. The first was Harold Brown's staff meeting for key people in the DDR&E. The other was the secretary of defense's (SecDef) staff meeting, which Harold, I, and his other deputies would attend. Curiously, despite its name, Secretary McNamara did not routinely preside. Usually, Roswell L. Gilpatric, the deputy secretary, or John H. Rubel chaired the so-called SecDef staff meetings. (Although Rubel was formally listed as Brown's deputy, he was also designated as special assistant to the SecDef, and as such carried extra authority.) Decisions on major OSD issues were arrived at by an inner circle composed of McNamara, Gilpatric, Brown, Hitch, and key assistant secretaries such as Alain Enthoven (system analysis) and Paul H. Nitze (international security). Others included Sol Horowitz (director of organization and management), Cyrus R. Vance, and John T. McNaughton (successive general counsels), and—to a lesser extent—the three service secretaries. Across the board, McNamara probably relied most upon Hitch and Enthoven because they were the keepers of the flame for using analytical techniques in decision making. On technological issues, Harold Brown was his closest confidant, but in many of the interoffice battles, Hitch's troops took advantage of their better access to ensure their views prevailed.[3]

For the most part, we in the DDR&E tried to manage by exception, or even more effectively, by setting up general criteria whereby we could see if an overall approach made sense. We had special rules to deal with the practice of "concurrency"—the over-lapping of research, development, production, and testing that in the 1950s had successfully permitted the rapid albeit costly deployment of ballistic missiles.[4] In our realm of early R&D, concurrency meant going ahead with design and fabrication of a system before having confirmed the feasibility of all its components, so long as these subsystems used proven technologies.

Since World War II, use of "cost plus fixed fee" contracts had been DOD's preferred way of doing business. Under McNamara, the OSD and the services devised a new contracting philosophy, which incorporated many aspects of concurrency, called Total Package Procurement. Rather than negotiating contracts for successive phases of the acquisition process—such as advanced development, limited production, full production, and support—McNamara and his disciples hoped to both speed up the process and hold down costs of major systems by contracting up front for most or all of these activities. The new policy delegated wide responsibilities to prime contractors while centralizing oversight authority in the OSD. The selection of a contractor would rely largely on a detailed "paper competition" conducted early in the acquisition process, resulting in more fixed price or "cost plus incentive fee" contracts. As evidenced by the few major programs that tried Total Package Procurement, the practice invited unrealistically low bids to win the extended contracts. Although not yet understanding all the implications of this policy, I felt uneasy about the way it did away with the time-honored practice of building prototypes or conducting thorough tests before contracting for full production.

Other McNamara reforms proved more successful. Requiring thorough early documentation helped determine if there was a real need for a proposed system, if it was technically feasible, if reasonable trade-offs had been made between cost and performance, if the system duplicated something already being done, and if it would be compatible with other systems. While considered somewhat revolutionary at the time, one would be hard-pressed today to find a request for proposal (RFP) for a major new program that had not addressed these issues. As the impact of McNamara's reforms began to hit home, the services discovered that they had lost a large degree of latitude in what they considered their own prerogatives. Yet no one I talked to at that time questioned the right of the secretary of defense to make final policy decisions. Nevertheless, as McNamara himself later admitted, "We had to make sweeping changes to achieve these goals. It meant moving senior civilian officials much deeper into the management of defense programs. That made a lot of people, both in and out of uniform, uncomfortable."[5]

If asked today if I thought Robert S. McNamara did a good or bad job, I'd have to give a very mixed appraisal. In some respects, he merely implemented authorities that had been vested in the secretary by the Defense Reorganization Act of 1958. I'd give him credit for instilling a greater sense of discipline into the budget process and for quantifying factors that needed hard data for making informed decisions. He also helped standardize procedures and reduce some of the unnecessary parochialism among the services. As one minor but telling example that has endured to the present day, the OSD under McNamara devised a logical new system of alphanumeric designations to identify current and future weapon systems, replacing individual schemes long used by each service (such as McDonnell's Phantom II fighter being called the F4H by the Navy and the F-110 by the Air Force). On the other hand, I would fault McNamara for not paying nearly enough attention to the morale of those most affected by his changes. I think that deficiency cost the nation the services of a lot of valuable people who more or less retired in place when they found their knowledge and experience being ignored. With better salesmanship and more humility, McNamara could have been much less divisive. After all, it was a time of increasing defense budgets and new opportunities for the military.

Then Vietnam changed everything. Trying to deal with the conflict there posed challenges very different from administering the internal affairs of the Pentagon. Much later, McNamara sort of apologized for helping lead our "slide down a tragic and slippery slope," but too late to undo much harm to his reputation.[6] His role in the war still overshadows most other aspects of his time as secretary, but I am getting ahead of the story.

Within the OSD, I think the DDR&E was one of the better directorates, and Tactical Warfare Programs one of its better offices. TWP was made up of about 40 people in three categories: career civil servants, political appointees such as myself, and military officers. (In those days, most OSD personnel were males except for personal secretaries and clerical employees.) Besides providing continuity, many of the career civilians were experts in one or more areas of military technology and knowledgeable about the defense science and technology (S&T) infrastructure. In TWP we had about a half dozen military action officers, mostly majors and lieutenant colonels, who helped interface with the services while also providing expertise in selected programs. The most senior uniformed officer was my deputy, a one-star flag officer. Three men held that position during my tenure. The first was Air Force brigadier general John O'Neill, who was heavily involved with the NATO Air Defense Ground Environment program. He was followed by Army brigadier generals Robert York and William Beverly. These three men were tremendous assets to the smooth operation of the office, and I would work again with O'Neill later in the decade.

Promoting Innovations in Tactical Weapons

Each of the professionals in TWP had a portfolio of programs which, on average, represented about $40 million in R&D dollars. Their judgments were particularly important in drawing up the next year's program objective memorandum (POM)—a key document in the PPBS process. In some ways, we viewed ourselves as technology entrepreneurs. Two good examples of this were night-vision devices and counterinsurgency aircraft.

Shortly after Kennedy took office, the President's Scientific Advisory Committee (PSAC) chartered a study, led by Dr. Luis Alvarez, to look at technologies for dealing with insurgencies threatening friendly governments around the world. The committee concluded three technologies could have a significant impact: (1) night-vision devices, (2) small, simple aircraft capable of operating from unimproved airfields, and (3) a means of precision location.[7] Harold Brown asked one of my best people, Al Blackburn (known as "Blackie"), to review this study and see what might be done to support its recommendations.

The major existing night-vision program was built around the use of large infrared spotlights—an easily targeted technology dating back to World War II—so Blackie started digging deeper.

He found what he was looking for in an obscure office buried deep inside the Army staff. With a relatively small budget, this office had been sponsoring some promising research with low-light television technology. When Blackie asked the officer in charge if the program needed additional funding, he was told that another $250,000 would give it a needed boost. After getting a few other people, such as Gene Fubini, interested in the program, Blackie was able to get $1.5 million. The night-vision program took off from there, leading first to the famous "star-light scope" introduced in Vietnam and eventually to even more advanced devices, such as night-vision goggles.

Another program that got its start in similar fashion was a counterinsurgency (COIN) aircraft. Not enthusiastic about investing in new aircraft of relatively low performance, the services deemed refurbished T-28 Trojans, A-1Es, and other old planes as a suitable response to the COIN requirement. Blackie was unconvinced and soon found a kindred soul at China Lake, the Navy's innovative weapons test center in California, where a Marine lieutenant colonel named K. P. Rice was attempting to build his own experimental COIN aircraft with negligible support from higher headquarters. In 1963, thanks to the efforts of Blackie, a line item for a new COIN aircraft appeared in the Navy budget. This was soon followed by an RFP to build the Light Armed Reconnaissance Aircraft (LARA). The end result was the North American-Rockwell OV-10 Bronco, which made its maiden flight in 1965. Deployed to Vietnam three years later, the Bronco was adopted by the Air Force as a forward air control platform (marking targets for faster jet aircraft). The OV-10 has most recently gained notice for its role in the war on drugs in Colombia. Without the DDR&E's initiative and support, this cost-effective plane (built for less than $.5 million per unit) would most likely have never made it into the inventory.

Our office was also responsible for a variety of R&D initiatives for vertical or short takeoff and landing fighters and transports. V/STOL technology had begun to come into its own in the mid to late 1950s.[8] At that time, the Tri-Service Assault Transport Program was launched, spawning several innovative transport designs that we examined. (I was on a trip to the Cornell Aeronautical Labs to look at its work on tilt-rotor technology when I heard the traumatic news of President Kennedy's assassination.)

38

I felt we were homing in on a design that the USAF could support: the LTV-Hiller-Ryan XC-142A cargo and assault aircraft. With its four propellers churning, the XC-142 made its maiden flight in 1964. Five of the boxy-looking, tilt-wing aircraft were ultimately built. Ultimately, vibration, mechanical problems, and high pilot workloads during transition between vertical and horizontal flight doomed the program—some of the same ills that three decades later would plague the tilt-wing V-22 Osprey.

Given a lack of interest in developing V/STOL technology for jet fighters on the part of the Air Force and Navy R&D establishments, it is not surprising that the focus of V/STOL fighter development shifted to Europe. The British government was developing the most promising such aircraft: the Hawker P.1127 Kestrel, a vectored-thrust fighter that evolved into the Harrier. It cost us only about $1.5 million per year to support the Army's participation in Harrier development. I repeatedly resisted advice to back out by arguing this was the cheapest way to get some valuable R&D accomplished, even if at times I felt like a voice crying in the wilderness. Eventually, the V/STOL banner was picked up by the Marines as a way of meeting their requirement for organic close air support (CAS), resulting in the AV-8 Harrier. Today, the concept lives on with the V/STOL version of the Joint Strike Fighter.

It is not generally acknowledged that Secretary McNamara and the DDR&E played an important role in the restructuring of Army aviation. After the Korean War, the Army saw the Air Force's fighter development community—focused on supersonic interceptors and nuclear strike aircraft—turn its back on close air support. As a result, the Army began to look seriously at performing CAS and meeting other air support requirements, such as battlefield airlift, with its own resources. Two of the more influential Army officers promoting this concept were members of the DDR&E: Brig Gen Robert Williams and Col Edwin "Spec" Powell, the latter in my office.[9] Working behind the scenes, these two advocates got Brown, McNamara, and other key players in the OSD to take a close look at where the Army was going with its aviation programs. Pressure from McNamara helped lead to the creation of the US Army Tactical Requirements Board, better known as the "Howze Board" after its president, Lt Gen Hamilton H. Howze. The work of this group, which evaluated operational concepts for air mobility, became the engine for sweeping changes in Army aviation (and a point of some contention with the Air Force).

I followed these developments closely. Nominally, Spec Powell worked for Sam Perry in my Combat Systems Division, but he operated fairly independently while keeping Bob York and me in the loop. Eventually we lost him to Fort Benning, Georgia, as the Army Staff's representative for the field trials of a full-strength air-assault division modeled on the findings of the Howze Board. Not long after, an airmobile division organized around these precepts was sent into combat in Vietnam.[10]

Although we didn't fully realize it at the time, our work on improving Army aviation and counterinsurgency capabilities helped pave the way for a growing involvement by the DDR&E in the Vietnam War. In some respects, Vietnam was becoming a giant warfare laboratory. In fact, the Advanced Research Projects Agency (ARPA—then an element of the DDR&E) established two combat development and test centers (CDTC) in Southeast Asia. Their mission was to improve host nations' counterinsurgency capabilities while offering a way to conduct field trials of new American equipment, such as the Army's Caribou light cargo plane. In early 1963, I observed the CDTCs in action while on an extended visit to Thailand and Vietnam. I also had the opportunity to meet briefly in Hue, South Vietnam, with South Vietnamese president Ngo Dinh Diem and members of his family. My general impression was that the overall American effort in Vietnam was reasonably well coordinated—but that solving the basic problems of the country would take several years before showing any real progress.

My Role in the Controversial TFX Program

The Tactical Fighter Experimental embodied several of McNamara's reform objectives. He wanted any expensive new weapons to surpass the capabilities of existing systems by a significant margin and thereby avoid early obsolescence. On the other side of the coin, he was not enamored with buying the best possible technology but only what was needed to satisfy stated requirements. Finally, he preferred systems that could be shared by two or more services. The TFX seemed to meet these criteria. The USAF was looking for a long-range attack aircraft capable of Mach 2.5 at high altitude and a low-level dash capability of Mach 1.2, while the Navy required a two-seat, carrier-based fleet defense fighter that could protect carrier battle groups at a much greater distance than existing aircraft. In February 1961 McNamara directed the Air Force and Navy to study development of a single aircraft to satisfy both of their needs. He even thought that such an aircraft might meet the Army's and Marine Corps' requirements for CAS aircraft. The TFX's promised combination of combat radius, loiter time, payload, low-altitude speed, and advanced avionics would make it markedly superior to any existing tactical aircraft.[11]

When I arrived on the scene, I was led to believe that the TFX issue was pretty well settled. Sensibly, the Army and Marine Corps had been allowed to opt out of the program, while the Air Force and the Navy appeared to be working together to achieve McNamara's goal of at least 65 percent commonality in the core components of their two versions. Boeing and General Dynamics (GD) (in partnership with Grumman) were locked in a tight competition. Boeing's approach was an aircraft (or rather two aircraft with substantial commonality) that would feature titanium structures, thrust reversers, and top-mounted inlet ducts. GD proposed an aircraft with greater commonality and more use of proven technologies. With many billions of dollars at stake, both sides lobbied hard for their candidates.

In November 1962 the source selection committee, with Air Force and Navy members, found both companies' proposals acceptable but voted unanimously that the Boeing proposal had "a clear and substantial advantage." Nevertheless, McNamara and his closest associates in this endeavor—Air Force secretary Eugene Zuckert, Navy secretary Fred Korth, Harold Brown, and perhaps Charlie Hitch and Alain Enthoven—decided to overrule the committee's choice and award the contract to GD. They cited greater commonality, less technical risk, and more realistic cost and schedule estimates as major factors in their decision. The fact that Boeing had not built a fighter aircraft since the 1930s may have been a contributing factor.

With the exception of Secretary Zuckert, the decision did not go down well with the Air Force's senior leadership, some of whom had been members of the source selection committee.[12] Gen Bernard Schriever, commander of Air Force Systems Command (AFSC), was among them. One of the alleged drawbacks of the Boeing concept was its use of titanium to reduce weight. Boeing's detractors claimed that industry didn't know how to work with titanium, which meant too much technological risk. This led to a phone call I received from General Schriever shortly after McNamara's decision was announced. He said something to the effect of, "You know, John, there is a big travesty going on. We are being asked to go with General Dynamics on the TFX. Boeing's airplane is obviously much better, but we can't seem to sell anybody over there on it. Now that you've come on the job, I'd like to have you look at another new airplane." When I asked for some details, he said, "I can't talk about it on the phone, but if you're at Andrews [AFB, Maryland,] tomorrow morning at 0730, I'll fly you to a place where you can see it."

Intrigued, I went out to Andrews, and in several hours we landed at a base out west. There in a hangar sat an almost unbelievably futuristic airplane built mostly of titanium in great secrecy by the Lockheed Skunk Works. I was duly impressed. It was an A-12 Oxcart, which would soon lead to the celebrated SR-71 Blackbird. I still don't know whether the argument about not being able to machine titanium was a key factor in the TFX selection or if it got invented later. As Zuckert claimed, using titanium in the complicated swing-wing mechanism might have been problematic at that time. But after learning about the A-12, I was glad that I did not have to go before Congress and defend the TFX decision on that basis.

The TFX became the highest visibility project in the Tactical Warfare Programs office. While others handled the politics of the decision and resulting hearings on Capitol Hill, Al Blackburn and I worked with the Air Force, Navy, and the contractors to try to set up a management structure and a myriad of other details needed to keep the program moving forward. Despite my reservations about the source selection process, I knew it was too late to turn back. In view of the controversy, I got Brown, Hitch, Enthoven, and several others to take an oath (metaphorically speaking) that they would not try to scuttle the program. Eventually the Air Force people involved and, to a lesser extent, their Navy counterparts, seemed to acknowledge support for the new aircraft, but articles critical of the TFX continued to appear in the press with disturbing regularity. Even so, Australia joined the program in 1963 by signing up to purchase 24 aircraft "off the drawing board." So, for the rest of my tenure, I did what I could to ensure that it would be a successful procurement. The first USAF version of the TFX aircraft (the F-111A) rolled out of GD's Fort Worth, Texas, plant on 15 October 1964.

There are some who still believe that the procurement decision was driven purely by unethical political considerations at the highest levels of government. Be that as it may, the TFX decision clearly signaled that the Kennedy administration was serious about asserting civilian control over major defense programs. It also reaffirmed McNamara's principle that any new system should be a significant departure from its predecessors—and the swing-wing F-111 certainly was that—yet not push the envelope too far. Although the Navy later opted out of the F-111B program, they did buy another swing-wing airplane with Grumman's F-14 Tomcat. Significantly, the Soviets copied its variable geometry with their MiG-23 Flogger. Now, 40 years later, the desire for commonality that gave birth to the TFX is being played out with the Joint Strike Fighter (JSF). This time, however, everyone seemed to be on board from the start. Does this to some degree vindicate McNamara's philosophy? The circumstances are very different, as is the diversity of JSF variants, but there is a striking similarity between the arguments used then to justify the TFX program and what we hear now in praise of the JSF.

Progress and Problems with Other Systems

While the TFX continued to dominate the news, I was trying to get the Navy's light attack aircraft (LAV)—needed to replace the A-4 Skyhawk—started as soon as possible. It was supposed to be an offshoot of an existing airplane so that the Navy wouldn't have to absorb another completely new design. Rumors began to spread that we also wanted to interest the Air Force in the LAV. In reality, my position was to get the new program firmly under contract before talking about possible Air Force interest. On the other hand, USAF officials rightly believed that if they would later have to buy the airplane, they probably ought to have a hand in deciding what it looks like. Because of timing, however, I could not agree to any changes that might compromise the program for its prime customer. In February 1964 Ling Temco Vought (LTV) Aerospace Corporation won the LAV competition. Designated the A-7 Corsair II, it was a subsonic derivative of Vought's F-8 Crusader. Shortly after Harold Brown became secretary of the Air Force in late 1965, he decided to buy a modified version of the A-7 as the Air Force's first specialized ground-attack aircraft in a generation.

I also devoted considerable time to the CX-4 heavy lift transport, intended to replace the big but obsolete C-133 Cargomaster and complement the new but smaller jet-powered C-141 Starlifter. The Air Staff was of the opinion that the CX-4 should be a relatively modest upgrading of current technology. At Systems Command, however, Bennie Schriever wanted to make this new program (now called the CX-X) a real step forward. He argued that it would be 20 years before we would have another such opportunity. We finally got Brown, Hitch, and Enthoven all on the same wavelength, agreeing that a really large airplane using advanced engines would provide a significant increase in efficiency (e.g., cost per ton-mile), better meet the Army's requirement for carrying armored vehicles and bulky cargo, and also offer commercial possibilities. To this end, I had been putting money into product improvement programs for more powerful and efficient high-bypass turbofan engines. RFPs for the CX-HLS (Heavy Logistics System—its latest name) went out to industry in April 1964. Among the responses were the wide-body designs that would become the Lockheed C-5A Galaxy and the Boeing 747. Unfortunately, the C-5A program would serve as the main test case for Total Package Procurement, and as with the F-111, would come back to haunt me in the future.

I think my time in TWP saw considerable progress on many nonaviation systems. The Army was well along in developing Lance, a class of surface-to-surface missile accurate enough to be useful without nuclear warheads. All the services were increasing the effectiveness of conventional ordnance by a variety of means, principally in fragmentation techniques, and had programs under way for improved air-to-ground weapons, including Shrike, Walleye, Bullpup, and Condor. New antitank weapons such as the tube launched, optically tracked, wire guided (TOW) and Shillelagh missiles were also moving along, while the man-portable Redeye antiaircraft weapon was ready for production. Airborne electronics for navigation and weapon release were being greatly improved, and new technologies for conventional ordnance delivery had some promise for eventually approaching the all-weather capability that had long eluded airpower.

The most ambitious of the missile programs was the Navy's Typhon, a system designed to meet the threat of mass attacks by Soviet antiship missiles. It featured long- and medium-range ramjet-powered missiles capable of flying at well over 5,000 mph and controlled by a revolutionary phased-array radar for engaging multiple targets. Unfortunately, these requirements were too far ahead of the available technology, and even if the serious technical problems we were learning about could be overcome, I came to believe that the cost and difficulty of retrofitting existing ships with the new system might be prohibitive. Based on these factors, McNamara canceled the program in January 1964, but some of the lessons learned from the Typhon eventually contributed to development of the highly sophisticated Aegis air defense system.

An Interlude with NATO in Paris

In early 1964, as the end of my planned two-year stint in the DDR&E began to approach, I was told that Singer now had no intention of letting me return to State College or of cashing in my stock options.[13] I therefore became more interested in other opportunities in both the defense industry and the Navy, where Paul Nitze, who was now secretary, asked me about being his assistant for R&D. When this was delayed, I chose to leave the Pentagon for Paris and become NATO's assistant secretary general for scientific affairs—a move abetted by Harold Brown. The die was cast when the NATO Council confirmed my nomination at a meeting in April.

My family and I left for France on 3 July 1964 aboard the luxury liner SS *United States*—a mode of transport rapidly being made obsolete by the first generation of jet airliners. On this voyage I was fortunate to become friends with Air Force brigadier general Russell E. Dougherty, who was on his way to become deputy director for plans and operations (J-3) in the US European Command (EUCOM). The Doughertys had two teenage boys about the same age as our girls, and we all got along famously. Russ and I would become colleagues when I returned to the Pentagon and have remained close friends ever since.[14] Upon

45

landing in France, my eldest daughter Pam served as our inter-
preter. After six weeks in a nice Parisian hotel, we moved into an
apartment spacious enough to house all six of us comfortably
and still have room for guests. The three younger kids learned
French by attending local schools, while I had a charming private tutor courtesy of the American Embassy.

In my NATO position, I wore three hats: principal advisor to
Secretary General Dirk U. Stikker on scientific and technical
matters, chairman of the Science Committee, and chairman
of the newly approved Defense Research Directors' Committee
(DRDC). I was classified as a State Department science officer,
Foreign Service Reserve, level one (FSR-1), earning a base salary
of $19,600, which was later raised to $26,000. My predecessors had been distinguished scientists with limited managerial
skills, so I was the first technocrat to run the Scientific Affairs
Division. Its professional positions were filled by civil servants
nominated by their respective governments. My immediate staff
consisted of a deputy, five or six senior scientists, several administrative people and personal secretaries, and an executive-
type secretary to help with the committee work. The division
administered three major activities: the Science Fellowships
Program, which provided funds for several hundred young scientists to conduct research and study in other countries; the
Advanced Study Institutes Program, which conducted a series
of high-level symposia on selected scientific topics; and the
Research Grant Program, which funded collaborative international research and a few national projects of special interest.[15]
This research normally involved a field of physics, electronics,
or chemistry, and tended to lean more in the direction of basic
science than military technology.

NATO had created the Science Committee in 1957 to provide
a forum in which interested member nations could be represented by a distinguished scientist. The new DRDC was intended to have high-ranking national defense officials focus
more on military requirements. It was constituted as a formal
NATO body upon my arrival, with Harold Brown having been
its chief proponent. He and Secretary General Stikker wanted
the DRDC to sponsor research that could mature into truly
cooperative weapon system development programs. I convened
its first meeting in October 1964. Thereafter the directors met

each year in April and October, while their deputies met in January and June. The NATO Secretariat had given the DRDC very general terms of reference, leaving the members to map out a more detailed charter. Unfortunately, I discovered that they held widely differing views as to exactly what kinds of R&D the DRDC should sponsor and how it fit into the overall NATO planning process. In view of entrenched national interests, these issues remained largely unresolved.

I soon realized that my existing staff would not be able to handle the additional responsibilities that came with the DRDC, so I began seeking more help. The first of my new people came from an unexpected source: the Advisory Group for Aeronautical Research and Development (AGARD). Also headquartered in Paris, AGARD was roughly modeled on the USAF Scientific Advisory Board. Both were creations of Theodore von Kármán, the Hungarian-born scientist who helped infuse an emphasis on science and technology into the US Air Force after World War II.[16] Von Kármán's right-hand man was Dr. Frank Wattendorf, who became AGARD's first director. When the time finally arrived for Frank to move on, I was fortunate to acquire him as my special assistant. He stayed with me for about eight months, helping launch the DRDC. Thanks to Harold Brown, I then obtained the services of another AGARD official, Air Force colonel George Munroe. In July 1965, we were joined by Spec Powell, the highly competent Army officer who had worked for me in the DDR&E. He subsequently made important contributions in getting NATO to adopt satellite communications and to support what later became the Roland short-range air defense system.

By the time Spec arrived, my confidence in the viability of the DRDC was waning. The concept that prompted its formation—fostering cooperative development and procurement programs under NATO auspices—appeared to have been too optimistic, or at least premature. The status of my new committee was typical of many NATO endeavors. Like some other international bodies, NATO existed as much for diplomatic reasons as to accomplish anything concrete. Its 15 members had to reach a consensus before it could make a decision. The alliance was, of course, first and foremost meant to be a deterrent to Soviet aggression. Yet, it had other less tangible attributes. There was a lot of truth in the cynical saying that NATO's true purpose as regards to Europe was to keep the Russians out, the Americans in, and the Germans divided. It also provided a polite rationale for maintaining large allied forces in West Germany long after the occupation. Another diplomatic benefit of NATO was to keep the Greeks and Turks from fighting each other too violently over issues like the Aegean Sea and Cyprus.

I had the misfortune to be with NATO when it faced its worst internal crisis up to that time: Charles de Gaulle's decision in 1965 to withdraw French military forces from the NATO command structure, close allied facilities in France, and supposedly rely on its independent *force de frappe* for nuclear deterrence. Ostensibly, this was France's response to what its leaders interpreted as a decoupling of US security concerns from those of Europe under the guise of flexible response.[17] (I'm just glad de Gaulle didn't force NATO headquarters to move to Brussels while I was there.) The disruption that France's action imposed on NATO was a cause for alarm among policy makers and military planners. In the hallways and conference rooms of NATO headquarters, de Gaulle's decision created an almost intolerable situation for me. I was forced to remain polite to the French representatives, even as they worked to undermine our efforts to create a more rational R&D system for the alliance.

In the excitement of our early days in Paris, I had entertained the notion of staying on for four years. After Dr. Stikker resigned as secretary general (because of ill health), support for what we were trying to accomplish with the DRDC began to fade. He was replaced by Manlio Brosio, an Italian diplomat of the old school. Brosio was a pleasure to work with (and had a nephew who dated my daughter Pam), but it eventually became clear that my goals for the DRDC were no longer realistic. Thus, not long into my second year, I decided that two years out of the American R&D community would be long enough.

In the fall of 1965, I began renewing contacts in some US defense companies. I also had a couple of visitors who presented me with another job opportunity. One was Maj Gen Jack O'Neill, a colleague from my early days at the DDR&E, who was now head of the Air Force Electronics Systems Division (ESD) at Hanscom AFB, Massachusetts, near Boston. Another was the renowned scientist, Dr. James R. Killian, who was chairman of the Corporation of the Massachusetts Institute of Technology (MIT) and a trustee of the MITRE Corporation, a federal contract research center (FCRC), whose main customer was the Air Force. MITRE's president was reaching retirement age, and both General O'Neill and Dr. Killian were looking for a replacement. General Schriever later phoned to also tell me I ought to consider this opening.

On my next visit to Washington, I stopped by to see Harold Brown, who told me I'd make a good choice for the MITRE job. As regards my questions about working in an FCRC like MITRE as compared to the government or a private defense company, he said words to the effect that "the in-house laboratories [manned by military and civil service people] are like loyal but rather dowdy old wives. The nonprofit research centers are more like longtime mistresses, and the defense contractors that you claim you want to go to are more like expensive whores, who will do whatever someone pays them to do." Harold followed up with a letter in late November 1965 stating, "I would have no cause for concern about MITRE's prospects given the strong public-minded leadership that I know you would provide."[18] It was becoming clear that I was part of an orchestrated campaign to provide new leadership at MITRE. Notwithstanding all this encouragement, I also took Pat to Los Angeles to follow up on previous offers from Dr. Allen Puckett at Hughes Aircraft and Donald Douglas Jr. at Douglas Aircraft. Despite their hospitality and enticements, she and I then voted for returning to Boston.

Except for my frustrations with the DRDC and de Gaulle's government, our tour in Paris had been quite pleasant and very educational for the entire family. A few years after I left, the DRDC was disbanded as such and merged with an organization called the Defense Research Group under the Conference of National Armaments Directors. The old Science Committee still exists and now reports to the assistant secretary general for scientific and environmental affairs.[19] Oddly enough, with the strong support of the French government, AGARD remained a fixture in Paris long after most components of NATO headquarters moved to Belgium. In 1997 it was absorbed by the new NATO Research and Technology Organization.

Closing a Circle at the MITRE Corporation

When I arrived at MITRE on 1 July 1966, my journey as a technocrat had come full circle. Twenty years earlier—two

49

years after Navy radar training at MIT—I had started my professional career as a junior engineer at the nearby Cambridge Field Station developing analog radar systems. I was now returning to the Boston area to run an organization that owed its very existence to some pioneering work by my old employer and MIT's Digital Computer Laboratory. This time I would be entering the new realm of digital electronics and data processing, which MITRE and its forebears in MIT's Lincoln Laboratory had pioneered with development of the Semi-Automatic Ground Environment (SAGE) air defense network for the Air Force. MITRE's technical director, Robert R. Everett, was the most prominent of the 450 employees absorbed from Lincoln Lab, while its first president, Clair W. "Hap" Halligan, had been brought in from Bell Labs.[20] Seven years later, these were the two gentlemen who greeted me at the door of MITRE's headquarters in Bedford, Massachusetts, on the outskirts of Boston, for my first day on the job.

From the very beginning, I knew I was going to like working there. When members of its board of trustees first interviewed me, we had agreed on my position's responsibilities and salary ($60,000 a year). MIT's vice president for finance even advanced me $20,000, which allowed me to make the down payment on a lovely five-bedroom home on the outskirts of Concord, Massachusetts. It looked out on a small lake, which reminded us of our farmhouse in Pennsylvania with its pond in the backyard.

The fundamental product MITRE delivered to its clients was systems engineering, to include defining requirements, conducting design and engineering studies, providing technical support, and assisting with tests and evaluations. These tasks were being applied to a category of emerging military programs that transcended individual weapons, equipment, or facilities. Loosely identified as command and control (C^2)—or "L-systems" in Air Force parlance—these systems tended to be extensive and diverse, involving multiple functions performed by numerous people. They not only broke new ground in several technologies but also in the field of engineering known as human factors.[21] By virtue of its work on SAGE and related programs, MITRE was at the forefront of these new developments, although MIT and its Lincoln Lab would continue to perform some of the more interesting and innovative work. The growing importance of C^2 and related areas, such as communications and radars, had also led to establishment at Hanscom AFB of AFSC's Electronic Systems Division.[22] Concurrently, MITRE became its principal source of technical advice and engineering support. MITRE's relationship with ESD was both synergistic and symbiotic. MITRE also worked closely with the defense contractors who ultimately competed for and built the systems the ESD/MITRE team initiated.

Being president and CEO of a federal contract research center posed a set of issues not found in either government or industry. The FCRCs' special strengths lay in their relative independence from political influence or profit motivations, their flexibility, and their body of specialized knowledge.[23] Yet, it sometimes seemed that we were everybody's whipping boy. Congress was unhappy with FCRCs' privileged status and was capping their share of defense budgets. The General Accounting Office (GAO) was investigating our business practices and questioning our right to collect fees and acquire property. Many in the defense industry were unhappy because they saw us as unfair competition. In the case of MITRE, the Air Force was unhappy because they were not getting as much help as they thought they needed.

By the time I arrived, MITRE had evolved from a company with essentially a single focus (the SAGE program) and client (the USAF) to an organization with several other customers. During most of my tenure, we operated under two annual revenue limitations: a $27 million Air Force ceiling within an OSD ceiling of about $32.5 million. Because of inflation, these caps threatened us with negative growth and having to do more with less. I thought our best solution was to seek more revenue from agencies unaffected by these caps.[24] This diversification strategy was either tolerated or frowned upon by our Air Force masters, depending on their confidence in our ability to remain dedicated primarily to their interests. For example, Gen James Ferguson, commander of AFSC, concerned that "diversification would dilute the quality of Air Force support," warned me that "all work for DoD and non-defense agencies must have a close relationship to MITRE's basic work for the Air Force." He cautioned that if other activities exceeded approximately 25 percent of total funding, "the Air Force would have to reconsider the basis for MITRE's FCRC status."[25]

With this sword hanging over our heads, I spent much of my time obtaining as many new customers as possible while trying to keep our Air Force sponsors happy with, or at least tolerant of, these efforts. In addition to work for existing clients such as the Federal Aviation Administration (FAA) and the National Aeronautics and Space Administration (NASA), we examined command and control systems for the recently created Arms Control and Disarmament Agency, did postattack planning for the Office of Civil Defense, and even branched out to examine data processing issues for the Department of Health, Education, and Welfare and the Internal Revenue Service. We also supported air quality studies for the National Weather Service and a precursor of the Environmental Protection Agency. On my own initiative, we looked for projects that yielded benefits to society at large and even helped various local government institutions in New England. I was encouraged indirectly in these efforts by the new secretary of defense, Clark Clifford, who urged DOD and, by extension, its contractors, to "do more for the common good." I also felt strongly that engineers had an innate responsibility to use some of their skills for the betterment of society and the environment, expounding on this philosophy in a speech to the Institute of Electrical and Electronics Engineers (IEEE) in 1967.[26]

When I took over at MITRE for what I said would be a five-year commitment, I knew I was facing some major internal challenges. Just prior to my arrival, General Schriever signed a report that characterized MITRE as a solid company but one that was "unaggressive and lacking in initiative."[27] It was still doing good work on many projects, but these were generally evolutionary in nature, as opposed to the more dynamic missile and space programs supported by Aerospace Corporation, MITRE's counterpart in California.[28] My marching orders from the board of trustees were to "bolster the military's confidence in MITRE and to broaden the company's base of operations."[29]

My initial impression was that MITRE, as a technical organization, was okay, but, as a business enterprise, it needed better marketing with the upper echelons of the Air Force and other clients. I therefore encouraged the senior staff to get out and talk more with people whose portfolios coincided with our interests and to participate more actively in advisory and committee work beyond the company's immediate projects. As for me, I agreed to serve on OSD's Defense Science Board (DSB) as a member of its Tactical Aircraft Committee and headed a DSB study on infiltration and interdiction (a hot issue in Southeast Asia). I also served on the Air Force Scientific Advisory Board (SAB) and the Defense Intelligence Agency (DIA) Advisory Committee. This approach, which everyone nowadays calls *networking*, had worked well at HRB, and I saw no reason why it would not serve us well at MITRE. On the civic side, I renewed my membership with the Young Presidents' Organization and did a lot of volunteer work in the Boston area, especially with a black-owned bank and small businesses.

Although I made some personnel changes at MITRE, Bob Everett remained as the vice president for technical operations, with John Jacobs as his assistant. Some of the key players at the division level were Charles A. "CAZ" Zraket (Defense Communications Agency and Air Traffic Systems), David Israel (Systems Engineering), and F. Robert Naka (Applied Science Laboratories). Their three divisions were responsible for about 60–70 percent of our revenue. I did not get to know Bob Naka as well as I would have liked until a few years later. For most of my time at MITRE, he was out in the West Coast leading an important project to improve the surveillance of objects in space. Everett and Jacobs spent most of their time on internal operations, while I concentrated on external affairs and relationships with our controlling body, the MITRE board of trustees and its executive committee.

By and large, the board comprised the same group that had established MITRE a decade earlier and needed some changes. Probably the most significant change occurred in July 1967, when none other than James Killian, a trustee since 1960, stepped in as chairman. Dr. Killian, who was a key figure in the mobilization of American science and technology during World War II, had overseen the creation of MITRE when he was president of MIT. As Eisenhower's trusted special assistant for science and technology, Killian formed the President's Scientific Advisory Committee, which was instrumental in initiating post-Sputnik reforms in science and mathematics education and in establishing NASA. Since 1959 he had served as chairman of the MIT Corporation. By also assuming chairmanship of

MITRE's board, Killian demonstrated the importance he placed in the future of the company.[30]

One month after becoming chairman, Killian named Henry Loomis and Courtland D. Perkins to the board. I had strongly recommended Dr. Perkins, who was chairman of the aeronautical engineering department at Princeton. He had served as Air Force chief scientist in the mid-1950s and its assistant secretary for R&D in 1960. On the other hand, I was sorry to see the departure in September 1967 of Dr. Luis Alvarez, who had been one of the board's most distinguished members.[31] In September 1968 five more members of the old guard were replaced by new faces. One of those who left was former secretary of the Air Force James H. Douglas Jr., who I thought was a good man. According to MITRE's first official history, "Probably the most effective instrument for greater board involvement in the conduct of MITRE's work was the formation of the Technical Advisory Committee at the board meeting of September 1967."[32] I strongly supported the formation of this group, the purpose of which was to evaluate the company's technical work and provide an interface with middle management. During my time at MITRE, I think the board of trustees became a highly effective oversight body that compares favorably to the other boards with which I later become associated.

MITRE Goes to War—Igloo White and Other Projects

Many of my frequent trips to Washington were both a cause and a consequence of our expanding roles in support of the war in Vietnam. MITRE first focused on an air defense system code-named Combat Lightning, which encompassed a network of ground-based radars, Air Force and Navy airborne early warning (AEW) aircraft, and sea-based radars on Navy ships—all linked to command centers at DaNang Air Base, South Vietnam, and Udorn Royal Thai Air Force Base (RTAFB) in Thailand. Combat Lightning was superseded in 1967 by a new program, Seek Data II, which benefited from work being done by our Systems Design Laboratory to automate many air control functions. We also helped with mobile radars and communications equipment of the gradually maturing 407-L Tactical Air Control System (TACS).

We began our most extensive project in Southeast Asia not long after my arrival. In mid-October 1966, Dr. John S. "Johnny" Foster, who had taken Harold Brown's former job as director of research and engineering, called to tell me that DOD was working on an urgent and highly classified project under an organization euphemistically called the Defense Communications Planning Group (DCPG). He explained it was a systems engineering type of project, so that's why he thought of MITRE. He asked if I had anyone who could quickly come to Washington and help run this project. I thought about it as we talked and said we had such a guy. His name was Dave Israel, the technical director of our Systems Engineering Division. Dave went down to the Pentagon to meet Johnny first thing the next morning. When Israel arrived at Foster's office, Johnny assumed that Dave was prepared to start work immediately, and that's what happened. He was given the job of pulling together all the DCPG's diverse technologies to provide an integrated picture of the infiltration of troops and supplies from North Vietnam, either across the demilitarized zone (DMZ) or down the Ho Chi Minh Trail into Cambodia and South Vietnam. The trail (actually many roads and paths) included long stretches through the Laotian mountains. The DCPG's concept relied on sophisticated sensors to detect the passage of troops and vehicles for interdiction, most likely by air strikes.

The barrier idea had been around since the early 1960s, but nothing much came of it until early 1966, when McNamara referred the concept to the Jason Committee, an influential science policy group under the auspices of the Institute of Defense Analyses (IDA). The scheme that resulted would be unofficially dubbed "the McNamara Line."[33] Lt Gen Alfred Dodd Starbird, US Army, headed the DCPG. His primary job was commander of the Defense Communications Agency (DCA), which was sponsoring MITRE's work on the National Military Command Center. It was General Starbird's desire for a systems engineering expert that resulted in Israel's sudden move to Washington. Dave became the DCPG deputy for engineering, and MITRE never did get him back. We also picked Jack Dominitz to lead a growing MITRE team supporting the DCPG, most of which was located at the Naval Observatory.

Meanwhile in Southeast Asia, the Army and Marines installed some sensors by hand, while the Navy and Air Force dropped others from the air to embed themselves in the ground or hang from the rain-forest canopy. Managing the collected data on the Ho Chi Minh Trail became the task of a dual IBM 360 computer that I later visited at Nakhon Phanom (NKP) RTAFB. The anti-infiltration system went by a number of code names, with Igloo White being the best known within the Air Force. A special unit at NKP, Task Force Alpha, served as the nerve center for the sensor network and related interdiction operations, which went by the code name Commando Hunt.

The Electronic Systems Division was given the job of procuring and managing the installation of facilities and equipment in Southeast Asia, with MITRE providing technical support. At the height of the effort, we had two teams working on what we called simply "the project": the 50-person Washington-based group under Jack Dominitz and another 50 at Bedford to support Task Force Alpha in the operation of Igloo White and to handle other related projects. Both rotated people to the war zone. We also set up a permanent office in Bangkok to manage all our activities in-theater and assigned some of our best people there (to the detriment of certain other work). We paid these people well and expected much in return. Some of them, such as Victor De-Marines (who would become president and CEO of MITRE from 1996 to 2000) stayed there for several years and provided much needed continuity and expertise.

The antipersonnel barrier along the DMZ was never completed, even though sensors diverted to the defense of Khe Sanh in early 1968 proved their effectiveness in helping lift the North Vietnamese siege. Igloo White, however, was fully operational from 1968 through 1972. At NKP a SAGE-like structure served as the hub of a dazzling array of technologies. This facility, the Infiltration Surveillance Center, was reputed to be the largest building in Southeast Asia. Spread along the Ho Chi Minh Trail were seismic and acoustic detectors of a dozen different makes and models. Beech QU-22 radio relay aircraft (code-named Pave Eagle) picked up the weak sensor signals, which were relayed through orbiting EC-121 Warning Star aircraft to NKP. Cued by this network were a variety of strike aircraft, from loitering propeller-driven gunships to "fast movers" (jet fighter-bombers), all operating under the watchful eyes of airborne command posts. This vast and highly classified effort, which probably cost at least a billion dollars per year, was supported by an amalgam of contractors, DOD civilians, and military personnel.

By the end of 1967, a quarter of MITRE's annual effort was devoted to the war in Southeast Asia, and this continued for several years. MITRE gained valuable knowledge of field operations and was able to observe firsthand how systems performed under actual combat conditions. This helped us accelerate development of the 407-L TACS, which was eventually deployed in Europe, and also in refining aspects of what became the Airborne Warning and Control System (AWACS) through our work with Combat Lightning and associated AEW aircraft. We also deployed a technical assistance team to Saigon to conduct an in-depth analysis of the Seventh Air Force's mission planning and execution process. As a result, operational planning time was significantly reduced, and there were fewer ineffective sorties blamed on late or inaccurate mission planning information (such as "frag" orders).

Unfortunately, despite the sophistication of the technology we brought to the interdiction campaign and other operations, the North Vietnamese were more resilient and resourceful than our leaders gave them credit for at the time. Even with American success in hampering their logistics and destroying much of their infrastructure, they prevailed in moving enough men and supplies into South Vietnam to keep fighting.

Accomplishments and Angst

Despite some remaining financial and management challenges, I was not displeased with the condition of MITRE as we entered 1969. We employed about 2,000 people and had a truly global presence, with offices in Japan, Thailand, Panama, the Netherlands, England, Italy, and Germany (three sites) as well as about a dozen locations throughout the United States, including a new building in McLean, Virginia, for most of our 450 employees in the Washington area.

As for our primary customer, MITRE continued to contribute to a wide variety of Air Force programs. We were actively involved in range support and instrumentation. We performed system engineering studies for ARIS III, an advanced intelligence collection ship designed to provide coverage of multiple reentry vehicles. Of great future significance, we assumed systems engineering responsibility for the AWACS. Also for air defense, we prepared a plan for using over-the-horizon backscatter (OTH-B) radars for long-range, low-altitude coverage around the United States. For the Strategic Air Command (SAC), we responded to an urgent requirement to design an interim high altitude radiation detection system to identify exoatmospheric nuclear bursts that could threaten Minuteman missiles. We developed and installed an interim capability in SAC's Looking Glass airborne command posts and its underground center in record time. All told, these and other DOD programs, such as our work on the World-Wide Military Command and Control System (WWMCCS) and the National Military Command Center, accounted for some 88 percent of our total revenue. On the civil side, we were gradually expanding work for the Departments of Commerce; Housing and Urban Development; Health, Education, and Welfare; Transportation (primarily the FAA); NASA; and even the US Postal Service. We also continued to seek opportunities to work with state agencies and local communities.

From the foregoing, one might assume that 1968 had been a great year for me, but this was not necessarily the case. I was having a difficult time keeping my professional life compartmented from our mounting national crisis. Although enjoying life in Concord, I certainly was not enjoying the worsening civil unrest and political polarization brought about in large part by America's involvement in Vietnam. My older children, like many of their contemporaries, were strongly opposed to the Vietnam War and knew something about the work MITRE was doing to support it. Pat shared their sentiments. With its adverse effects at home and abroad, the war was causing considerable heartache and anxiety for me and many colleagues.

With hundreds of our employees working on projects in Southeast Asia, one could argue that the war was good business for MITRE. We were seeking to provide our Airmen, soldiers, sailors, and marines with the best possible technology—

technology to allow them to do their wartime jobs more safely and efficiently. As time went on, however, I could not escape the conclusion that Vietnam had become a big black hole into which the US government could throw as much effort and treasure as it wanted without achieving victory, and all the while alienating much of world opinion. After attending a briefing on the war by President Johnson and Secretary of State Dean Rusk, a fellow Davidson alumnus, I even wrote the latter a long letter outlining my distress with the administration's bankrupt Vietnam policy. Before I had the audacity to put it in the mail, I got to thinking about the consequences. Although signing as a private citizen, the fact remained that my views would reflect on the MITRE Corporation and the Air Force. Then, while still pondering what to do, Lyndon B. Johnson's sudden announcement on 31 March 1968, that he would not seek a second term gave me an excuse not to mail it. The letter remained in my desk as, late in the year, I began to consider an offer to join the new administration of President Richard M. Nixon, who had campaigned on a platform of "peace with honor."

Leaving MITRE would not be easy. My job there was challenging and rewarding, despite frustrations caused by its FCRC status and the war in Vietnam. Yet I now had an opportunity to return to the Pentagon in a very important job for the Air Force and the Department of Defense. The conflict in Southeast Asia had shown that the United States was not as well prepared to fight limited wars as we had once thought. For the Air Force, the Vietnam War reinforced its need to develop better systems for continuous surveillance over extended combat areas and to acquire a more accurate arsenal of weapons to deal with elusive targets without killing so many innocent bystanders. At the same time, there was a need to maintain the effectiveness of our strategic forces, which the Soviets now seemed intent on matching or surpassing. Although these were difficult problems, new technology, such as we were exploring at MITRE, had much to offer towards their solution. I believe that when citizens are offered the chance to serve in positions that can deal effectively with major issues facing the country, they should make themselves available. It was also my sincere hope that Mr. Nixon could resolve the situation in Vietnam. In front of my colleagues at MITRE, I publicly pledged to do all I could to help manage the resources of the Air Force as efficiently as possible to give the nation the forces it needed.[34]

Notes

1. The acronym DDR&E was used to identify both the position and the organization that it supervised.

2. For good impartial summaries of Robert McNamara's policies and management style, see Roger R. Trask, *The Secretaries of Defense: A Brief History, 1947–1985* (Washington, D.C.: Historical Office, OSD, 1985), 28–34; and Douglas Kinnard, *The Secretary of Defense* (Lexington: University of Kentucky, 1981), 109–12. When the PPBS was later extended to six years, the FYDP was renamed the "Future Years Defense Plan."

3. At the end of his tour with the DDR&E, McLucas wrote an essay published in HRB-Singer's monthly journal, the *Short Circuit*, "My Two Years in the Pentagon," July 1964, 7–10, which has been helpful in preparing parts of this section.

4. See Jacob Neufeld, *Ballistic Missiles in the United States Air Force, 1945–1960* (Washington, D.C.: Air Force Office of History, 1989), 93–244. The early missile programs also had the luxury of reducing risk through parallel development of comparable systems (e.g., the Air Force Thor and Army Jupiter IRBMs and the Air Force's Atlas and Titan ICBMs).

5. Robert S. McNamara with Brian Van DeMark, *In Retrospect: The Tragedy and Lessons of Vietnam* (New York: Random House, 1995), 23.

6. Ibid., 107. One year after McLucas died, McNamara expounded on many of the themes of his book in the Academy Award–winning documentary, *The Fog of War*.

7. For a summary of the long struggle to develop a precision location capability that ultimately led to the global positioning system (GPS), see chapters 5 and 8.

8. See Michael J. Hirshberg, "V/STOL: The First Half Century," 1997, http://www.aiaa.org/tc/vstol/VSTOL.html.

9. Edwin "Spec" Powell provided some of these details in a telephone interview with McLucas and Alnwick on 21 November 2001.

10. The development of airmobile operations would be summarized in a best-selling book by Lt Gen Harold Moore and Joseph L. Galloway, *We Were Soldiers Once . . . and Young* (New York: Random House, 1992) and depicted for an even larger audience in the 2001 motion picture based on this book and starring Mel Gibson.

11. For a good academic study, see Robert J. Art, *The TFX Decision: McNamara and the Military* (Boston: Little, Brown, & Co., 1968). For a comprehensive technical review of the F-111's evolution, see Joe Baugher, "General Dynamics F-111 History," 9 August 1999, http://www.f-111.net/JoeBaugher.html; and Bill Gunston, "F-111," chapter 5 in Bill Sweetman et al., *The Great Book of Modern Warplanes* (New York: Portland House [for Salamander Books Ltd.], 1987), 265–329.

12. Zuckert, who had enjoyed great influence as assistant secretary of the Air Force under Stuart Symington in the late 1940s, was at first so unhappy with McNamara's centralization of power that he thought of resigning. He decided, however, to try to help the Air Force adapt to life under McNamara by serving out his term. George Watson, *The Office of the Secretary of the Air Force, 1947–1965* (Washington, D.C.: Center for Air Force History, 1993), 205.

13. Typically, stock options have restrictions that prohibit an employee from exercising them until a certain period of employment has been completed.

14. When John McLucas died, Russ Doughtery would be influential in arranging an impressive interment ceremony for him at Arlington National Cemetery in January 2003.

15. NATO, *The NATO Handbook*, 16th ed. (Paris: NATO Information Office, 1963), 50–52; and *NATO and Science: Facts about the Activities of the Science Committee of the North Atlantic Treaty Organization, 1959–1966* (Paris: NATO Scientific Affairs Committee, 1967), 12–15.

16. Jan van der Bliek, ed., *AGARD: The History, 1952–1997* (Paris: NATO Research and Technology Office, 1999), 2–3. For more about von Kármán and Wattendorf, see Michael H. Gorn, *The Universal Man: Theodore von Kármán's Life in Aeronautics* (Washington, D.C.: Smithsonian, 1992).

17. David R. Mets, *NATO: Alliance for Peace* (New York: J. Messner, 1981), 98–102.

18. Harold Brown, SecAF, to John McLucas, NATO Scientific Affairs Div., letter, ca. 20 November 1965, in McLucas personal files.

19. NATO, *The NATO Handbook*, August 2001, http:www.nato.intl/hand book/2001/hb130125.html.

20. MITRE Corp., *MITRE: the First Twenty Years: A History of the MITRE Corporation (1958–1978)* (Bedford, Mass.: MITRE Corporation, 1979); John F. Jacobs, *The Sage Air Defense System: A Personal History* (Bedford, Mass.: The MITRE Corporation, 1986); and Robert Everett, telephone interview by McLucas and Alnwick, 24 June 2002. Bob Everett, a former president and CEO of MITRE Corp., shared his memories throughout this section.

21. Davis Dyer and Michael Aaron Dennis, *Architects of Information Advantage: The MITRE Corporation since 1958* (Montgomery, Ala.: Community Communications, 1998), 35. To fill in some details, the rest of this section draws upon this book and the previously cited *MITRE: The First Twenty Years*, as well as documents from the McLucas papers on file in the company's archive at its Bedford facility (hereinafter cited as MITRE Archive).

22. For background on the early history of both MITRE and ESD as milestones in the development of systems engineering, see Stephen B. Johnson, *The United States Air Force and the Culture of Innovation* (Washington, D.C.: Air Force History and Museums Program, 2002), 180–223.

23. For a contemporary look at FCRCs, see James Hessman et al., "Federal Contract Research Centers: DoD's Cerebral Reserve," *Armed Forces Journal*, 28 September 1968, 6.

24. MITRE Corporate Planning, "Updated and Revised Corporate Three Year Plan (Draft)," 9 November 1967, 20–23, in MITRE Archive.

25. Gen James Ferguson, commander, Air Force Systems Command, to John McLucas, president, MITRE Corporation, letter, 17 March 1967, in MITRE Archive.

26. John McLucas, "The Social Responsibility of the Engineer" (speech at an IEEE meeting, 11 October 1967, in MITRE Archive).

27. Gen Bernard A. Schriever, "Air Force Relations with the Not-for-Profit Corporations," HQ AFSC, 14 April 1966, part 2, 22, in the MITRE Archive.

28. For a detailed comparison of Aerospace and MITRE through 1965, see Johnson, *The United States Air Force and the Culture of Innovation.*

29. *MITRE: The First Twenty Years*, 105.

30. Killian served as MITRE's chairman until 1969 and remained on the board of trustees until 1982.

31. An extraordinarily versatile scientist, Alvarez developed several radars and the triggering device for one of the atomic bombs during World War II, was a pioneer in understanding cosmic rays and discovering subatomic particles, and became one of the early proponents for the theory that an extraterrestrial body helped cause extinction of the dinosaurs. In 1968 he received the Nobel Prize for physics.

32. *MITRE: The First Twenty Years*, 104.

33. Peter Bush, "The Story behind the McNamara Line," www.chss.montclair .edu/english/furr/pbmcnamara.html, July 2002. A version of this article appeared in *Vietnam* magazine, February 1996, 18–24.

34. Information Services Office of the MITRE Corporation, "Goodbye and Good Luck: Dr. McLucas goes to Washington," *MITRE MATRIX*, March/April 1969, 9.

Chapter 3

My Air Force Years
People and Politics

Without doubt, 1968 was one of the most traumatic years in American history—certainly within my lifetime. Opposition to the Vietnam War, emergence of a "counterculture" of rebellious youth, the assassinations of Robert Kennedy and Martin Luther King Jr., urban riots, and an angry backlash against social changes seemed to be tearing the nation apart. After the chaotic Democratic Convention in Chicago and the third-party candidacy of George Wallace, Vice Pres. Hubert H. Humphrey did not seem to be a very credible candidate.

Like most Southerners, I had been raised a Democrat, but after moving north I became a moderate Republican who believed in a bipartisan foreign and defense policy. In the context of the times, Richard Nixon's platform of peace with honor was appealing. He got my vote.

Joining the Laird-Packard Team

With Nixon's victory, I soon became involved in helping recommend people to fill several midlevel jobs in the new administration's defense establishment. By now I belonged to a loosely organized "technology mafia"—one of the old-boy networks that helped staff the military-industrial complex. At first, however, I had no intention of returning to government service myself, at least not for several more years.

Despite the bitter political climate in the nation at large, the transition from the Johnson to the Nixon administration seemed to go fairly smoothly. From my perspective, Nixon's best cabinet selection was his last-minute choice of veteran Republican congressman Melvin R. Laird as secretary of defense (SecDef). A long-time member of the Defense Subcommittee of the powerful House Appropriations Committee, Laird was respected on both sides of the aisle. From this political base, he came to the Pentagon having obtained full authority from Nixon to select his own staff without the intimate White House participation that shaped the key personnel rosters of other agencies. Laird in turn selected the now-legendary David Packard, cofounder of Hewlett Packard (H-P) and a pioneer of what became Silicon Valley, as his deputy. Like many others who knew him, I consider Dave one of the greatest combinations of technologist, business leader, and public servant in American history.

Laird and Packard perfectly complemented each other, with Mel keeping his eyes on the big picture of strategy and politics and Dave focusing on technology and the details of management. For the third-ranking position in the Office of the Secretary of Defense, they decided to keep Dr. John S. Foster as director of Defense Research and Engineering. A handsome and articulate scientist, Johnny Foster was a strong advocate of keeping ahead of the Soviets in all areas of technology, including nuclear weapons (which he had helped develop in his previous job as director of the Lawrence Livermore National Laboratory). His personality was such that one could not help but like him. I had become quite friendly with Johnny when I was at MITRE, and he was popular with much of the Air Force's military leadership.

Unlike some secretaries of defense who have preferred to work through a handful of appointees in the OSD, Mel Laird intended to pay close attention to his service secretaries and military leaders. To be his secretary of the Air Force (SecAF), Laird selected Robert C. "Bob" Seamans Jr., formerly a top executive at NASA and currently on the faculty at MIT, to replace Harold Brown. I had recently invited Dr. Seamans to join the MITRE board of trustees, but he declined because of already being on the board of the Aerospace Corporation. Harold, who graciously stayed in office until Bob was confirmed in mid-February, gave him a list of people with strong R&D backgrounds that he thought well qualified to be the new undersecretary. My name was on his list, but since Bob knew me only by reputation, he asked James Killian about my suitability. Bob later wrote that when Killian realized he was serious about trying to recruit me away from MITRE, some of the color drained from Jim's face. After checking with Mel Laird and Central Intelligence Agency (CIA) director Richard Helms (for reasons that will become apparent in two more paragraphs), Bob quickly arranged to meet with me while on a stopover at Hanscom AFB, where he made his job offer.[1]

My wife, Pat, was adamant that returning to the Pentagon was a very bad idea. She liked living in the Boston area and remembered all the income we sacrificed in our last move to Washington. She was also troubled about Vietnam and thought I was already too close to "those people who run the war." That I might become one of them was not a pleasant prospect for her. I saw the issue from a rather different perspective, shared by my eldest daughter Pamela, a senior at Wellesley at the time. While Pam would have preferred that I not get more deeply involved in the war, she also believed that I might do a better job of helping the nation out of the quagmire than someone else. My younger daughter, Susan, was far away at Reed College in Oregon, and my older son, John, was a junior in high school. When told that I had decided for us to leave Concord, he arranged to move into a neighbor's house as a live-in babysitter so he could graduate with his class. Rod was still a youngster of 13.

During the early weeks of 1969, I made more than one trip to Washington to talk with Laird, Packard, Seamans, Foster, and others. Secretary Laird assured me about his and the administration's resolve to end the war in Vietnam. From my membership on the USAF Scientific Advisory Board as well as my work at MITRE, I was familiar with many of the technical challenges and opportunities facing the Air Force. Bob Seamans made the prospect of joining him irresistible by having the undersecretary's position responsible for managing the National Reconnaissance Office (NRO)—one of the most secret and technologically advanced organizations in the entire government. As director of the NRO, I would report to the secretary of defense and work closely with the director of central intelligence. I had been interested in remote sensing since my early days at HRB, and the NRO dealt with the world's most advanced long-range sensors and imaging technology. With this tipping the scales, I told Dr. Seamans, "I'll take the job even if my wife thinks I'm nuts." She did; but notwithstanding, I prepared to begin work as undersecretary of the Air Force. The Senate Armed Services Committee approved my nomination along with those of several other new Pentagon appointees on 26 February 1969.

My Roles as Undersecretary

A changeover of presidential administrations always means extra business for Washington-area real estate agents. Pat and

65

I soon selected a house on Lake Barcroft near Falls Church, Virginia. I thought its view of the water would make it easier for her to leave our home on the pond in Concord. Pat and Rod moved down from Massachusetts to join me in June. In the intervening months, I stayed in a one-room apartment and spent most of my waking hours at the Pentagon.

I reported for duty on 17 March 1969. In addition to my special role in the "black world" of national intelligence, I had three major responsibilities as undersecretary: (1) to advise Secretary Seamans in making decisions on Air Force issues and programs; (2) to serve as acting secretary in his absence (which amounted to about one-third of the time) by handling issues in the way I thought he would if present; and (3) to administer those areas of Air Force business he specifically delegated to me, such as acquisition programs involving electronic systems, foreign military sales and other international issues, and certain personnel matters. With Bob being an aeronautical engineer and me being more or less an electronics engineer, we had most of the bases covered when it came to understanding the basic technologies of new Air Force systems.

Bob and I very quickly established a way of working together that allowed me to serve as his alter ego but still have enough flexibility to meet my NRO responsibilities (covered in a later chapter). Throughout our four years together, I never failed to be impressed by his strong leadership, good judgment, and integrity. Bob and I had a lot in common professionally, although our personalities were somewhat different. In general, I was probably just a bit more easygoing in my management style, while he was somewhat more of an activist. In general, I think we made a great team. I had the impression that our immediate predecessors did not, probably because Harold Brown did not delegate as much authority to his undersecretary, Townsend Hoopes (who had been assigned to him by the White House).

Instead of sitting behind our desks, Bob Seamans and I both liked to go out and make personal contact with as many people as possible. The job of secretary required extensive worldwide travel—sometimes for a week or more at a time—to give speeches, visit installations, attend meetings, confer with allied government officials, perform various protocol duties, and stay informed. At that time the Air Force had many more bases scattered around the world than it does today, and Bob Sea-mans probably traveled more than most secretaries. As part of his busy schedule, he made sure to visit our forces in South-east Asia at least every six months. Seeing situations in person gives a service secretary extra credibility with OSD officials and Congress. (Just before Bob's resignation, someone on his staff calculated he had taken 175 trips in four years.)

Besides me, the SecAF's primary staff included four assistant secretaries (each of whom had several deputies of his own), a general counsel, and two military assistants holding the rank of colonel or brigadier general. These latter two served as liaison to the "blue suit" Air Force (i.e., its uniformed personnel), keeping us educated on military protocol and up-to-date on issues requiring actions by the secretary. They also dealt with areas not in the portfolios of the assistant secretaries, such as current operations and intelligence, and helped arrange trips, especially those going overseas. Most mornings would start with a short meeting to confirm schedules, needed actions, and other daily business. One of the military assistants might also give Bob and me a summary of ongoing operations, based on briefings given earlier in the morning to the Air Staff leadership. On Mondays, we would hold a staff meeting attended by the assistant secre-taries and other key personnel, who would report anything of general interest and receive guidance from Dr. Seamans.

Assistant secretaries were presidential appointees usually recruited from outside the civil service. Of the four assistant secretaries, I worked most closely with Grant L. Hansen, the assistant secretary for research and development, whose of-fice dealt with most acquisition programs. Grant came to the Air Force with many years of experience in the air and space industry. The others were Phillip N. Whittaker, assistant sec-retary for installations and logistics; Spencer J. Shedler, as-sistant secretary for financial management; and Dr. Curtis W. Tarr, assistant secretary for manpower and reserve affairs (including personnel matters). Richard J. Borda replaced him in 1970, while all the others served out full four-year terms. Senior military officers on the secretary's staff included Maj Gen John R. Murphy and later John C. Giraudo as director of legislative liaison (responsible for congressional relations) and Maj Gen Henry L. Hogan as director of information (i.e., public affairs). In view of the growing criticism and oversight by Congress and the antimilitary tone of the media during the period, both of these jobs were particularly challenging.

Two of the first people I met after arriving on the job immediately become very important in my daily life: Kathy Rizzardi and Mike White. Kathy had served my predecessor, and I soon found her to be one of the best secretaries in the Pentagon. When Dr. Seamans' secretary retired a year or so later, Kathy was offered the opportunity to move into his office. When I asked why she had turned down the promotion that went with the job, she said she would stick with me until I moved over there myself. And that's what she eventually did. (She even stayed with me when I went to the FAA.) My driver, Mike White, was a combination of loyal employee, good friend, and someone who knew every shortcut to keep his boss on schedule no matter how bad the traffic. Like most government executives who had a driver pick them up, I spent the commute to work skimming morning newspapers and material from my briefcase.

Another previous member of the undersecretary's staff whom I was happy to keep on my team was Phillip F. Hilbert, the deputy undersecretary for international affairs. Extremely well informed, Phil ran our own little "State Department," monitoring base rights agreements for overseas units and keeping an eye on the stream of Air Force hardware supplied to nations around the world under the foreign military sales program and various foreign aid arrangements. I created another deputy undersecretary position for systems review in which to place Harry Davis, a long-time civil servant. Formerly the deputy for special programs to Alexander Flax, assistant secretary for R&D under Harold Brown, Harry was very knowledgeable about the NRO and—when it came to almost anything having to do with electronics—he was a genius.

The Air Force's top leaders occupy a suite of offices on the fourth floor of the Pentagon's outer E-Ring between corridors eight and nine. I also spent some of my time in a less attractive office complex, hidden behind a heavy vaulted door a few minutes' walk away on the C-Ring, which served as NRO headquarters. My undersecretary's office was adjacent to Dr. Seamans'. Both of us enjoyed a beautiful view of the Potomac River, with many of Washington's famous buildings and memorials on the skyline. I often watched airliners making their final approach to Washington's National Airport less than a mile downriver. During the Cold

War years, Pentagon residents joked morbidly about the center courtyard being "ground zero"—a perfect bull's eye for Soviet ballistic missiles. But I don't remember ever worrying much about an airplane hitting the Pentagon, either by accident or on purpose, a possibility that became all too real three decades later.

In those days, Headquarters Air Force (HAF) had separate and somewhat parallel civilian and military staff organizations: the Office of the Secretary of the Air Force (OSAF) and the Air Staff. The OSAF or "Secretariat" (as the secretary's staff was also known) was only a fraction the size of the workforce reporting to the chief of staff of the Air Force (CSAF), known collectively as the Air Staff. Even so, most functions other than operational matters were also represented in the Secretariat. In such a dual-tracked bureaucracy, with considerable duplication and overlap of functions, the relationship between the leaders at the top was key to effectiveness. Although service secretaries perform a constitutional role by assuring civilian control over the military, their senior uniformed leaders exercise considerable authority by virtue of their experience, professional training, and web of personal relationships spun over a career of rising through the ranks. Furthermore, the service secretaries were removed from the operational chain of command, while the chiefs—by being members of the Joint Chiefs of Staff (JCS)—were closely involved in operations.[2] Service secretaries are basically responsible for organizing, equipping, training, and sustaining their departments, but implementation of these functions is usually delegated to others. According to Jack Stempler, the general counsel during most of my Air Force years, the only thing that could not be delegated was the authority to accept gifts on behalf of the Air Force. (Such donations can range from real estate to museum artifacts.)

When it comes to credibility with Congress and other outside observers, the very cultural identity of the uniformed leaders that enhances their status within the services can be a disadvantage in the sense that loyalty to the institution might be seen as limiting their objectivity and openness to new ideas. Witness a cliché (popular at the time) that a general never met a new weapon he didn't like. The thin layer of appointed civilian leaders, especially the service secretaries, can provide valuable "top cover" for the generals and admirals—explaining things that go wrong in a way that they cannot. But civilian leaders can do this credibly only when their military staffs keep them informed. Since the service secretaries are not in the operational chain of command, there is no guarantee this will always happen. In 1973, after I became acting secretary of the Air Force, the Department of Defense suffered considerable embarrassment when it was revealed that Secretary Seamans and the rest of us had been fed misleading reports on bombing missions in Southeast Asia (see chap. 6). On 21 December 1972, about six months before this story broke, I wrote down the following thoughts about the civilian-military relationship in a memorandum for record.

> Proper management of the DoD requires good relations all around, but also demands a healthy skepticism on the part of civilians even though, with respect to military matters, they frequently are amateurs being taught by the professionals. (On the other hand, the civilians are more likely to be professional in areas where the senior military are sometimes amateurs; e.g., in legal, business, public relations, diplomatic issues.) There is a tendency of many people to look with awe on military braid and medals and not analyze what is being proposed. Similarly, or conversely, there have been times when not enough weight was given to military experience.[3]

In July 1969, just four months after my arrival, Gen John P. McConnell, who had been CSAF throughout the buildup in Southeast Asia, reached his time for mandatory retirement. Bob Seamans and I found him very competent but somewhat protective of what he considered blue-suit matters. McConnell, who could be quite charming, had maintained good rapport with administration officials and key members of Congress, but the frustrating course of the war had obviously been an ordeal that had taken a toll on him. Bob had begun considering which general should be the next chief soon after he became secretary. Although vice-chiefs had traditionally been elevated to chief, we looked closely at several other candidates, most notably Gen Bruce Holloway (then heading the Strategic Air Command). In the end, however, Dr. Seamans—with the approval of Mel Laird—chose the current vice-chief, Gen John D. Ryan, to serve for the next four years as the Air Force's seventh chief of staff. Jack Ryan had been a bomber pilot in World War II and later served as commander of SAC. Bob never regretted his choice, and I too found General Ryan a real professional as well as being easy to work with. He absorbed information quickly and made sound decisions. Outwardly, he was a shy man of few words, but when he did say something, it was always worthwhile, and he always seemed to "keep his cool" no matter what the crisis.[4] He was also a devoted father, whose two sons graduated from the Air Force Academy in the Class of 1965 and became Air Force fighter pilots. His oldest son, John Jr., was tragically killed in an aircraft accident in 1970. His other son, Mike, would rise through the ranks to become the 16th Air Force chief of staff. I once remarked to his father that I thought the Air Force secretary knew a lot more about what the Air Staff was doing than the Army and Navy secretaries did about their military staffs. Although always open to our ideas, Ryan replied, probably only half in jest, that such a situation wasn't necessarily all for the better.

In my role as undersecretary, I most often dealt directly with my counterpart on the Air Staff, the vice-chief of staff. During my first few years, this was Gen John C. Meyer, a highly decorated fighter pilot with 26 aerial victories in World War II and Korea. I got along very well with Meyer, who was more articulate and colorful than Ryan. At an award presentation in the Pentagon just after his retirement ceremony in Omaha (where he headed SAC), I recall him giving a short speech, saying with typical humor that he viewed his retirement with mixed feelings: "half joy and half pure ecstasy."

The offices of the chief and vice-chief were just down the hallway from our offices, so it was convenient to meet face-to-face when issues required. Yet tight schedules and diverse commitments orchestrated by a phalanx of executive officers, secretaries, and administrators did not routinely allow for ad hoc contacts, although we often ate lunches together. The best time to just walk in on the chief or vice versa seemed to be Saturday mornings, a quiet time when only the key people normally came in to catch up on work left over from the previous week. Unlike the practice that began in the late 1990s, the secretary and chief did not routinely hold combined staff meetings. Both Bob and I liked to gather information and develop opinions through conversations and other informal contacts, both inside and outside the Pentagon. For the record, however, we had to document most decisions with initials or signatures on correspondence,

"staff packages," and other paperwork that had already been reviewed, revised, and coordinated by the interested offices in the
Air Staff and Secretariat. The more complicated or controversial
matters, including those involving major acquisition programs
or policy changes, were often accompanied by briefings—highly
scripted presentations featuring slides or vu-graphs.

Although some friction is inevitable in any large organization
with a host of often-competing factions and agendas, I felt that
HAF operated in a very collegial and orderly manner both during
Dr. Seamans' tour as secretary and mine that followed. William R.
"Bill" Usher, who served both of us as military assistant, thinks
Bob tended to deal more directly with the ranking generals on
the Air Staff and at the major commands (MAJCOM), while I delegated more responsibility to the assistant secretaries to handle
issues with their uniformed counterparts. The good people working for Bob and me undoubtedly made our jobs easier.

Defense Management in the
First Nixon Administration

My satisfaction with the Air Force leadership was matched by
my respect and appreciation for the people heading the Office
of the Secretary of Defense during the first Nixon administration, starting at the top with Melvin Laird. As the largest federal
agency, DOD has the reputation of being one of the most un-
manageable organizations in Washington. The Pentagon cannot
be run like a large private corporation. As Norman R. Augustine,
a Pentagon colleague who became one of the nation's greatest
defense industry executives, told me in June 1999 during my
research for this book, "There is a tendency for people in the
private sector to say, 'Well if you would just do business like we
do . . . , everything would be okay.' There are fundamental differences. Having a board of directors of 535 people is just one
example."[5] Norm was, of course, referring to Congress.

Several secretaries have tried to run DOD like a business, but
most have fallen far short of the mark. Mel Laird, using a different approach, has in my opinion come closer than anyone
else. Although he had never supervised anything larger than a
congressional office, Mel had some qualities important to mak-

ing a good manager: he knew the territory, he was intelligent, and he knew how to choose and motivate people. His nine terms in the House of Representatives, where he served on its powerful appropriations committee, gave him a head start on dealing with the complicated defense bureaucracy. He also knew that, to work effectively together, people need a common understanding of how their own goals fit into an overall policy. Mel could be highly critical when something didn't go the way he thought it should, but I always found him very fair and supportive.

Although Laird had not run a big organization before becoming SecDef, he had the good sense to hire David Packard, knowing he needed a deputy with proven management experience and technical knowledge. Mel and Dave made a wonderful team. Mel also established personal relationships with people who worked one or two layers below him. Many secretaries of defense, either because of their authoritarian management styles or because of fear of spreading themselves too thin, have tended to isolate themselves in their offices and work closely with only their immediate staff and a minimum of others. Perhaps because Mel Laird was used to sitting through congressional hearings and participating in the many social gatherings required of a successful politician, he patiently spent a large portion of his time holding meetings and staying in close touch with all his key people. As for Packard, he had been a great practitioner of "management by walking around" in building his company. Although not able to do this in the huge Pentagon, his door was usually open to those of us who needed to see him. He was also a proponent of DOD and its components following the "management by objectives" technique—setting needed goals and measuring progress toward meeting them. I considered Dave Packard as more than a good manager; to me he was a mentor. In a Pentagon ceremony in January 1973, I said to those present, "I have to look back on my association with Dave Packard as one of the fine experiences of my life." Thirty years later, I still do.

Laird and Packard had a regular schedule of meetings to which they adhered as often as possible. On Mondays they held a large staff meeting with 30 or 40 people in attendance, including OSD undersecretaries, the service secretaries, and the JCS. With the Vietnam War at the top of his agenda, Laird spent many hours each week in meetings with the Joint Chiefs, who were led by Gen Earle Wheeler until July 1970 and Adm Thomas Moorer for the next four years. On most Wednesdays, Mel and Dave hosted a weekly lunch with the service secretaries as a group, which I attended whenever Bob Seamans was away. The two of them also met separately with the secretary and undersecretary of each service once a week, usually Tuesday afternoons for Bob and me. On Fridays, Dave conducted a breakfast meeting of all the intelligence chiefs who reported to the secretary of defense. For much of my time as head of the NRO, these included Lt Gen Donald Bennett of the DIA; Vice Adm Noel Gayler of the NSA; and Dr. Robert A. Frosch, representing the underwater reconnaissance function. In August 1972 Air Force lieutenant general Samuel Phillips took over the NSA. In view of Dave Packard's technical expertise, Mel usually had him handle NRO business. All this personal contact allowed Laird and Packard to stay up to speed on the major activities of their key people. The meetings also provided good liaison among several key constituencies, promoted "bonding" among the defense leadership, and helped to suppress the misunderstandings and mixed signals that inevitably arise in the multifaceted and often compartmented defense establishment.

The three service secretaries and their undersecretaries also got together for separate lunches about once a month and otherwise kept in touch. The secretaries of the Army during my years with the Air Force were Stanley Resor, Robert Froehlke, and Howard "Bo" Callaway. I enjoyed working with Bo (a former Georgia congressman) when I was secretary, although we didn't do as much business with the Army as with the Navy. The Air Force shared more weapon systems with the Navy, whose secretaries were John Chafee (former governor and future senator from Rhode Island), John Warner (future senator from Virginia), and William Middendorf. Warner and I served together as undersecretaries and then secretaries, so we got to know each other pretty well. I later met John's second wife, Elizabeth Taylor, while celebrating the birthday of another well-known Elizabeth at the British ambassador's residence. After Warner left the Navy to head the Bicentennial Commission in June 1974, I got along admirably with Bill Middendorf, who replaced him.

In looking back at service secretaries since World War II, I find it interesting that many of the Air Force secretaries have been scientists or engineers, perhaps because of the special importance of technology to the Air Force, while most of the other service secretaries have been businessmen and lawyers, including some with successful careers in politics. (Of Air Force secretaries, only Stuart Symington ever achieved high elected office.) I was especially pleased that John Warner, who appreciated the role of service secretary, became a key member of the Senate Armed Services Committee. In the second Clinton administration, he successfully opposed the nomination of an individual that he (and I) did not think worthy of being appointed Air Force secretary.

As regards defense policy, Mel Laird had a number of priorities, which he did not hesitate to make known. Some of these supported the objectives of the Nixon administration; others reflected his own ideas. He wanted to solve the country's two most divisive defense issues by reducing American military involvement in Vietnam as quickly as possible and replacing selective service (the "draft") with an all-volunteer force. He also began upgrading the status and capability of the Reserve and National Guard components, applying the term *Total Force* to this policy. Other goals included making better assessments of the threat from the USSR; reinvigorating NATO and rebuilding US military resources in Europe (which had been drawn down for the war in Southeast Asia); fostering a cooperative style he called "participatory management" within DOD; and establishing realistic fiscal guidance compatible with what he believed Congress could be convinced to approve. Although keeping intact most of the PPBS process, Mel resumed the use of service budget ceilings, with the services given more latitude to "reprogram" resources within these ceilings.

I quickly found working in the Pentagon under Melvin Laird a lot different from my previous experience under Robert McNamara. A comparison of their management styles and effectiveness must, however, take some account of the different circumstances prevailing during their respective periods in office. McNamara came in at a time when the services had still not fully accommodated themselves to the legislation of 1947 and 1958 that created and strengthened the OSD. By Mel's time, open defiance against centralization had diminished, and his smooth way of handling people problems avoided most of the acrimony triggered by McNamara. Laird's philosophy was to draw on the strengths of the services while giving the military leadership credit for the value of their experience and traditions. I was one of many people who found the Laird-Packard team's attitude toward the military more constructive than that of McNamara (and some recent SecDefs as well). While friction between the military departments and the OSD did not disappear during Laird's tenure, it was kept at a level that was not too disruptive.

When it came to dealing with the White House, Laird served as a buffer between Nixon's people and the services. At first our natural tendency in the Air Force was to respond promptly and directly when we had a call from anyone in the presidential mansion. At an early staff meeting, however, Mel told us not to do anything except refer the caller to Carl Wallace, his personal assistant. Laird wanted both to control DOD–White House relations and protect us from being caught up in the administration's hidden agendas. I think his normal practice when dealing with the White House was to talk only with Nixon and National Security Advisor Henry Kissinger, at least on matters of substance. Despite frequent frustration at the White House with Laird's independence, he was almost invulnerable to pressure or intimidation. As Nixon's confidant John Ehrlichman ruefully acknowledged, "Laird was so effective with his old congressional cronies that everyone realized he was irreplaceable."[6]

Laird's years on Capitol Hill served him well when it came to the political aspects of running the Pentagon. He could walk into most congressional offices and be greeted as a friend or respected colleague. This gave him a head start in gaining a sympathetic hearing if not outright support for whatever issues were on the table. When Mel left the Pentagon, he was proud of never having lost a key vote—quite an accomplishment for a Republican cabinet officer dealing with a Democratic Congress during a very contentious period. Although Laird put a lot of effort into congressional relations, he did not try to restrict or micromanage the services' contacts on the Hill—so long as they didn't go behind the OSD's back and try to get support for their own pet programs. When it came to congressional testimony, Mel wanted to review what we planned to say about general policy or "big-picture" type issues, but let us testify as we saw fit on technical matters or more routine issues.

My congressional contacts as undersecretary were fairly routine, primarily involving specific acquisition programs. Both the House and Senate of the 91st and 92nd Congresses (1969–72) were firmly under the control of the Democrats. Yet, despite the investigations and hearings on Pentagon mismanagement and efforts to cut "wasteful" defense programs, enough Democrats sided with most Republicans to generally support defense measures. In part, that reflected residual effects of the bipartisan foreign policy consensus that had generally prevailed since World War II, and in part, the Nixon administration's preemptive strategy to reduce defense budget requests as it drew down forces in Southeast Asia. Despite Nixon's huge electoral margin over George McGovern in November 1972, the 93d Congress (1973–75) saw a relatively minor adjustment in the strength of the two parties and a noticeable shift to the left in political philosophy. Yet even some of the more liberal members seemed to gain a renewed appreciation for maintaining defense capabilities after the Middle East War of October 1973. Furthermore, the chairs of most influential congressional bodies were held by promilitary southern Democrats throughout this period. When Pat and I had moved into our new house on Lake Barcroft in June 1969, Harry Davis hosted a neighborhood party to welcome us. Among the guests was Representative Robert F. "Bob" Leggett of California, a member of the House Armed Services Committee. He told me the most important thing I had to remember when dealing with Congress is that the power of a committee chairman is greater than that of all the other members combined. I kept this in mind in my congressional contacts over the next eight years.

Within the Pentagon, Mel Laird delegated most administrative details to Dave Packard, who thought the OSD and the service headquarters were "overorganized." Dave did his best to streamline the bureaucracy and reduce the size of headquarters. Some of his structural changes in DOD came as the result of the Presidential Blue Ribbon Defense Panel, chaired by Gilbert W. Fitzhugh. Its final report in July 1970 made a total of 113 recommendations to improve organization and procedures, ranging from closer command and control of combat operations to a merger of logistics functions. Many of the panel's recommendations, however, would have further strengthened the OSD at the expense of the services, and I advised against a

77

number of them in a long memo on 10 August 1970. The most influential of its recommendations were probably in the area of procurement (described in the next chapter).

One of the panel's recommendations was to create an assistant SecDef to oversee telecommunications and automatic data processing, which in light of my experience at MITRE, I agreed were becoming more interdependent. In August of 1971 Mel and Dave decided to establish this new assistant secretary for telecommunications position and offered it to me. I was happy with my Air Force and NRO duties and didn't feel it was yet time to leave if they could find someone else for the new job. I was grateful that they instead chose Eberhardt Rechtin to be the new assistant secretary for telecommunications in January 1972. Two years later, my eventual successor as secretary of the Air Force, Thomas Reed, assumed that position.

Dave Packard, who probably sacrificed more financially than almost any person of that era to become a public servant, came under increasing financial pressure to return to the private sector. Whereas most other appointees had to sell their stock in any company doing business with the Pentagon, Dave negotiated a special arrangement with Senator John Stennis, longtime chairman of the Senate Armed Services Committee. Rather than sell his 30 percent stake in the company he had cofounded (which would hurt William "Bill" Hewlett and smaller investors), he would assure that he did not profit from his holdings while in office. If Hewlett-Packard's stock were to rise, he promised to sell enough to cancel the gain and contribute the proceeds to charity. In early 1971 H-P stock started to run up in price, and every day he stayed on the job began costing him big money. Mel begged him to stay as long as possible, and by the time Dave actually left in mid-December 1971, I recall his stock's gain as about $22 million (a considerable sum in those days). He gave a lot of it to Stanford, but the Wolf Trap Foundation for the Performing Arts, which I later chaired (see chap. 7), benefited from his generosity by receiving about $2 million.

In the antidefense atmosphere of the 1970s, Pentagon perks came under attack from "watchdogs" in Congress and the media. Most senior officials, for example, had long enjoyed the privilege of eating free lunches in executive dining rooms. By the time I left the Pentagon, we were being assessed four or five dollars a day for this fringe benefit. I remember once when Dave Packard was asked if he didn't think it was appropriate for him to pay for lunch out of his own pocket rather than eat at taxpayers' expense. He answered that he was already paying for his own lunch—at a rate of many thousands of dollars per day (the amount of stock value he was losing by staying on the job). This anecdote serves to illustrate one of the problems of recruiting successful business executives to work in government. (I myself took a much more modest cut in salary to leave MITRE for the Air Force: from $65,000 to $38,000.) The situation has gotten worse in subsequent decades. Rapidly escalating corporate compensation—combined with onerous background investigations, intrusive oversight, and occasional bad publicity—has made government service seen even less attractive to leaders from the private sector than during the Cold War era.[7] There were indeed some genuine scandals during my time in the Pentagon, including senior officials using government resources for personal reasons and accepting questionable favors from defense contractors. I tried hard to maintain my integrity and avoid even the appearance of impropriety and am pleased to have completed my years of public service with my reputation untarnished.

As will be described in subsequent chapters, I think the Air Force made significant progress in many areas under the Laird-Packard team, despite the difficult environment for national defense at the time. Dave's replacement as deputy SecDef, Kenneth Rush, proved to be an interim appointment. One of Nixon's favorite professors at the Duke Law School, Rush was confirmed as deputy secretary in February 1972 after having served as ambassador to Germany. Less than a year later, he moved to Foggy Bottom as deputy secretary of state, having been assured by Nixon he would replace William Rogers as secretary.[8] I became a friend of Ken's during his short Pentagon tour, but when it came to overseeing technology and weapons procurement, he was no Dave Packard. He also was less involved with NRO matters than Dave, and Laird began presiding at our Friday intelligence meetings. As for Mel's tenure, he remained firm in his well-publicized commitment to serve only four years.

Right after Nixon's landslide victory over George McGovern in November 1972, people sitting in presidential appointee slots received an unpleasant surprise. Nixon was determined to clean house in the Executive Branch, and an unsigned letter written on White House stationery was circulated to all noncareer employees directing us to submit our resignations immediately to the president. It was as if McGovern had won the election. Mel Laird called a staff meeting to address the topic of the letter. We all felt it was an insulting document, but none resented the unexpected directive more than Mel, who considered it a breach of his contract as well as bad etiquette. He told us not to submit our resignations to the White House but to sign letters and give them to him. He would submit them as a package in his own good time. In his memoirs, Nixon admitted his call for mass resignations had been a mistake because he "did not take into account the chilling effect this action would have on the morale of people who had worked so hard."[9]

Shortly thereafter Mel Laird left the Pentagon, as scheduled. At his final staff meeting on 15 January 1973, I jotted down his farewell remarks. He expressed satisfaction for the progress achieved in Vietnamization (see chap. 6), dealing with people problems, and improving management effectiveness, in both defense agencies and the services. He was concerned, however, that DOD and the services seemed too often on the defensive in the public affairs arena and said we needed to tell more about the improvements being made. After recalling praise from Dave Packard about the cooperative spirit in the Pentagon, Mel ended by saying that, when he departed on Monday morning, he would leave feeling that he had received the best possible support from both military and civilian personnel. As a postscript: most of the officials on the Laird-Packard team continued to hold periodic reunions for 20 years thereafter—an indication of camaraderie probably unequalled under any other secretary of defense.

Serving as Secretary

In addition to Mel Laird, other good people were leaving the Pentagon in the early months of 1973, including most of our assistant secretaries. Aware that Bob Seamans's days as secretary were also numbered—but being in no big hurry myself to move on—I thought it would be worth some psychic income to serve for a while as acting secretary, thereby providing some continuity until a new secretary was on board.

Laird's replacement was Elliott Richardson, a versatile public servant who had held several high-level positions in government, most recently secretary of health, education, and welfare. Bob Seamans and I both liked Elliott, who had earned an excellent reputation as lieutenant governor of Massachusetts in the mid-1960s. We were just getting him broken in on Air Force issues when Bob decided it was his time to leave. When being recruited, Mel had indicated that he would be in line to replace Dave Packard in the Pentagon's number two job. Obviously, the appointment of Ken Rush (at the behest of the White House) had temporarily postponed this opportunity. Even before William Clements was sworn in as the next deputy secretary on 30 January 1973, it was evident that Bob had lost all chance of higher office in the Nixon administration. In addition to the president's desire to make partisan appointments during his second term, Bob believed an "off-the-record" comment to several reporters in late 1972 had got him on the White House's "blacklist." In answer to a parting question on whether the Air Force might still have a role in Southeast Asia three years hence, his honest response was that this might be a possibility. The White House was not amused when this leaked to some media outlets as contradicting the administration's promise of Vietnamization. When Bob was nominated to be president of the National Academy of Engineering, he decided to depart the Air Force in mid-May.[10]

In late March, with Bob now scheduled to leave, Elliott Richardson invited me down for lunch. He began by saying, "This is the time I wanted to ask you to be the next secretary of the Air Force." I started to protest that I wasn't sure I could afford to stay in government another four years when he continued, "but I don't have that authority." Elliott explained that he had wanted the same deal that Mel Laird had made to pick his own people, but the White House was insisting that all service secretaries be true political appointees, each with some kind of constituency. He divulged the leading candidate for the next Air Force secretary was Gov. John Love of Colorado. I told Elliott that I appreciated his faith in me, but I also welcomed the chance to go out and look for a better paying job. (Sending four children to private colleges had just about used up all our savings). Twenty years later, the personal diary of H. R. Haldeman, Nixon's very efficient chief of staff, was published posthumously. It revealed the following entry for 2 April 1974: "I had a bunch of miscellaneous phone calls. Elliot Richardson [is] all concerned about personnel in the Defense Department and things being held up there, especially on the Air Force secretary where we vetoed his man."[11]

Soon the disarray in the White House caused by the accelerating Watergate investigation began to interfere with its desired personnel moves. Four months into Nixon's second term, at least 14 of the 34 top civilian positions in the Pentagon were either vacant or had incumbents scheduled to leave without replacements being named. I became acting secretary on 15 May but had already found a permanent job. Not being eager to leave the Washington area, which I had come to consider as home, I jumped at a chance to join the Communications Satellite Corporation (Comsat), headed by Joseph V. Charyk (another former Air Force undersecretary). We agreed for me to start in July as vice president of research and engineering at Comsat headquarters in downtown Washington. On 17 April 1973, I submitted resignation letters to Secretary Richardson and President Nixon. In the latter, I acknowledged "my great affection and respect for Mel Laird and Dave Packard" and further wrote:

> These four years have been a rewarding time for me, and I take pride in the accomplishments of the Department of Defense and the Air Force during that time. I have enjoyed especially the opportunity to act as a focal point for the Air Force's increased interest in broadening the representation in its ranks of all our citizens. In addition, my special responsibility to the Secretary of Defense in operating a program outside normal Air Force management [i.e., the NRO] has been equally rewarding to me.

So I began winding up my affairs at the Pentagon, while the Watergate affair continued to take its toll on the administration. Elliot Richardson was called upon to take the sensitive job of attorney general in place of Richard Kleindienst, who in turn had recently replaced John Mitchell.[12] For a while, there was some wishful thinking about David Packard returning to be secretary of defense. Ironically, in an effort to shore up his rapidly deteriorating relations with Congress, Nixon prevailed on Mel Laird to return to his administration and replace the recently fired John Ehrlichman as his domestic policy advisor.

To serve as the new secretary of defense, Nixon decided to move Dr. James R. Schlesinger from the CIA. I had gotten to know Jim fairly well when he oversaw the NRO's secret budget while he was deputy at the Office of Management and Budget (OMB). He was sworn in as defense secretary on Monday, 2 July, the same day I was scheduled to start at Comsat. On the preceding weekend, I had a phone call from Schlesinger. He asked, "John, what was all this I've heard about you leaving the Air Force?" I admitted it was true. He said no, I had to stay on as secretary. Obviously, recent events at the White House had strengthened his bargaining position as compared to Elliot's, and he wanted to hire experienced people, not politicians. So I said in jest, "Why don't you call my new boss and tell him I'm not coming to work next week?"

With some trepidation, I went over to Comsat on Monday morning while Charyk and his top people were having their staff meeting. I had to tell them I was there to apologize because Dr. Schlesinger had persuaded me to stay on as secretary of the Air Force. I didn't stick around long enough to hear what was said about my capriciousness. But Joe, having once been undersecretary himself, understood why being promoted to SecAF was an offer I could not refuse. Considering I had started my civilian career in one of the Air Force's lowest ranking civilian positions, I felt humbled that I was now going to have its top job. I went back to the Pentagon, and we set up the necessary meetings in the Senate for being confirmed as secretary.

Several months earlier, Bob Seamans, Jack Ryan, and I had been looking at who should be the next chief of staff of the Air Force. We had agreed that Gen George S. Brown, commander of Air Force Systems Command, was the right choice. Independently from me, Jim Schlesinger came to the same conclusion based on advice from Donald Rice, a former colleague of his at OMB and president of RAND from 1972 until 1989 (when he became secretary of the Air Force in the first Bush administration). As luck would have it, the Senate was now vetting Brown for CSAF and me for SecAF at the same time. Senator Stuart Symington of Missouri was acting chairman of the Armed Services Committee, while John Stennis of Mississippi was at Walter Reed Army Hospital (recovering from a gunshot wound suffered when he was mugged in front of his apartment). As a result, I found it noteworthy that the man who would chair our joint confirmation hearing had been the first secretary of the Air Force back in 1947. Symington opened the session by saying that John McLucas's experience "eminently qualifies him for this post" and later closed by telling me "the Air Force will be fortunate to have you on the job as secretary."[13] In between, most of the proceedings focused on recent revelations about the unreported bombing in Southeast Asia raised by Senator Harold Hughes of Iowa and a recent delay in development of the B-1 bomber (both subjects discussed in subsequent chapters). Despite Senator Hughes's probing, George and I thought the overall hearing was quite friendly, and we were soon installed in our jobs.

Jim Schlesinger swore me in as the 10th secretary of the Air Force on 18 July 1973. Former secretaries Gene Zuckert and Bob Seamans were there, as was Senator Strom Thurmond from my old home state of South Carolina, a good selection of current government officials, and most of my family. In my remarks, I reminisced, "When I came here four years ago, it was to join . . . a very professional group, and I've enjoyed very much that association. The time was not all that pleasant; we had our good days and our bad ones. . . . When it was all over, though, and we brought our men home from Vietnam, most of us . . . felt it had been a worthwhile effort. As we move into the next four years, we face a whole new set of challenges." Among them I identified maintaining morale, recruiting and retaining personnel in the all-volunteer force, improving equal opportunity for minorities and for the majority of American citizens who are women, making more progress in procurement of new systems, and improving efficiency, including our base structure.[14]

Working for James Schlesinger was quite different from working for Melvin Laird. Jim was a brilliant intellectual and disciplined thinker—the first PhD to become secretary of defense. As a long-time analyst for the RAND Corporation, chairman of the Atomic Energy Commission, and briefly director of central intelligence, he brought unrivaled national security credentials to the Pentagon's top job, despite being, at age 44, the youngest secretary of defense up to that time. Philosophically, he was more concerned with international security issues and less interested in domestic politics than Mel Laird. Apprehensive about Soviet advances in weaponry, Schlesinger was unwilling to preside over a continued decline in the defense budget. Indeed, he pushed for significant enhancements in both conventional and nuclear forces. On a personal level, he was not as patient with the give-and-take of personal relations as Laird. More in the manner of an elite university professor, Schlesinger did not suffer fools gladly. As regards those of us in the Pentagon who worked for him, he felt we were all adults who didn't need his regular attention. Like Laird, however, he tended to delegate the bureaucratic details of defense management to his deputy.[15]

Shortly after becoming secretary, I mentioned to Jim that Mel had held a couple of regular meetings that I thought were helpful, such as his weekly sessions with the service secretaries. He responded by asking, "Whatever would we all talk about?" When he had reason to communicate with subordinates, he said he could just call us on the phone or ask us to come to his office. He didn't seem to appreciate the synergy that might result from sharing ideas or the sense of teamwork that such regular group contacts can engender. He did hold a weekly staff meeting, usually on Monday, but it was a formal affair with numerous attendees and limited opportunity for conversation. In a profile of Schlesinger in 1975, his public affairs officer, Joseph Laitin (who had known him since his days at OMB), was quoted as follows: "Jim . . . is a people oriented Bob McNamara. Strip through that layer of Jim's intellectual arrogance, and do you know what you'll find? Another layer of intellectual arrogance. I say that knowing he's one of the best and nicest men I've ever met in government."[16] Having the advantage of already being acquainted with Jim, I too found him affable and unpretentious. I think I had better access to him than the other service secretaries, but he certainly kept certain zones of privacy that could not be penetrated.

About the time Elliott Richardson took over from Laird, the White House replaced Deputy Secretary Ken Rush with William Clements, a Texas oil executive who would later become that state's governor. When Schlesinger arrived several months later, Clements did not bond with him as part of a congenial team, as Packard had with Laird.[17] As for myself, I had a businesslike relationship with Clements, without any overt problems. He was certainly a good administrator. I had the impression, however, that he was willing to suffer me up to a point, so long as I was useful to him. There was a joke at the time that if you didn't like the answer you got from Clements you could go to Schlesinger, and he would overturn it every time. If you thought you'd get a no from Schlesinger, you could wait until he was out of town and then ask Clements.

When Johnny Foster (who once hoped to replace Seamans as SecAF) left DDR&E in June 1973, his replacement was Dr. Malcolm R. Currie, a former Hughes manager who had been a vice president of Beckman Instruments since 1969. I worked very closely with Mal Currie during my last two years with the Air Force and found him of great help with several key acquisition programs, such as the Lightweight Fighter and the global positioning system. (Currie rejoined Hughes in 1977, eventually retiring as CEO of Hughes Electronics in 1992.) I encouraged the Air Force assistant secretaries to meet directly with Currie and their counterparts in the OSD without having to go through me unless they desired. I believed in delegating as much authority as possible for day-to-day business, never being what might be called "a control freak."

Within the Air Force, at least at the headquarters, I like to think there was a general sense of relief when I unexpectedly stayed on as secretary. There had been growing concern about how long it was taking to name a replacement for Seamans. With all the uncertainties overseas and turmoil in Washington, I at least was a known quantity, and my appointment represented continuity. I was suspicious of making changes just for the sake of change, or reorganizing in an attempt to improve operations unless something was truly screwed up. Too many leaders, when they can't think of anything else positive to assert their authority, decide to reorganize. Of course with the post–Vietnam era under way, we did have to continue decreasing the overall size of the Air Force, to include closing installations and cutting back on headquarters.

As soon as I was selected to be secretary, I began deciding on a staff. For the sake of continuity, I retained Col William R. Usher, who had already served Bob Seamans for the previous two years, as my military assistant. (The more than four years he remained in that position is longer than anyone before or since.) I also acquired Col John S. Pustay to be my executive assistant. Both Bill and John became general officers. Col Keith McCartney replaced Pustay in March 1974, and Keith remained my trusted executive assistant for the rest of my time with the Air Force, soon thereafter becoming a general officer as well. I really appreciated his positive attitude and ability to get things done. Jack Stempler continued to be a great asset as the Air Force general counsel. As a former legislative liaison for Laird, he could give me sage political advice as well as legal opinions, and he enjoyed excellent relations with key members of Congress. William B. Robinson had replaced Phil Hilbert as deputy undersecretary for international affairs in 1972. I raised his position to report directly to me as the assistant to the secretary of the Air Force for international affairs.

Of the other four assistant secretaries, only William W. "Bill" Woodruff in financial management remained from the end of Bob Seamans's tenure. I offered to make Norm Augustine, then the assistant DDR&E, my assistant secretary for R&D, but as luck would have it, just that morning he had accepted Bo Calloway's offer to become Army undersecretary. I was then fortunate to obtain Walter B. LaBerge, technical director of the prestigious Naval Weapons Center at China Lake, California, as my R&D assistant. I also hired David P. Taylor as assistant secretary for manpower and personnel and Frank A. Shrontz as assistant secretary for installations and logistics. They were all good people and solid performers. Frank came from Boeing, which was grooming him for the chief executive job. Up to that time, he had worked on the commercial side of the company, and Boeing's chairman, Albert "T." Wilson, encouraged Frank to get some defense experience. His time with the Air Force was a great experience for both of us. As for my military deputies, Maj Gen Marion L. "Boz" Boswell served as director of legislative liaison until being replaced by Maj Gen Ralph J. Maglione in August 1974, and Maj Gen Robert N. Ginsburgh continued as director of information until his deputy, Brig Gen Guy E. Hairston, was promoted to replace him in June 1974. Also in 1974, Dennis J. Doolin replaced Bill Robinson as my international affairs assistant.

Until I could get a suitable replacement for myself confirmed as undersecretary, I continued to manage the National Reconnaissance Office. Eventually I picked James W. Plummer to be undersecretary and NRO director. Jim was the vice president and general manager of Lockheed Missiles and Space Company, where he had worked on satellites for the Air Force, NASA, and the NRO for many years. The White House announced his nomination in late November 1973. On my way to work on 2 December I opened the *Washington Post* to see an article about Senator William Proxmire's concern that appointing Plummer might pose "a serious possibility of conflict of interest" because of his air and space industry ties. Thankfully, Proxmire's press release was not one of his notorious "golden fleece" awards about government waste. I quickly called Senator Stennis about how to handle this apparent roadblock in the confirmation process. A few minutes later he called back: "Mr. Secretary," he said in his characteristic southern drawl, "do you think you could have Jim Plummer in my office tomorrow morning."

After an overnight flight from California, Jim was there on schedule. Stennis then had Proxmire come down to his office and, after several minutes of chitchat, Stennis left the two of them together. When he returned, Proxmire stood up, turned to Jim, and said, "Congratulations Mr. Undersecretary." Plummer had obviously impressed the Wisconsin senator with his qualifications and that he had no intention of returning to Lockheed. Stennis then moved the process along, and by mid-December, Jim Plummer was confirmed.

Of the four Air Force chiefs of staff I worked with, George Brown was the one with whom I had the closest relationship. We shared mutual respect, and I really appreciated his candor. George's swearing-in ceremony on 1 August 1973, marked the 30th anniversary of the day he had piloted a B-24 on the famous but deadly low-level bombing raid against the Ploesti oil fields in Romania. When I joined the OSD in 1962, George was McNamara's military assistant, having previously served Thomas S. Gates in the same capacity. He later served as assistant to Gen Earle Wheeler, chairman of the JCS. In addition to knowing his way around the Pentagon, George was a proven combat leader. When I became undersecretary in 1969, he was commander of Seventh Air Force in Southeast Asia, the Air Force's most important operational command. To get his fourth star, he was soon promoted to commander of Air Force Systems Command at Andrews AFB, Maryland, where he impressed me in our frequent contacts involving acquisition programs. All previous CSAFs had served as vice-chief, but in view of the emphasis we were placing on modernizing the Air Force after Vietnam, we thought George would be the best choice. Seamans recommended him to Richardson in March 1973, a choice that Schlesinger later endorsed as well.

Brown became chief of staff a couple weeks after I became secretary. Our partnership was cut short in July 1974, when

Schlesinger picked him to become chairman of the JCS after less than a year as CSAF. On military matters, his judgment and decisions were superb. Unfortunately, in this high-profile position his lack of discretion in public forums got him into hot water on several occasions. Yet I continued to hold him in highest regard and was relieved that he was allowed to remain chairman despite making a misleading remark to some Duke law students in October 1974 about Jewish influence in the press and banking industry and a couple other ill-chosen statements later in his term.[18]

When George Brown was selected to be chairman of the JCS, he, Jim Schlesinger, and I agreed that the right man to replace him was Gen David C. Jones. Jim told me many years later he had planned from the beginning to make General Brown chairman as soon as Admiral Moorer retired, and that he already had his eye on Jones as Brown's replacement. Probably the last military leader who will ever attain such high status without a college degree, Dave Jones had left his studies to join the Army Air Forces in early 1942. His innate intelligence, imagination, and management abilities served him well. He was George Brown's vice-commander at Seventh Air Force before becoming the commander of United States Air Forces in Europe, where he streamlined its command structure while convincing NATO to create a unified air headquarters for Central Europe. Dave was not as personable as George or as inspirational a leader, but he seemed more creative and open to new ideas. He was also more judicious in dealing with the press and politicians. I respected Dave and worked well with him, although at times his desire to impress our superiors at the OSD and the White House may have trumped our partnership. (In any case, we certainly got along infinitely better than my Navy counterpart, John Warner, did with his CNO, Adm Elmo Zumwalt.) Dave was intent on succeeding George Brown as chairman of the JCS, which he eventually did under Pres. Jimmy Carter in 1978—the only time yet that two successive Air Force generals have held the nation's top military post.

My becoming secretary meant more official duties for Pat. In addition to entertaining visitors, going to luncheons, teas, banquets, and other social functions, and continuing an active calendar of church and volunteer work in Arlington, keeping up with my travel itinerary made her life even busier. "I try to go with him whenever I can," Pat said in an interview. "Last fall we discovered we had been away seven weekends out of eight." Reflecting the often hectic schedule of official travel, she found that "it's amazing what a strong impression you can get of a place when you visit less than an hour."[19] When we stopped at Air Force installations, she would spend time with military wives, base schools, volunteer groups, host-nation agencies, and other community-oriented organizations—giving me valuable feedback on factors affecting morale and living conditions. Her commitments and volunteer work left little time for the domestic activities she cherished. Our children continued to enter adulthood while I was in the Pentagon. Pam graduated from Wellesley and got a doctorate at Rutgers University after marrying Jeffrey Byers. John graduated from Wesleyan College in Connecticut, while Rod later graduated from Columbia University in New York. Susan took a sabbatical to serve in Volunteers in Service to America (best known as VISTA) before graduating from Reed College.

Although the Vietnam War was now over (at least for American forces), my first year as secretary was marked by international tension, economic disruptions, and—most of all—political turmoil. The Senate began its Watergate hearings in May 1973. Revelations about the incident and subsequent cover-up as well as related scandals kept undermining the authority of the White House. Henry Kissinger, Nixon's long-time national security advisor, assumed the additional title of secretary of state in August 1973. The resignation of Spiro Agnew in October led to the appointment of Rep. Gerald Ford of Michigan as vice president. Ford provided the nation with an honest and respected head of state after Nixon resigned in August 1974. Although the Department of Defense was still deeply scarred from the Vietnam War, its reputation emerged largely unscathed from the nation's constitutional crisis. At his staff meeting three days after the change in presidents, Secretary Schlesinger read us a message from Gerald Ford praising the department for its steadfast service during the recent time of turmoil.

Schlesinger wanted to reeducate the American public on the importance of a strong defense. He believed the services should publicize their successes, but that all DOD components should speak with one voice on critical issues. At his first staff meeting on 3 July 1973, he made a point that we had one military establishment supporting national objectives, and that we must put away parochialism while maintaining pride of service. Jim gave us his philosophy on dealing with Congress on 14 August. He noted that Laird had run "a one-man show"—admittedly a very effective one—but that other DOD leaders needed to have more contact with key people on the Hill. He also believed that the department needed a better institutional memory to deal with congressional and private critiques of defense requirements, such as those being effectively presented by the Brookings Institution, the Center for Defense Information, and the Federation of American Scientists. In Congress, Jim was respected for his intelligence and honesty, but he did not benefit from the broad base of support with moderates and liberals that Laird had cultivated.

Even more so than his predecessor, President Ford depended on Henry Kissinger to run foreign affairs and set national security policy. The manipulative Kissinger and the combative Schlesinger, who had both graduated from Harvard in the class of 1950, feuded openly and behind the scenes on both strategy and tactics. In view of Ford's reliance on Kissinger, Schlesinger was at a distinct disadvantage in this contest. Ford later wrote that when he became president, he respected Jim's intellect but thought he was too patronizing in dealing with Congress.[20] Although Schlesinger established reasonably good relations with most congressional leaders as time went on, he continued to have run-ins with the White House and antagonize the president.

My experience from dealing with Congress while I was undersecretary made it easy to represent the Air Force when I became secretary. I remember how George Brown and I used to go to the Hill together. One day we were scheduled to testify before George Mahon's House Armed Services Appropriations Subcommittee. When we arrived, he asked us, "Where are all these background and backup people you guys normally have?" I responded, "Well, we thought we'd just come as we are" or something to that effect. "I'm used to seeing a typical crowd of people walk in and try to overpower me with details," he said. "Mr. Chairman," I explained, "it seems to me that the kind of questions that should be asked at your level is the kind that General Brown and I ought to be able to answer." He liked that.

During my final year as secretary, we had to work with the post-Watergate 94th Congress (1975–76). Distinguished by nu-

91

merous liberal freshmen, it continued the trend of its recent predecessors to give increasing scrutiny to defense programs and push for reforms. Yet I found the fears by many in the defense establishment of a drastic swing to the left somewhat exaggerated. No doubt the new Congress was different. It featured many younger members, but the occupational structure stayed about the same, with 288 lawyers and 379 military veterans among the 535 members. In a detail interesting to me as a technocrat, my legislative liaison office counted only four engineers and two scientists in both houses. Reflecting the impatience of the new Democrats to gain influence, they began an assault on the House's seniority system (much as a large crop of new Republicans would do 20 years later under Newt Gingrich). One of the first victims was F. Edward Hebert, who was deposed as chairman of the House Armed Services Committee. Fortunately for the Air Force, Melvin Price of Illinois (whose district included Scott AFB) replaced him. During the early months of 1975, General Jones and I appeared before all four major defense-related committees to defend the Air Force's budget for fiscal year 1976. I had interesting conversations with many of the freshmen congressmen, only a few of whom in my opinion could be truly characterized as "radical." They asked critical but fair questions and showed great interest in our programs, both informally and in the committee hearings.

During my years with the Air Force, we basically fought a holding action against even more drastic cuts in the budget than what occurred. Between fiscal year 1969 and fiscal year 1975, DOD spending sank from almost $149 billion to about $93 billion (in 1976 dollars). In the Air Force, our budget outlays (also in 1976 dollars) declined from $49.8 billion in fiscal year 1968 to $28.7 billion in fiscal year 1976. Force structure was reduced accordingly. In 1968 the Air Force had 856,000 active duty military personnel and 322,000 full-time civilian employees—a total of 1,178,000. By 1975 these numbers had shrunk to 612,000 active duty military personnel and 278,000 civilians—a total of 890,000, or a decline of 24.4 percent. Despite shrinking budgets, we were able to begin some necessary steps toward modernizing weapon systems.

In defending the Air Force request for $30 billion in fiscal year 1976, I answered a question before the House Armed Services Committee by saying, "I would hope we would not take any more continuing cuts. I think we're down about where we ought to be. But if next year it comes down to a choice between taking a cut [in force structure] or one of our weapon systems, we'd take another cut [in force structure]."[21] At the time, I believed strongly that modernizing our equipment was the most pressing need. In view of the antiwar movement, accelerating inflation, the worsening economy, the growing deficit, the expansion of entitlement spending, and the collapse of the Nixon presidency, I am satisfied that we in the Pentagon did about the best we could to obtain the resources necessary during a very difficult period in our nation's political and military history.

From the very beginning of his presidency, Gerald Ford was handicapped in recruiting new cabinet members and other senior officials by having less than a full term ahead of him, not to mention the unpleasant atmosphere in post–Watergate Washington. President Ford therefore relied heavily on reassigning incumbents to fill key vacancies in his administration, as I learned myself in October when he asked me to become head of the Federal Aviation Administration. By the end of that month, Ford and his White House staff were completing plans for several other major personnel changes, an action that became briefly known as the "Halloween massacre."[22] To replace Schlesinger as secretary of defense, Ford chose Donald Rumsfeld, a 43-year-old former congressman from Illinois and ambassador to NATO who had been working as his chief of staff in the White House. (In a trip report after visiting Rumsfeld in Brussels during October 1973, I had written of him: "He is a man who has had a lot of political experience at a young age, and I predict he will have a great future unless something goes wrong somewhere.") Many in the Congress and the press interpreted Schlesinger's firing as not just a personality conflict, but as the price he paid for being such a staunch advocate of defense spending and outspoken skeptic about arms negotiations with the Soviets.

As one of Schelsinger's last official functions as secretary of defense, he graciously presided over my farewell ceremony at Andrews AFB on 7 November 1975, a few days before he too left the Pentagon.[23] Rumsfeld was sworn in as SecDef on 20 November, three days before Jim Plummer became acting SecAF upon my move to the FAA.[24] It had been an interesting six years and eight months in both my life and the history of the Air Force, some episodes of which I will describe for posterity in the next three chapters.

Notes

1. Robert C. Seamans Jr., *Aiming at Targets* (Washington, D.C.: NASA, 1996), 157.

2. The Defense Reorganization Act of 1986 ("Goldwater-Nichols"), as implemented in Title X of the *US Code*, reduced the operational roles of the service chiefs by strengthening the authority of the JCS chairman and establishing a deputy chairman.

3. Copies of many memoranda and other correspondence while McLucas was at Headquarters Air Force, as well as transcripts of speeches, agendas, meeting summaries, trip reports, calendars and other documents, were filed as personal papers. These have been invaluable in helping us reconstruct the thoughts and events described in this chapter and the three that follow. Because of the large number of primary documents used, they are not individually referenced in these endnotes, but sources used for direct quotes are alluded to in the text. Articles and other published material used for quotes are cited in these notes. Microfilms of the McLucas papers are available at the USAF Historical Research Agency, Maxwell AFB, Alabama, and the original paper files have been donated to the US Air Force Academy Library in Colorado Springs.

4. One of Ryan's colleagues on the JCS, Adm Elmo Zumwalt, wrote: "As for Jack Ryan, he was not an easy man to figure out, being laconic by temperament or habit, and evidently more at home as an operational commander. . . . However, I always found what he said, when he said it, to be sensible, honest, and, mercifully, brief." *On Watch* (New York: Quadrangle, 1976), 277.

5. From transcript of an interview of Norm Augustine by John McLucas, 21 June 1999. After serving as assistant DDR&E and undersecretary of the Army, Augustine went on to become CEO of Martin-Marietta and Lockheed-Martin.

6. John Ehrlichman, *Witness to Power: The Nixon Years* (New York: Simon and Schuster, 1982), 97.

7. In one example of special concern to McLucas, the second Bush administration took 10 months to find someone both qualified and willing to be undersecretary of the Air Force and director of the NRO. Peter Teets was finally confirmed for these positions in December 2001.

8. Statement by Kenneth Rush in Gerald S. and Deborah H. Strober, *Nixon: An Oral History of His Presidency* (New York: HarperCollins, 1994), 272. William Bundy in *A Tangled Web: The Making of Foreign Policy in the Nixon Presidency* (New York: Hill and Wang, 1998), 513, contends Rush "would almost certainly have become secretary of state in mid-1973 if it had not been for Nixon's need for Kissinger's prestige to hold off the impact of the Watergate scandal."

9. Richard Nixon, *RN: The Memoirs of Richard Nixon* (New York: Grosset and Dunlap, 1978), 769. Jerry Jones, director of the White House personnel office, later admitted that "almost the entire Department of Defense [i.e., political appointees] was wiped out. On a substantive basis, that was the wrong thing to do. . . . [T]he problem was that the president and Secretary Laird simply didn't get along very well." Strober and Strober, *Nixon: An Oral History of his Presidency*, 371.

10. In December 1974 President Ford hired Bob Seamans to be the first chief of the new Energy Research and Development Administration, which became a principal element in the Department of Energy later created by President Carter.

11. H. R. Haldeman, *The Haldeman Diaries: Inside the Nixon White House* (New York: Putnam's, 1994), 627. Love was named the White House's director of energy policy in June 1973.

12. Richardson served as SecDef only from 30 January until 24 May 1973. He then lost his new job of attorney general in the "Saturday Night Massacre" on 20 October 1973. The White House—in the person of Alexander Haig—asked him to fire Archibald Cox as special prosecutor. Richardson said that he could not do that in good conscience and resigned. Haig turned to the deputy attorney general, Bill Ruckelshaus, who also resigned. The third in line was Robert Bork, the solicitor general, who fired Cox.

13. Senate, *Nomination of McLucas and Brown: Hearing before the Committee on Armed Services*, 93d Cong., 1st sess., 13 July 1973, 1–18.

14. McLucas's papers include a transcript of the speech, which was also summarized by Claude Witze, "USAF's New Leaders," *Air Force Magazine*, September 1973, 58–59.

15. For profiles of Laird, Schlesinger, and others, see Trask, *The Secretaries of Defense: A Brief History*; and Kinnard, *The Secretary of Defense*.

16. As quoted by Nick Thimmesch, "Schlesinger: The Early Years," *Air Force Times*, 5 March 1975, 14.

17. Gerald R. Ford, in his book, *A Time to Heal* (New York: Harper & Row, 1979), 321, wrote that "Schlesinger didn't get along with William P. Clements . . . ; in effect, he kept telling me that I should get rid of Clements."

18. For a biography, see Edgar F. Puryear, *George S. Brown, General, U.S. Air Force: Destined for Stars* (Novato, California: Presidio, 1983). After each incident, Schlesinger lobbied effectively to keep him on board as chairman.

19. Sue Toma, "Pat McLucas: An Interview," *Air Force Times*, 13 March 1974, 33, 35.

20. Ford, *A Time to Heal*, 132.

21. "McLucas Firmly against Force Cuts," *Air Force Times*, 12 March 1975, 5.

22. Ford describes his difficult relationship with Schlesinger in *A Time to Heal*, 320–30. Other changes included replacement of William E. Colby by George H. W. Bush as director of central intelligence, selection of Lt Gen Brent Scowcroft as national security advisor to allow Kissinger to be a full-time secretary of state, replacement of Rogers Morton with Elliott Richardson as secretary of commerce, and withdrawal of Nelson Rockefeller from renomi-

nation as vice president. Schlesinger later served as the first secretary of energy under Carter, and his postgovernment career would include becoming chairman of the MITRE board of trustees.

23. Most of Schlesinger's immediate staff at the OSD also had to leave, and McLucas was happy to get Joe Laitin to be his public affairs director at the FAA. For more on his busy final weeks as SecAF, see the end of chapter 6 and the beginning of chapter 7.

24. Thomas C. Reed, the OSD's director of telecommunications, was sworn in as the next secretary of the Air Force on 2 January 1976. See his book, *At the Abyss: An Insider's History of the Cold War* (New York: Ballantine, 2004), 192–210, for a chapter on his 15 months as SecAF, during which he generally continued on the course set by McLucas.

Chapter 4

Modernizing the Force
New Systems for Future Air Supremacy

"As secretary of the Air Force," I stated several weeks after being confirmed, "my main job is to see that the Air Force does not become obsolete."[1] This referred both to the effectiveness of its equipment and the skills of its people, but at that time, the future capability of the USAF's aircraft and other hardware seemed more at risk than its human resources.

While the Soviet Union was embarked on a concerted effort to modernize its weapon systems, the Johnson administration had been taking money from research, development, and procurement programs to help fund the Vietnam War. In effect, our armed services deferred investing in the future to meet current operating expenses. Despite continued cutbacks in overall defense spending by the Nixon administration, Bob Seamans and I wanted the Air Force to acquire a new generation of weapon systems and other needed equipment.

Reforming Acquisition Management

Since the mid-1960s, the Air Force had turned much of its attention away from developing new systems to modifying existing ones for meeting requirements in Southeast Asia. The "Century series" of nuclear-armed fighter-bombers and interceptors developed in the 1950s had proved less than desirable for conventional combat and counterinsurgency operations. So, somewhat to its embarrassment, the Air Force had to adapt the Navy's A-1 Skyraider, A-7 Corsair II, F-4 Phantom II, and the Marines' OV-10 Bronco for employment in Southeast Asia. The USAF also relied heavily on Navy-developed missiles, namely the Air Intercept Missile (AIM)-7 Sparrow and AIM-9 Sidewinder, for air-to-air combat. As the challenges faced in Vietnam became better understood, the Air Force was successful in modifying other systems and developing some new weapons for combat there. Notable innovations included side-firing gunships, various sensors, remotely piloted vehicles (RPV), electronic jamming pods, precision-guided munitions (PGM), and computerized command and control systems (such as those described in my MITRE section).

At the same time the Vietnam War drained funds for pursuing new systems, some of the major programs initiated during the early McNamara years encountered embarrassing cost overruns and performance problems. Two were of special concern when I became undersecretary: the F-111 fighter-bomber, known as the TFX when I began my association with it in 1962, and the giant C-5A Galaxy transport, which had its beginnings as the CX-4 during my previous Pentagon tour. Mel Laird reportedly once said, probably only half in jest, that his three major problems as secretary of defense were the Vietnam War, the F-111, and the C-5A.[2]

The C-5 was developed and built under the new total package procurement procedures described in chapter 2. Much to the frustration of Gen Bernard Schriever, commander of AFSC from 1961 to 1966, the inflexibility of the C-5 contract inhibited Air Force officials overseeing Lockheed from taking corrective action when they discovered problems. The F-111 and C-5 also highlighted some downsides of "concurrency," that is, overlapping the development, testing, production, and deployment of new systems. Although concurrency had worked with early ballistic missiles, this strategy was less successful when applied to aircraft programs and tactical weapons.

In response to cost overruns, performance deficiencies, congressional criticism, and unfavorable media attention, Richard Nixon promised to reform defense acquisition.[3] Deputy Secretary of Defense David Packard led these efforts, which were just getting under way when I arrived on the scene. Bob Seamans and I shared a common philosophy with Packard. We all thought McNamara and his team (of which I had once been a small part) had gone badly off course with some of their ideas, especially the total package concept and not building prototypes or demonstrators. Although in the short term these items might seem unnecessarily time-consuming and expensive, I came to believe that we could not afford *not* to do prototyping, which I considered a form of insurance against future disasters. As I wrote at the end of 1969, "Analysis must eventually be backed up by hardware. Much of the notorious overruns on systems comes from postponing the cost of verifying ideas with hardware into the production phase, where we pay for it ten or a hundred fold."[4] I also thought that the accelerating pace of scientific progress was leading to instability in program management. In a memorandum for Secretary Laird on 8 January 1970, I suggested that "part of our difficulty is caused by technological advances which make yesterday's modern systems seem obsolete (and in need of upgrading) today. In part, the problem is one of [in]adequate control at the top."

During his first several months in office, Dave Packard initiated a number of changes in acquisition policies, often called the Packard Reforms. His new rules included preparing "selected acquisition reports" to keep Congress better informed about major programs, making more realistic cost estimates, more precisely defining operational requirements, conducting technical risk analyses, decreasing concurrency in favor of sequential schedules, and returning to the practice of building prototypes. Packard also established the Defense System Acquisition Review Council (DSARC), forerunner of today's Defense Acquisition Board (DAB), to review program status and recommend milestone decisions to the SecDef after each phase of a major program—from concept definition through development and into production.

I gave Dave Packard my wholehearted support and did my best to sell his initiatives within the Air Force. His new management philosophy, as I favorably described it at the time, was "to decentralize authority, pinpoint responsibility, and optimize innovation and flexibility." I especially liked his "fly before buy" philosophy. "Our management procedures," I further wrote, "must concentrate on maintaining competition as long as possible through the various stages of R&D and procurement. Competitive development allows one to choose between alternatives, weighing performance and cost."[5] We later put this policy into practice with competitive "fly-offs" and demonstration/validation (dem/val) projects, such as the Lightweight Fighter (YF-16 versus YF-17), A-X Close Air Support (YA-9 versus YA-10), and Short Takeoff and Landing (STOL) Transport (YC-14 versus YC-15) programs. We also used more prototypes for subsystems and experimental equipment. In the decades since, I have been glad to see this basic concept continued with the Advanced Tactical Fighter (YF-22 and YF-23) and the Joint Strike Fighter (X-32 and

X-35). I must admit, however, that quantum leaps in computer modeling and simulation have allowed early evaluations without the need for actual hardware. And for really large aircraft, such as the C-17 Globemaster III and B-2 Spirit, building complete prototypes would have been too costly.

As part of their new participatory management philosophy, Laird and Packard restored to the service secretaries more independent responsibility for source selection decisions. In view of the major new systems being developed, Bob Seamans made a goodly number of such decisions, having set up an impartial procedure based on the merits of contractor submissions, not political influence, which helped deter protests by losing contractors. In each case, the process culminated in a source selection decision memorandum (drafted by his military assistant in coordination with the chairman of a source selection advisory council) documenting the reasons. When I became secretary, I continued Seamans's procedure, although Jim Schlesinger was more involved in the process than his immediate predecessors. After reviewing the detailed findings of a source selection evaluation board, I would go over the factors favoring the competing contractors with Schlesinger or Deputy Secretary of Defense Clements and inform them of my tentative decision. In every case, they agreed with my choice. I would then make the announcement for large contracts at a press conference or issue a press release for less newsworthy contracts.

While I was undersecretary, Grant Hansen was our main focal point for acquisition matters in his capacity as assistant secretary of the Air Force for research and development (SAF/ RD). Under him were several deputies with offices that dealt with requirements, research, engineering, development, and laboratories. We later dropped the deputy for engineering and added a deputy for technical and information systems. When I became secretary, I hired Walter B. LaBerge as my assistant secretary for R&D. Coming from the Naval Weapons Center at China Lake, California, he knew a lot about developing new technologies. We realigned his R&D organization into deputies for advanced technology; tactical warfare systems; strategic systems; space programs; and command, control, and communications. This arrangement, I believed, interfaced better with actual Air Force programs and organizations in the field.

Although the Office of the Secretary had great authority over acquisition matters, the Air Staff—specifically the deputy chief of staff (DCS) for Research and Development (AF/RD)—was heavily involved as well. It interfaced directly with AFSC and the operational commands on acquisition matters, from defining requirements to implementing policies. Four lieutenant generals headed AF/RD during my Air Force career: Marvin L. McNickle from February 1969 to January 1970, Otto J. Glasser from February 1970 to June 1973, William J. " Bill" Evans from August 1973 to August 1975, and Alton D. Slay from September 1975 to March 1978. I had no real problems with any of these gentlemen, although Otto Glasser considered me something of an enemy because of my NRO role. He was one of the Air Force's early space pioneers, and he later became frustrated by the Air Force's lack of authority over the National Reconnaissance Program. I really liked Bill Evans. He looked like the Hollywood ideal of an Air Force general, but his good looks were backed up by a strong intellect. Walt LaBerge also enjoyed a smooth relationship with Evans, who I always expected to become chief of staff. I think he would have made a good one.

Even though there was some duplication between SAF/RD and AF/RD, I believe that sharing acquisition responsibilities between the Secretariat and the Air Staff was advantageous. Unlike today, it gave the chief of staff a major stake in the acquisition process. The two sides seemed to work amicably together, and I can't recall the secretary and the chief having to mediate many disputes between their staffs. So-called reforms in the late 1980s removed most acquisition functions except requirements and test and evaluation (T&E) from the Air Staff.[6] My discussions with recent chiefs of staff and other general officers have revealed a degree of frustration at the lack of blue suit influence on the acquisition process.

In addition to improving procurement policies and procedures, Dave Packard also emphasized improving the quality of DOD personnel involved in acquisition work. For example, Dave was the main force behind establishing the Defense Systems Management School at Fort Belvoir, Virginia, in 1971. In line with Packard's DOD-wide initiatives, Bob and I looked at ways to enhance the acquisition career field within the Air Force, both for military and civilian personnel. To fill acquisition-related jobs, we needed a good

101

mix of civil service employees for their continuity and specialized knowledge and military personnel for their operational experience and assignment flexibility. Occupations included scientists, engineers, contract administrators, quality assurance specialists, auditors, and project managers. On aircraft and weapons projects, many of the military officers also had to be "rated" (i.e., flying) personnel. Because the traditional officer assignment system encouraged them to get a wide variety of experiences to qualify for promotions, many managers did not remain long enough with acquisition programs to make much of an impact. We encouraged extending the acquisition assignments of senior officers and raising the rank of major program managers. For example, when I came on board, the $4 billion C-5 program, which was getting more congressional attention than just about anything else the Air Force was doing outside of Vietnam, was being run by a colonel. Meanwhile, the Air Force had numerous general officer billets scattered around obscure air divisions and numbered air forces where nothing much ever happened.

The need to develop a professional acquisition workforce with its own career ladder versus the desire to get acquisition personnel with "real-world" experience continues to be an issue. In my day, we had Air Force Systems Command to take care of most acquisition professionals. As I understand the situation today, there are special Air Force and DOD personnel management programs to do this. Some top Air Force generals have told me, however, that the blue suiters in the acquisition corps spend most of their time becoming experts in acquisition but no longer know enough about the operational side of the Air Force.

As for the entrenched defense acquisition structure of the late 1960s, implementing reforms did not come easy. I must admit to having changed some of my own philosophy since the time I worked in the OSD, but the trend toward centralization was hard to stop. In my first year as undersecretary, I made the following critique of the Pentagon bureaucracy we inherited:

> Excessive layering and centralization have been serious problems throughout the Department of Defense. The layering has been a product of adding staff for years, without streamlining. The addition of OSD offices led to growth of military staffs rather than to their reduction. And because good people were usually chosen for these jobs, they needed something to do. Therefore, they frequently usurped the authority [that]

lower levels used to have. There were two, three, and four-star generals who had to go to the Pentagon for very minor decisions. Even in those cases where some authority was delegated, many senior officers had a psychological barrier, a feeling of little authority and a fear of relentless staff interference.[7]

Dave Packard was making internal changes to defense acquisition at the same time the somewhat influential Presidential Blue Ribbon Defense Panel, chaired by Gilbert W. Fitzhugh, studied the entire military structure (see chap. 3). Most of its reorganization proposals were not implemented, but the panel's conclusions on acquisition were generally consistent with much of Packard's philosophy. In my analysis of the Fitzhugh report for Dave and Mel Laird, submitted on 10 August 1970, I wrote that "I agree with its basic recommendations to do a better job of operational test, with most of its recommendations in procurement, and with a need for better long-range planning both in R&D and in general."

As regards to operational testing, the panel's call for establishing independent operational test and evaluation (OT&E) organizations to help ensure that complex weapon systems really worked in the field provoked opposition within the Air Force.[8] As early as February 1971, I suggested in a memo to Grant Hansen, "We ought to take a look . . . to see whether we think our OT&E function is independent enough." In April 1971 Dave Packard revamped DOD testing practices in this direction. He defined the engineering phases of testing (from preliminary laboratory experiments through flight test) as development test and evaluation (DT&E). He then instituted the initial operational test and evaluation (IOT&E) to verify a weapon system's performance under realistic field conditions before deciding on full-scale production. Congress soon incorporated Packard's policy into public law, which effectively banned concurrency from major aircraft programs. Then, in 1972, the Commission on Government Procurement reported that all but one of 22 weapon systems deployed to Southeast Asia from 1965 to 1970 had suffered major deficiencies in the field. Nevertheless, many within the Air Force, especially Air Force Systems Command, opposed creation of a separate operational test agency. After more outside pressure, and with my concurrence, General Brown (formerly an AFSC commander) in late 1973 ordered establishment of the Air Force Test and Evaluation Center, later renamed the Air Force Operational Test and Evaluation Center (AFOTEC).

While Dave Packard was immersed in the details of management, Mel Laird stayed on top of the overall acquisition process. As I saw it, his most influential role was to determine how best to satisfy congressional concerns and help us sell our programs on the Hill. He also had some specific ideas on fixing total package procurement that I thought were very constructive. For example, he agreed that it was good to plan for all the different expenses that accompany an airplane buy—items like spare parts, simulators to train pilots, and test equipment. So we kept the best parts of the total package concept. I think I accurately expressed this balanced philosophy in early 1970 when I wrote, "Total Package Procurement is an excellent cost control method for procuring certain types of military material. Such procedures make sense for buying systems that do not involve a great deal of new technology." For example, a fixed price total package contract worked well with the C-9A Nightingale medical evacuation aircraft, which was based on the DC-9 airliner and used mostly off-the-shelf equipment.[9]

In most areas, Headquarters Air Force and Air Force Systems Command embraced steps to adapt acquisition management practices to the Packard policies. The early 1970s brought administrative changes, such as a comprehensive series of formal program reviews to keep AFSC and Air Force headquarters, the OSD, and Congress informed of major programs while still hoping to delegate authority to the product divisions and system program offices (SPO) that did most of the work.[10] In retrospect, however, the momentum toward centralization proved inexorable except in very special circumstances (such as "black" programs). The commander and his staff at Headquarters AFSC, we at Headquarters Air Force, and our superiors in the OSD all wanted to be kept informed of progress and to be able to take action to solve problems. This led to various reporting requirements and review panels, up to and including the DSARC.

Although Dave Packard made a lot of changes during his three years as deputy SecDef, he encountered the same challenge that frustrates almost all DOD leaders when it comes to acquisition. The Pentagon is a huge beast. You can tilt its head, but it takes a long time to get it to go where you want. Even Packard, despite his changes in the bureaucratic processes, could only do so much in the time allotted. Interestingly enough, Dave was able to revisit defense acquisition a decade and a half later when Pres. Ronald Reagan named him to head his Blue Ribbon Commission on Defense Management. Benefiting from Packard's prestige, the commission's final report in 1986 helped usher in even more sweeping changes in acquisition procedures and organization than Packard had implemented from inside the Pentagon.

Defense acquisition policies continued to evolve in the mid-1970s after the departure of Dave Packard and Mel Laird. With the twin threats of inflation and declining defense budgets as a backdrop, I disclosed some of these new policy directions to defense industry leaders at a world air and space conference in San Francisco in October 1974. I announced there that "design to cost" would now be the most important criterion in selecting new weapon systems, and that contractors would need to make tradeoffs between cost and performance, so long as the resulting design could satisfy requirements.[11] As analysis techniques improved, we would look more closely at life-cycle or ownership costs, including logistical support after delivery. This in turn could mean better reliability and maintainability, which was becoming an ever-greater problem as electronics and software took an increasing share of equipment costs. To promote improved electronics dependability, the DOD began a comprehensive program for such components to be standardized as much as possible with multiplex buses and modular "black boxes."

As time went on, I became increasingly unhappy with the resumed drift toward centralization and micromanagement, often through what I saw as abuse of the Defense Systems Acquisition Review Council. When the DSARC was first implemented, decisions on only three of eight phases in the life cycle of a new system were to be made by the secretary of defense. The other five were supposed to be a service prerogative. In July 1975 I was so bold as to make the following complaint in a speech at the Defense Systems Management School. "We have seen in the recent past a proliferation of pre-DSARC and post-DSARC activities, which is generating excessive workloads for our program managers and their staffs and inhibiting timely decisions. In some ways, the responsibilities that properly belong to the [s]ervices are being weakened by direct OSD staff involvement."

105

Judging from the general trend toward even more centralized DOD management over the next quarter century, I was just another voice shouting into the wind.

In most cases, however, people are more important than process. Getting knowledgeable and experienced individuals from the private sector to serve as political appointees in the acquisition field was not easy in the early 1970s, and I believe it is even more difficult in today's environment. In that regard, I agree with something Johnny Foster said when he was the DDR&E. He expected good appointees to have conflicts of interest and accepted this—just so, he knew what those conflicts were. That way, they could be watched to ensure they did not take advantage of their positions. In essence, a balance of conflicting interests would more realistically protect the government than trying to purge connections with the defense industry.

Some service secretaries and undersecretaries have been reticent about working too closely with corporate leaders for fear of being perceived as too friendly. I always felt that I could deal with any such impressions, and I liked to work directly with industry people, getting to know what they saw as their biggest problems and trying to help if possible. I also wanted them to hear firsthand whether we were happy or unhappy with their work. In many cases, it was the latter. I recall once asking a glad-handing company executive visiting my office, "If you were grading yourself on the job you're doing for us, what would it be?" Taken aback, he finally admitted a "C-minus." He must have taken this message back to his company, as its performance began noticeably to improve. Among the industry executives I enjoyed working with most (on either Air Force or NRO business) were T. Wilson at Boeing, Thomas Jones at Northrop, Sanford "Sandy" McDonnell at McDonnell Douglas, James M. Beggs at General Dynamics, Gerhard Neumann at General Electric, Allen Puckett and Albert "Bud" Wheelon at Hughes, Edwin "Din" Land at Polaroid, Irv Kessler at RCA, and Simon "Si" Ramo and Rube Mettler at TRW Corporation.

Sustaining Basic and Applied Research

In addition to the immediate pressures of acquisition programs, I focused attention on the longer-term need to reinvigorate research activities. Air Force funds for R&D had fallen from $2.06 billion in fiscal year 1964 to only $1.56 billion in fiscal year 1968. By fiscal year 1974, the amount had climbed to $2.4 billion, and my last budget with the Air Force for fiscal year 1976 obtained $3.3 billion for R&D.[12] Unfortunately, these totals do not reflect the corrosive effects of inflation, which accelerated relentlessly during those years. As a scientist, I was especially concerned about a reduction of DOD funding for basic research by 50 percent from 1967 to 1975, from the equivalent of $139 million to only $78 million (adjusted for inflation).

Since World War II, the Air Force has used a combination of "in-house" government laboratories, university science and engineering departments, nonprofit corporations—including federally funded research centers—and defense contractors to perform basic and applied research as well as analytical studies. Each has advantages and disadvantages. With my background at HRB-Singer and MITRE Corporation, I felt fairly well qualified to deal in this arena. In addition to MITRE, the Air Force's other principal FCRCs, both headquartered in California, were the RAND Corporation, which had been founded in 1946 as the Air Force's "think tank," and the Aerospace Corporation, which had been created in 1960 to do systems engineering for space programs.[13] I thought the Air Force's in-house laboratories had come a long way since I first began working with some of them in the late 1940s and early 1950s. Although they lacked the special expertise of the FCRCs, I felt very comfortable defending our organic science and technology accomplishments on the Hill. The contributions of universities, however, were becoming more problematic. One of the more disheartening effects of the Vietnam War was the growing antimilitary sentiment on college campuses, with its negative impact on defense-related research. What happened at the Massachusetts Institute of Technology serves as an especially notorious example. It was also very distressing to both Bob Seamans and me on a personal basis, in view of our past connections with MIT.

Ever since World War II, MIT had been DOD's top university contractor. In the late 1960s, its support of defense research, especially by the prestigious Instrumentation Laboratory at Cambridge and Lincoln Laboratory at Lexington, Massachusetts, came under criticism from antimilitary activists in the faculty and student body. A chief victim of this crusade was Dr.

Charles Stark Draper, who had founded the Instrumentation Lab in the 1930s and thereafter developed inertial guidance and many other navigation technologies, including that used by NASA for the Apollo moon mission. After a veritable inquisition by a university-appointed review panel, Draper was forced to resign in October 1969. At the same time, MIT's leadership announced their intention to reorient the lab's projects from military technology to civilian projects. To help soften the blow to its founder, they renamed the Instrumentation Laboratory the Charles Stark Draper Laboratory. Some militants were still not satisfied, and in January 1970 MIT's executive offices were occupied and vandalized by demonstrators, including that of my recent overseer, James Killian. In response to all this, I actively supported an effort involving the DOD, NASA, and FAA to accommodate MIT's desire to diversify these labs into more civilian-oriented work. In 1973 the Draper Lab was divorced from MIT and became an independent entity.[14]

In addition to similar incidents at other universities, an amendment sponsored by Senate Majority Leader Mike Mansfield in 1971 curtailed the DOD from supporting "exploratory" research in general—anything without a direct military benefit. Dr. Jerome B. Wiesner, president of MIT and a former White House science advisor, testified to a Senate subcommittee in 1975 that this "has had a very negative effect on both basic and applied research activities in our country."[15] Although in the long run it was probably more appropriate for agencies such as the National Science Foundation to sponsor most such research, the Mansfield amendment's immediate impact was unnecessarily disruptive. Several years later, this unfortunate legislation was repealed.

With university research under siege, the status of our federal contract research centers became even more important. Yet as I knew from first hand experience, the FCRCs were subject to special scrutiny and constraints. In November 1969, for example, the House Military Appropriations Subcommittee criticized RAND for doing studies not directly related to the Air Force mission. One of the FCRC issues I dealt with was ensuring their costs were commensurate with their benefits. Among the advantages of FCRCs was their ability to hire experts at higher salaries than the government, yet retain them in specific disciplines longer than normal defense contractors. FCRCs could also be given access to a wide range of proprietary and competition-sensitive information without causing a conflict of interest. Some in Congress, however, criticized the salaries and overhead costs of our FCRCs. In February 1970 President Nixon placed a cap of $45,000 on most FCRC salaries in response to congressional complaints that some FCRC managers were making more than they were.

On top of the overall scrutiny of FCRCs, the RAND Corporation soon received some unwelcome publicity for one of the most notorious leaks ever to embarrass the executive branch. In June 1971 the *New York Times* began publishing the celebrated "Pentagon Papers," the top secret OSD review of US policy toward Vietnam commissioned by McNamara. The source of the leak was Dr. Daniel Ellsberg, a one-time DOD official working for RAND who had surreptitiously duplicated RAND's copy of the multivolume study. The contents of the study (which Mel Laird liked to refer to as the "McNamara Papers") covered events during the Johnson administration. Even so, the Nixon White House went after the press for publishing the documents and created the ill-fated "plumbers" unit to discredit Ellsberg and plug other leaks—a scheme that eventually led to the Watergate break-in. The administration angrily ordered the Air Force, as RAND's primary sponsor, to launch a tough security crackdown on the corporation. We found some laxness in procedures, but in retrospect, it seemed like a case of locking the barn door after the horse escaped. In the defense budget for fiscal year 1972, Congress dictated a cut in money for RAND as well as for the Institute for Defense Analyses, OSD's primary FCRC.

Before he left, Dave Packard asked me to think about a long-range plan to prevent the gradual demise of the FCRCs. Since they were serving a vital role, I thought we either had to persuade Congress to stop crippling them or come up with some other way of performing their services. With this in mind, I called together a study group in early January 1972. It included Lt Gen Otto Glasser (AF/RD), Maj Gen Glenn Kent (assistant chief of staff for studies and analysis), Joseph Jones (deputy assistant SecAF for R&D), and Jack Stempler (general counsel). We concluded that while Project RAND (the Air Force component of the corporation) had tended to drift away from serving our needs in the past, establishment of an Air Force advisory group two years earlier by General Glasser had re-established a much closer working relationship and ensured that recent RAND work was relevant to Air Force needs. After looking at alternatives—ranging from making it into a typical contractor to chartering it as a government corporation—we decided to keep RAND and our other FCRCs as independent sources of specialized expertise and hope the quality of their work would eventually quiet the critics.

Developing Weapon Systems

Upon becoming undersecretary, I was impressed by the new systems the Air Force had in advanced stages of development but worried about how we could pay for them if they all entered full-scale development in the next year or so. In June 1969 I recommended that we "prioritize" our major programs to spread out peak contracting costs over several years. My recommended priorities were to (1) push procurement of the F-15 Eagle fighter, which had manageable technical risks and was probably most urgently needed; (2) stretch out the promising AWACS by focus-ing initially on a careful evaluation of the two competing radars; (3) slow down the A-X close air support program to conduct a full competition between prototypes; and (4) more carefully refine re-quirements for the Advanced Manned Strategic Aircraft (AMSA) to control its escalating costs before getting too far along in its design. Although I can't claim too much of the credit for what was very much a corporate process, this strategy (with some re-finements) was essentially what happened in the 1970s.

Early in my Pentagon tour, when Secretary Laird was taking a lot of heat from his congressional buddies on the F-111 and C-5 fiascos, I had a call to come down to his office. (Bob Seamans was out of town.) Mel had just learned about an especially trou-bling failure in the C-5's wing during a fatigue test. "I want you to tell me," he said in a frustrated tone of voice, "what programs the Air Force is doing right." Taken somewhat aback, I asked, "What exactly do you mean?" He replied, "I mean an airplane that did exactly what it was supposed to do at the price you said it was going to cost." I had to get with the staff to do some re-search. A couple hours later, I went back and told him I'd found one program from several years ago: the C-141 Starlifter. "No fighters?" he asked, and I had to admit that the C-141 transport was the only example we could find in the last 10 years. In March 1970 I was able to send him a more complete list of 18 programs we thought were successful, including modifications to existing aircraft as well as helicopters, remotely piloted vehicles, spacecraft, radars, and electronic systems. Almost three years later, just before he left the Pentagon, Mel was passing out some awards and recalled when he embarrassed me about our aircraft programs. He said that now, when he discussed Pentagon procurement with Congress, he was happy to praise the Air Force's management of aircraft programs. Specifically, he was referring to the F-15 air superiority fighter, A-10 attack aircraft, E-3 AWACS, and B-1 strategic bomber.

Looking back at the 1970s, it is apparent that we followed a conscious strategy of investing in the future by devoting as much money as possible to new systems. Because of inflation and declining budgets, we were unable to spend as much as desired on logistical support and other elements of operational readiness, and we lost a lot of experienced people to downsizing. Readiness was also hurt by the need to furnish South Vietnam and Israel with equipment and supplies taken from Air Force stocks. Nevertheless, because of the long lead times required to produce modern weapon systems, we thought it advisable to press ahead with modernization while postponing the purchase of more mundane items that could be obtained more quickly when funds became available in the future. In effect, we anticipated that congressional reluctance to approve foreign interventions after Vietnam, the Nixon Doctrine to rely on regional allies as surrogates for American forces, and Kissinger's desire for continued détente with the Soviet Union made it unlikely that the United States would be called upon to fight another war in the near term. In hindsight, we were either astute or lucky. The renewed chill in the Cold War and the Islamic revolution in Iran ushered in the big defense buildup of the first Reagan administration, which provided needed money for logistics and personnel. Most of the new USAF weapon systems fielded in the 1980s, however, had already been developed.

During my almost seven years at Headquarters Air Force, I was privileged to see real progress on a variety of aircraft, weapons, and other systems. To get firsthand knowledge of the programs that crossed my desk, I flew on almost all of the Air Force's aircraft, taking the controls myself whenever possible, and tried to get hands-on experience with other equipment. I found those interludes to be one of the most enjoyable parts of my job. I believe history has shown that most of the new systems we nurtured in the first half of the 1970s have served the nation well in the decades to come. They helped deter the Soviets for the rest of the Cold War and then proved themselves in combat during the Gulf War of 1991, the air war over Serbia in 1998, and the occupation of Afghanistan in 2001. Modified and updated, a number of these systems will be in operation far into the twenty-first century.

While I was his undersecretary, Bob Seamans generally took principal responsibility for major aircraft and missile programs and some of the less classified space programs such as launch vehicles. He delegated oversight to me for remotely piloted vehicles, precision-guided munitions, and systems involving electronics and command, control, communications, and intelligence (C^3I). I was also responsible for the classified systems developed by the NRO. Except for the latter, this division of labor was not hard and fast. We both filled in for each other and, depending on the phase of a program, he might tell me to take over one of "his" aircraft for a while. As mentioned in the previous chapter, I had Harry Davis to provide expertise on electronic systems and Bob Naka as my deputy for national reconnaissance matters. When Dr. Seamans left, I assumed primary responsibility for all acquisition programs until Jim Plummer came on board to take over the NRO. In the following pages, I'll share memories about aircraft and weapons programs in progress during my watch, while the next chapter will deal with space systems.

Salvaging the F-111 Fighter-Bomber

One of our first big challenges was fixing the controversial F-111 program, which the Air Force was trying to groom into a specialized long-range strike aircraft. In the midst of a drawn-out flight test program, six F-111As deployed to Southeast Asia in March 1968 for an ill-fated combat evaluation. Then fatigue testing in August 1968 showed a fatal weakness in a wing carry-through box (the structure supporting its pivoted swing wings). Because most of the 125 F-111A models had already come off the production line, fixing this required expensive retrofits. Problems continued to crop up after I got to the Pentagon. On 22 December 1969, the Air Force lost its 15th F-111A when a wing broke off on a flight from Nellis AFB, Nevada, killing the two-man crew. We immediately grounded almost the entire fleet of 223 aircraft, and Congress soon began another round of hearings into the program.[16] Although (as Bob testified in March 1970) the F-111 actually suffered fewer accidents per flying hour than any of the Century series fighters a decade earlier, doubts about its structural integrity, as well as the costs of making the necessary corrections, were serious causes for concern.

We made General Dynamics conduct a rigorous inspection, cold-temperature stress testing, and a modification of 340 F-111s that was finally completed at the end of 1971. The extra cost was covered by canceling production of several airplanes. To be on the safe side, we also scheduled Air Force Logistics Command to conduct a follow-on inspection starting in 1973 for aircraft approaching a specified number of flying hours. This eventually solved the wing issue, but problems with the avionics (especially the "Mark II" package on the F-111D), engines, canopies, and other components continued to plague various F-111 models for many years.

Nevertheless, the F-111's ability to carry a heavy payload over long distances at high speeds during night and in bad weather at extremely low altitudes eventually gave the Air Force an unprecedented new strike capability. Flying low level in an F-111 was quite impressive. I had the opportunity to ride in one while it sped at Mach 1.2 only 250 feet above the Nevada desert. When I looked down to the side, I saw a V-shaped "shadow" caused by its supersonic shock wave dancing against the blur of sand and creosote bushes. General Dynamics ultimately built a total of 563 F-111s and FB-111s through 1976, with some of the original F-111As being converted to electronic jamming EF-111A models in the late 1970s and early 1980s.

During 1971, however, there was not much enthusiasm in either the Air Force or the DOD for continuing production of the troublesome and expensive aircraft. As early as June, I raised the issue of buying additional F-111s as well as A-7Ds with Mel Laird. I warned that even though we had not budgeted for con-

113

tinued production, perhaps we should be prepared for some re-programming of funds in case the White House wanted to keep the line open for political purposes, especially in an election year. According to H. R. Haldeman's diary, OMB director George Schultz called on 8 January 1972 to ask the White House about our plans to cancel F-111 production later in the year. "The Air Force doesn't want the plane and doesn't want to continue it," wrote Haldeman, "but Laird says Defense won't make the decision, that this is a White House political matter." Haldeman checked with President Nixon, who "said they should go ahead with it."[17] As for the A-7D, Congress kept adding unsolicited production funds for it (as well as for Lockheed's C-130 transport) to the Air Force's budget for many years to come.

After Britain cancelled its planned purchase of F-111s in 1968, the Royal Australian Air Force (RAAF) remained the aircraft's only foreign customer. The F-111's long range was especially attractive to the RAAF, which wanted to be able to cover Indonesia to the northwest. Incorporating longer wings and heavier landing gear (similar to that of the F-111B), the Australian model became the F-111C. By late 1969, after more than a year of delays and now costing over $300 million, delivery of the F-111Cs appeared to be drawing near. I was scheduled to visit Australia to work on arrangements for a secret satellite downlink station in December 1969 (see chap. 5), so Bob Seamans asked me to also do some public relations for the F-111C on my trip. I went to the RAAF's base at Amberley in southeast Queensland to be shown its new facilities. I also appeared on television with Defence Minister Malcolm Fraser (later elected prime minister), and we said nice things about our cooperative venture. A few days after I arrived home, the wing fell off the F-111A at Nellis, indefinitely postponing delivery of the Australian aircraft. I was glad only that the accident didn't happen while I had been "down under."

As a result of this latest disaster, we had to negotiate a new joint agreement, signed in April 1970 after some painful and delicate discussions with the Australian Ministry of Defence and Department of Air. Their 24 aircraft were kept in storage pending corrective action, and we had to extend a loan of F-4Es to the RAAF until the F-111Cs were modified with new wing carry-through boxes. That arrangement cost the USAF $100 million (as much as the original F-111C contract). General Dynamics eventually put more than a million man-hours into the modification and refurbishment program, which began in April 1972. The first of the renovated F-111Cs did not reach Australia until June 1973, five years later than originally promised. I trust the Aussies found the wait worthwhile. With retirement of the USAF's last EF-111s in 1998, the RAAF's F-111Cs along with several replacement F-111As and a later purchase of 15 F-111Gs (former SAC FB-111s) remained the only F-111s of 562 built still in active service.

Fixing the C-5 Galaxy Transport

Lockheed's wide-body C-5 transport and its total package procurement contract gave us another giant headache. With a length of 248 feet compared to 145 feet for the C-141A, the C-5 weighed more than two and one-half times and had five times the cubic volume of its little brother. Perhaps company engineers convinced themselves that they merely had to scale up the C-141 to make the C-5. In reality, trying to make a wing strong enough to carry its full payload but light enough to meet range requirements proved difficult, and pushing the giant fuselage through the air challenged the thrust of its engines. After delivery of the first C-5A in mid-1968, it was determined that the wings would only last for 7,000 flying hours versus the 30,000 called for in the contract. Trying to fix this and other discrepancies would be very expensive. By the time Bob and I arrived, the scope of the cost overrun was becoming apparent, and the Senate began hearings in May 1969 on what became known as the C-5 controversy.

Seamans oversaw a number of steps to identify the scope of the major problems, especially structural fatigue of the wing assembly. Coming on the heels of the F-111 problems, we took a lot of heat. Bob usually served as the Air Force's point man on the Hill, but I can remember Senator William Proxmire grilling me on how we could ever build aircraft with wings that break off. Our near-term solution for the C-5 was to limit the weight carried by C-5s except in emergencies. In May 1970, with the long-term fix to the wing problem still unsolved, Dave Packard discussed the program with Harold Brown, who said that he was appalled to learn now about problems that had been occurring during his watch. The only conclusion he could draw was that people had kept a lot of information from him. I told Dave it might also have been a case where the right questions had not been asked. Harold's assistant secretary for installations and logistics, Robert H. Charles, had been a champion of total package procurement and was against any renegotiation of the contract. Some of the military procurement officials also downplayed the seriousness of the C-5's problems. As late as the spring of 1969, Gen James Ferguson, commander of AFSC, publicly defended the use of total package procurement for the C-5 "without need for being protective or defensive about it."[18]

Lockheed's problems with the C-5 and other programs, such as its L-1011 Tristar, threatened it with bankruptcy if we held the company to its C-5 contract. Since Lockheed was also doing a lot of work for the NRO, Packard asked me what I thought about the company going broke. Because Lockheed Space and Missile Corporation was a separate subsidiary, I said its future was important to the NRO, but not that of the parent company. In view of Lockheed's place in the defense industry, Dave Packard went to Secretary of the Treasury John Connolly, who said Lockheed was too important economically to be put out of business. So Dave renegotiated the contract for the C-5. It held the company's losses to $200 million, considerably less than might have been justified. The early 1970s were an especially tough time for the entire air and space industry. Even Boeing was having problems. Once when Pres. T. Wilson of Boeing was visiting with Packard, he called me down to his office to join the conversation. Wilson said Boeing didn't have enough orders on its 747 to break even. Dave agreed the Air Force might use the 747 for the future airborne national command post if Boeing could offer a decent price. Boeing did, and the E-4 became the result.

The Air Force originally planned to buy 120 C-5As. Seamans was away when Mel Laird had to decide (probably at the behest of someone on the Hill) how to absorb the escalated cost of the aircraft, so General Ryan and I agreed to cut back our total buy to 81. The last of the C-5As were delivered in May 1973. Some years later the Air Force spent about $1.6 billion more to have the wings replaced to last 30,000 flying hours. Even so, maintaining the C-5A remains a major challenge to USAF airlift capabilities.

The manner in which the cost overruns on the C-5A were brought to public attention became an issue unto itself. While Harold Brown was still secretary, an Air Force civilian appointee named A. Ernest Fitzgerald, the deputy for management systems to the assistant secretary for financial management, took it upon himself to break the news of the C-5A's $2 billion cost increase to Congress sooner than planned. To those already suspicious of the Pentagon and the defense industry, the publicity-hungry Fitzgerald became a classic example of a "brave whistleblower," especially when his position was abolished after Bob Seamans became secretary. The upshot of this apparent firing of a conscientious bureaucrat was considerable criticism on Capitol Hill, especially from Senator Proxmire, a drawn-out appeal to the Civil Service Commission, more bad press for the Air Force, and minor celebrity status for Fitzgerald. The White House was even drawn in, with some of its embarrassing memos and discussions about what to do with Fitzgerald later revealed during congressional hearings.

In September 1973, several months after I succeeded Seamans, the Civil Service Commission finally ruled that the Air Force had to reinstate Fitzgerald with back pay. I had Bill Woodruff create a new position for Fitzgerald as deputy for productivity management in the Office of the Assistant Secretary for Financial Management, but I waited to rehire him until we got another opinion from the commission that this new job was comparable to the one he had once held. Mr. Fitzgerald returned to work in December 1973. Even so, he was unhappy with his new position, and I don't think he was entrusted with many substantive projects. Yet he continued to draw an Air Force paycheck, no matter how little he was able to accomplish.[19]

World events soon vindicated the C-5's operational utility if not its procurement record. When Egypt and Syria attacked Israel on 6 October 1973, I had just begun a two-week trip to Europe. As we visited various allied headquarters, embassies, and American bases, regular news reports and periodic intelligence updates revealed a surprisingly desperate struggle by Israel. By the time I reached US European Command Headquarters near Stuttgart on 13 October and received the latest intelligence, Israeli forces were in dire need of equipment and supplies. American air carriers refused to fly into the war zone, and all European nations except Portugal denied use of their airfields and airspace. Back in Washington, the president approved Jim Schlesinger's recommendation to use the Military Airlift Command for an emergency operation, one that George Brown had already set in motion. On 14 October while my party was flying west across the Atlantic, several C-5s and C-141s were flying east. After a refueling stop in the Azores, they began landing in Tel Aviv to the cheers of the awed populace. The sight of the massive Galaxies disgorging the heavy M-60 main battle tank and other outsize cargo was an impressive demonstration of their capability.

In the first 33 days of the famous Berlin airlift in 1948, 6,885 American transport sorties (mostly flown by C-47s) delivered 33,357 tons of supplies less than 250 miles to the besieged city. Our 33-day airlift to Israel, code-named Nickel Grass, required only 147 C-5 and 422 C-141 missions to move 22,395 tons an average of 6,400 miles. Based on lessons learned about the limitations of the C-141, we initiated a program to lengthen its fuselage to increase payload and add aerial refueling to increase range—a program that ultimately modified the Starlifter fleet to the C-141B model.

Ensuring Air Superiority with the F-15 Eagle

Motivated by the USAF's rather poor combat record against MiGs over North Vietnam and the threat posed by a new generation of Soviet aircraft, Air Force planners in the late 1960s developed the "fighter-experimental" (F-X) concept as a replacement for the F-4. A loose cadre of pilots, analysts, and engineers (sometimes called the "fighter mafia") focused on the F-X's need for air combat capabilities. Influenced especially by the energy-maneuverability theory of Maj John Boyd, they wanted the new plane to become the Air Force's first true air superiority fighter since the F-86 Sabre of Korean War fame. Lt Col Larry Welch, a very astute and persuasive officer in the Air Force Studies and Analysis Office (and more articulate than Boyd), redefined the F-X concept with realistic operational requirements and sold these to the Air Force leadership. F-X proponents then had to overcome the resistance in DDR&E, where some wanted the next fighter to also focus on air-to-ground missions.

Although the F-X project was too far along to require proto-types by the time the Laird-Packard-Seamans team took over, we felt the need to incorporate some lessons from the F-111 and C-5 programs. Bob and I got Dave Packard to veto long-standing OSD plans to use a fixed price total package contract (still desired by Johnny Foster) in favor of a cost plus incentive fee contract. We also added lots of demonstration milestones to the schedule. In July 1969 we began evaluating the final-ists in the design competition: Fairchild Hiller, North Ameri-can, and McDonnell Douglas. In December, after hearing from a source selection evaluation board and an advisory council, Bob announced McDonnell Douglas as the winner. We did use a competition between prototypes for the new fighter's engine and fire control system (specifically, Hughes and Westinghouse Corporation for its advanced Doppler radar). The resulting F-15 Eagle pushed the technology of its time, achieving an op-timum combination of airframe, engines, and avionics. Other jet airplanes could sometimes travel straight up, but the F-15 (with a thrust to weight ratio of 1.4:1) was the first able to ac-celerate while doing so.

Although the F-15 later experienced some growing pains, such as with its Pratt and Whitney F100 engines, its acqui-sition went more smoothly than the F-111. Starting in July 1972, the Air Force Flight Test Center at Edwards AFB, Califor-nia, began subjecting it to thorough testing before the approval for full production. To some degree, I think the success of the program can also be credited to our new personnel policies. Rather than assign only a colonel to run the system program office as in past projects, we gave the task to Brig Gen Benja-min Bellis (soon promoted to major general) with authority to run a "super SPO." Earlier in his career, he had been the Air Force manager for the phenomenal SR-71 Blackbird. He did another outstanding job with the F-15 Eagle. President Ford ceremonially accepted the Tactical Air Command's first F-15 at Luke AFB, Arizona, in November 1974. I'll never forget my first flight in the backseat of an F-15B. By the time it reached the end of the runway, we were at 15,000 feet!

One of my disappointments at the time was not being able to convince the Shah of Iran to buy F-15s instead of Navy-devel-oped F-14s. In addition to McDonnell Douglas, a lot of people in the Air Force, such as Ben Bellis, were very anxious for Iran to select the F-15. When the shah visited Washington in 1974, I escorted him out to Andrews AFB to show him an F-15 and an AWACS plane, both on display for that purpose. Since the shah was a pilot, he wanted to see the Eagle take off and fly. Unfortunately, a strong wind was blowing right up its tailpipes, and the engines couldn't be started until someone thought to turn the plane around. Although embarrassing, that was probably not the reason he opted to buy F-14s, with their long-range Phoenix missiles. We also took the shah up in the AWACS, which he found quite impressive. The United States was fortunate indeed that none of the shah's E-3s (unlike his F-14s) had been delivered before the Iranian revolution.

My only quibble with the F-15 was one that I've had with many weapon systems: the desire for excessive performance that adds to their cost. Even though the F-15 would carry air-to-air missiles and do most of its fighting under Mach 1.5 (where maneuverability was of primary importance), its contract called for a top burst speed of Mach 2.5. Many years later, Lt Gen George K. Muellner, one of the original F-15 test pilots who eventually became the Air Force's top military acquisition official in the 1990s, admitted to me that the only time an F-15 needed to fly that fast was to pass the qualification test called for in the contract. A bit later, I raised this issue with Gen Larry Welch, who became director of the Institute for Defense Analyses after retiring as Air Force chief of staff. Besides helping write the original F-X requirement, he commanded the 1st Tactical Fighter Wing at Langley AFB, Virginia, when it became the first operational F-15 combat unit in the mid-1970s. He explained that we did not really pay a lot just to reach Mach 2.5, and having that capability meant the F-15 could intercept an enemy more quickly and then have the excess thrust needed to accelerate during hard maneuvers. So in retrospect, some of my skepticism about the F-15's top speed may have been misplaced.

By the time I became secretary, I was getting tired of hearing how the Soviets were overtaking us in aircraft technology. So I pushed for a demonstration of the new F-15's incredible performance. Using a specially prepared airframe nicknamed the "Streak Eagle," this took place at Grand Forks AFB, North Dakota, in January 1975. The Streak Eagle broke eight existing records for reaching various altitudes then held by a US Navy F-4B and Soviet MiG-25 Foxbat. The Soviets soon reclaimed the overall time to height record with a souped-up Foxbat (the Ye-266M), but in 1976 the defection to Japan of Soviet lieutenant Viktor Belenko in his MiG-25 would show the Foxbat to be inferior to the Eagle in almost every respect except top speed at high altitude.

Progress on the F-15 was an early sign that Air Force aircraft procurement was getting back on track. One big reason was the leadership within Air Force Systems Command. In the previous chapter I described my high regard for Gen George S. Brown, who commanded AFSC from 1970 to 1973. Perhaps the most important contributor to modernizing the USAF for the post-Vietnam era was Lt Gen James T. Stewart, commander of the Aeronautical Systems Division (ASD) at Wright-Patterson AFB, Ohio, from 1970 to 1976. Jimmy was a brilliant, personable, and energetic officer—a real joy to work with.

Guaranteeing Close Air Support with the A-10 and AC-130

As indicated in chapter 2, one of the more bitter interservice controversies in the quarter century after World War II was the Army's complaint that the Air Force was not serious about providing CAS to soldiers engaged in ground combat. In 1966 an Air Force team responded with the A-X concept—the first jet plane designed to loiter at low speeds over the battlefield. Later evolution of the A-X also emphasized destroying enemy tanks and operating from austere airstrips near the front lines. The Air Force chief of staff, General McConnell, had supported this concept, but many in the Air Force and on the DDR&E staff opposed making the A-X such a simple, single-role fighter. Meanwhile, some in the Army still dreamed of flying their own CAS aircraft. Overcoming opposition to the A-X was not easy. After delays marked by hostility between the Army Staff and the Air Staff, Bob Seamans and Stan Resor stepped in and reached an agreement on the A-X program despite the misgivings of their respective chiefs of staff, Generals Jack Ryan and William Westmoreland. This is one of the few examples I recall when we exercised civilian control over the military leadership so overtly.

On a related matter, both Bob Seamans and I had to keep up pressure for the continued production and upgrading of the AC-130 Specter gunship. In an analysis I made of interdiction on the Ho Chi Minh Trail after visiting Southeast Asia, I called it "without question, our most effective truck killing weapon." Although eventually admitting the AC-130's value in the Vietnam War, much of the Air Force's uniformed leadership considered it a poor investment for the future because the lumbering plane clearly could not survive Soviet-style air defenses. In contrast, I thought it would still have a role to play. On 12 June 1970 I wrote in a memorandum for Bob, "In planning for the future, we will no doubt need an airplane with effective guns. . . . The gunship has the reputation for being vulnerable. Yet since 1 July [19]69 in Laos we have lost one AC-130 to enemy fire. During this period we had 55 jet fighter aircraft combat losses in Laos." Seamans and Dave Packard overcame the resistance of the Air Staff, Tactical Air Command headquarters, and the JCS to convert more C-130s to an upgraded AC-130 configuration known as "Surprise Package" that featured heavier weapons and improved sensors. In the decades since, the AC-130's usefulness in special operations and low-intensity conflicts has been vindicated many times over, most recently in Afghanistan.

As for the A-X, in September 1969 Bob and I got the agreement of several people, including General Ryan and Johnny Foster, that it would be a good program in which to try out some new management procedures: going for a prototype approach, eliminating the bulk of routine paperwork, and short-cutting many of the steps normally followed. The Air Force issued an RFP in May 1970 for two prototypes to compete in a classic "fly-before-buy" competition. Subsequently, Northrop built the YA-9, and Fairchild Republic built the YA-10. After an intensive and closely matched fly off during the second half of 1972, Bob Seamans selected the more rugged YA-10 in January 1973. We also held a competition at Eglin AFB, Florida, to select a contractor for the A-10's massive 30 mm Gatling gun. The Senate Armed Services Committee then made us conduct another fly-off to compare the A-10 with the faster and more sophisticated A-7D. I had nothing against the A-7D—I had flown on a two-seat version out of the Patuxent Naval Air Test Center and managed to drop a bomb on a sunken target ship—but it was not equivalent to the exceptionally robust A-10. After trials at Fort Riley, Kansas, in the spring of 1974, the A-10 was judged better for the close air support mission. The DSARC did not approve the final production go-ahead until preproduction models of the A-10 completed two more years of developmental and operational testing.

In service since 1977, the A-10 has proven itself cheap, dependable, and able to absorb more punishment than any other aircraft in history. Officially named the Thunderbolt II in memory of the P-47 of World War II fame, its pilots soon began calling it the "Warthog." Advances in surface-to-air missiles have called the survivability of all low-flying aircraft into question, and in the Gulf War and Kosovo operations, the A-10 operated mainly from midlevel altitudes. Despite some performance limitations that went with its low cost, I still consider the A-X program an exemplary case study on the effective use of prototypes. I hope the Air Force adds the necessary upgrades to keep it flying for many years to come.

Prototyping with the YF-16 and YF-17 Lightweight Fighters

The prototype competition that most engaged me was the Lightweight Fighter (LWF). On my first day on the job, when I was being introduced to other members of the OSAF, I entered Grant Hansen's office. On his desk I saw the drawing of a really sleek looking jet fighter. He had brought the picture with him from his previous job at Convair (soon to become part of General Dynamics). "That's a beautiful airplane—what is it?" I asked. It was an artist's concept of what became the YF-16. With encouragement from certain Air Force and DOD personnel, the company had already been preparing for the possibility of a small but high-performance jet fighter. Meanwhile, Northrop was working on a competing design that became the YF-17.

The idea of the LWF had originated with a splinter group of the Fighter Mafia that thought the Air Force needed a simpler, lower-cost fighter (initially dubbed the F-XX) to complement the F-15. This group included John Boyd (who in 1971 became a member of our new Air Force Prototype Study Group) and Pierre

Sprey, a civilian working for the assistant secretary of defense for systems analysis. I got to know Pierre fairly well, finding him to be bright but very opinionated. In the case of the LWF, I think his opinions were mostly correct. As for the now legendary Colonel Boyd, when presenting him an award in June 1975 for his contribution to the operational capability of the Air Force, I especially praised his persistence in promoting the innovative idea of energy maneuverability, "to the point where right now I don't think we would consider . . . a new weapon system without making this one of the dominant factors."

The F-XX proponents believed we needed quantity as well as quality to match the larger Soviet fighter fleet. This idea for a more affordable force structure, which became known as the "high-low" mix, also had potential for extensive foreign sales. Despite these arguments, which were strongly endorsed by Secretary Seamans and me, there was considerable opposition within the Air Force to pursuing the LWF because of fears that it might draw funding away from the more capable F-15. Dave Packard became a powerful proponent of the LWF, not the least because its development would involve the use of pure prototypes. He approved the program in late August 1971 as one of the advanced technology demonstrations he had solicited from the services earlier in the year.

As acting SecAF on the last day of 1971, I gladly approved release of a remarkably short and simple RFP for a small (approximately 20,000 pound), highly maneuverable aircraft capable of Mach 2 top speed and costing only about $3 million per copy if produced in quantity. The LWF competition began as an advanced technology demonstration, emphasizing aerodynamic performance without any commitment for a follow-on contract. Some of my critics later considered this a clever way to get the LWF's "foot in the door" without immediately provoking too much opposition from F-15 advocates. In the program's early stages, however, I was careful to put a damper on ideas about it being anything more than a great experiment.

On 1 April 1972 we awarded the prototype contracts to General Dynamics and Northrop. By limiting their financial and legal risks, these cost-plus-fixed-fee contracts with flexible schedules encouraged the contractors to take technical risks. In essence, we relied on carrots instead of sticks to obtain the desired results. Considering that their prototypes were supposed to reflect low-cost approaches to fighter design, both companies came up with innovative technologies that truly advanced the state of the art. GD's YF-16, for example, featured electric rather than hydraulic flight controls ("fly by wire") and innovative human factors engineering that allowed tight, high-speed turns. Reflecting the flexible nature of the LWF program, we reduced its specified top speed to Mach 1.6 to keep costs from escalating.

When some NATO allies established their requirement for a new fighter in February 1974, I became more eager to transition the LWF from demonstration models into an operational aircraft. The OSD leadership was amenable to this, but I continued to encounter opposition from most of the generals I worked with, up to and including George Brown. In essence, they didn't want anything to endanger procurement of 500 F-15s. We gradually won most of them over by pointing out the fiscal and political realities we were facing from Congress and influential defense analysts. I also had the support of Jim Schlesinger and Bill Clements, both of whom favored the high-low mix to allow a larger force structure. In addition to budgetary considerations, General Dynamics' big factory in Fort Worth might have added to Clement's enthusiasm.

On 7 March 1974 I advised the Senate Armed Services Committee about a new $36 million program element to begin work on an air combat fighter (ACF) for the 1980s. On 27 April Secretary Schlesinger formally notified Senator Stennis of our plan to transform evaluation of the two LWF designs from technology demonstrations into a full-blown competitive fly off to serve as the basis for the ACF. The YF-16 got a head start on 20 January 1974 when it made an unscheduled takeoff during a taxi test. Because of using new General Electric YJ101 engines, the YF-17 could not begin flight-testing until 9 June. By the end of our accelerated test program in January 1975, the two YF-16s had flown 347 sorties compared to 288 sorties for the two YF-17s. In accordance with our source selection plan, Col William E. Thurman, the ACF program director, chaired a source selection board comprised of personnel from various major commands and NASA that evaluated technical, logistics, and management factors. General Stewart chaired a source selection council to review and validate the board's findings. To complement the various American members, I added to the council a lieutenant general from the Royal Netherlands Air Force to represent the European multinational fighter program committee.

As secretary, I was the final source selection authority. The results of the fly off were close, but not agonizingly so. On 13 January 1975 I held a press conference to announce the YF-16 as the winner. Although both aircraft performed very well, I explained that the YF-16 demonstrated more agility, range, and acceleration. It also cost less, was more fuel efficient, and promised long-term logistical savings because its Pratt and Whitney F100 jet engine was the same basic model as used in the F-15. Later in the day, the Air Force awarded a $418 million full-scale development contract to General Dynamics Fort Worth Division to begin building F-16 Fighting Falcons.

In May 1974 even before we began the fly off at Edwards AFB, the four-nation NATO consortium (Belgium, Denmark, Norway, and the Netherlands) sent a delegation to discuss possibly selecting the LWF winner as a multinational jet fighter to replace their increasingly dangerous and obsolete F-104 Starfighters. In meetings with the multinational steering group during the next two months, Walt LaBerge, my assistant for R&D, assured the European officials that the winner of the LWF competition would enter the USAF inventory. In August Frank Shrontz, assistant secretary for installations and logistics, led an interagency American delegation to Brussels to discuss financial arrangements. The Europeans' desire for a quick decision caused us to compress our LWF test program and source selection by several months. General Dynamics submitted the formal F-16 proposal to the Europeans on 14 January, one day after my source selection announcement. At the same time, the consortium received offers from Sweden's SAAB-Scania for the Viggen 37E and from France's Dassault-Breguet for the Mirage F1.

On 21 May 1975, as the multinational decision approached, one of the YF-16s went on a European tour to demonstrate its capabilities. In addition to performance advantages, provisions for coproduction in Europe and other generous terms had already convinced all the nations to go with the F-16 except Belgium, which was under intense economic and political coercion from its neighbor to the south to accept the Mirage F1. Our side

126

also used some high-level sales pressure. In late May President Ford discussed the F-16 with Belgian premier Leo Tindemans, followed up by less promising discussions on 2 June between Jim Schlesinger and the Belgian defense minister, who American sources believed had probably been bribed by Dassault.

I was in France when, five days later, the European consortium (Belgium included) announced selection of the future F-16. That night General Dynamics threw a memorable party in Paris. As if to validate this decision, I had the pleasure of watching the visiting YF-16 put on a stunning performance at the Paris Air Show. I then talked about the selection on visits with defense officials in the four nations, all of whom cited its technical superiority. French foreign minister Jacques Chirac, however, lamented the rejection of the Mirage as a decision against the European air and space industry. The consortium's initial order of 348 aircraft marked the first of many foreign sales of the F-16. On the down side, complicated multinational manufacturing and financing arrangements led to an administrative and accounting nightmare for the Air Force and General Dynamics. GD deserves a lot of credit for making the idea into a reality, as does Frank Shrontz for his oversight of the process at Air Force headquarters. Within NATO, having the F-16 as the principal fighter for several air forces greatly improved logistics and interoperability.

When the YF-17 lost out to the YF-16, Northrop focused on the Navy's requirement to supplement the big and expensive F-14 Tomcat with a new fighter-bomber. Largely because of the Navy's traditional mistrust of single-engine jets for operations at sea, it was not interested in the F-16. Bill Clements was instrumental in working out a way for the Navy to accept a beefed up version of the YF-17. Northrop then teamed up as a junior partner with McDonnell Douglas to produce the F/A-18 Hornet, which has become the sea services' primary combat aircraft. So our Lightweight Fighter Program was a win-win proposition for the competing firms as well as the Air Force and Navy, which got the planes they wanted without having to compromise their own requirements.

Promoting the F-5E Tiger II

Some aviation experts consider Northrop's F-5 as an earlier example of a lightweight fighter. At just over 47 feet in length, 26 feet in wingspan, and 8,000 pounds when empty, the F-5A/B Freedom Fighter (derived from the T-38 Talon jet trainer) was truly small, lightweight, and well suited to the foreign air forces that flew it. In March 1969, the month I arrived at the Pentagon, Northrop began testing a slightly larger version of the F-5, with radar and other avionics for air-to-air combat. Shortly thereafter, we sponsored the International Fighter Aircraft (IFA) competition to pick a successor to the F-5A/B that could deal with the latest models of the MiG-21 Fishbed. In addition to Northrop, three other companies entered the IFA competition: Ling-Temco-Vought, Lockheed, and McDonnell Douglas. In November 1970 we declared Northrop's entry the winner, with an initial contract for 340 aircraft. One month later, Northrop's new aircraft was designated the F-5E and soon nicknamed the Tiger II. In June 1973, like the YF-16 two years later, the F-5E put on a spectacular display for potential customers at the Paris Air Show. Northrop would eventually build 792 F-5Es, 140 F-5Fs (two-seat trainers), and 12 RF-5Es, while factories in Switzerland, Korea, and Taiwan would build more than 500 additional variants of these models under license.

The Air Staff as a whole could not get very excited about the F-5E, which some blue suiters considered more of a toy than a real fighter. Bob Seamans and his staff, however, were more enthusiastic, mainly in view of its great foreign military sales potential. Grant Hansen, Phil Whittaker, Phil Hilbert, and later Bill Robinson were among the key advocates who helped make the F-5E/F program a success, and I gave it my full support when I became secretary. Even though the USAF had no interest in acquiring the F-5E as a frontline fighter, it did eventually adopt the agile little Tiger II (which was about the size of a MiG-21) for dissimilar air combat tactics (DACT) training. Based on the Navy's successful Top Gun program, the Tactical Air Command (TAC) in 1972 formed an aggressor squadron at Nellis AFB, Nevada, to provide F-4 pilots with a chance to fly against a smaller and more maneuverable adversary. At first TAC had to use less robust T-38s from Air Training Command as an interim measure. When Saigon collapsed in early 1975, a group of about 70 F-5Es originally destined for delivery to South Vietnam suddenly became available for Nellis. Additional aggressor squadrons were soon formed in Europe and the Pacific. Over the next 20 years, these specialized squadrons were very instrumental in improving the air combat capabilities of Air Force fighter pilots.

Demonstrating Advanced Transport Technologies

Besides the YF-16/17, the major prototype program during my watch began as a technology demonstration of jet-powered transport aircraft that could possibly replace the C-130 Hercules. For such a streamlined program, it had a rather clumsy official name: the Advanced Medium Range Short Takeoff and Landing Transport (AMST). Having been interested in STOL transports ever since the early 1960s, I wanted this program to be successful. In accordance with contracts awarded in November 1972, Boeing developed the YC-14 and McDonnell Douglas, the YC-15. The contracts allowed each company great latitude in their designs—specifying only critical requirements such as the size of the cargo bay, short takeoff distance (2,000 feet with a 27,000-pound payload), full payload size (53,000 pounds from a 3,500 foot runway), and ferry range (2,600 nautical miles). We even used two different types of contracts: a conventional cost-plus-fixed-fee contract for Boeing, and a cost-sharing limited government obligation contract for McDonnell Douglas, which allowed the latter company to retain ownership of one aircraft for possible commercial development.

The two companies responded with two very different and innovative aircraft. Boeing's YC-14 featured two high-bypass ratio General Electric (GE) turbofan engines extending above its supercritical wings, deflecting their exhaust over the upper surfaces to generate extra lift. McDonnell Douglas's YC-15 had four smaller Pratt and Whitney turbofan engines protruding under the wings, using slotted flaps that could be lowered into the engines' exhaust to achieve extra lift. The first YC-15 arrived at Edwards AFB in August 1975, and the even more revolutionary YC-14 arrived more than a year later. Testing on both experimental designs and their later modifications was very promising, although the Douglas entry was eventually judged as the best choice.

Yet even before I left the Air Force, it was becoming obvious that the AMST program lacked support for moving on to full-scale development. Lockheed had a lot of political influence for continuing C-130 production, and some in the Army feared a STOL transport would threaten its plans for a heavy-lift helicopter. Meanwhile, both the OSD and the Air Force were becoming more interested in a longer-range, dual-role transport that could meet strategic as well as tactical airlift requirements. With numerous other defense programs competing for money, the Carter administration dropped AMST funding from the defense budget in 1977. While it lasted, the program served as an interesting experiment in aerodynamics and propulsion options that later paid dividends when McDonnell Douglas developed the wide-bodied C-17 Globemaster III.

Meeting Technical and Political Challenges with the B-1 Bomber

One major new aircraft program that did not feature a "fly off" was the B-1 bomber. After five years of planning the AMSA—sometimes also called "America's most studied aircraft"—we issued an RFP for the B-1 in November 1969. North American Rockwell (soon renamed Rockwell International) won the cost-plus-incentive-fee contract award in June 1970 to build three prototypes. As with the F-15, these would be thoroughly tested to achieve various milestones before a production decision. Air Force Systems Command appointed an experienced general officer to run the SPO, Brig Gen Guy Townsend, who had learned a lot of lessons as program director for the C-5 in the mid-1960s.

Because of its strategic mission and projected cost, the B-1 was controversial even while still on the drawing board. Advocates of a strong nuclear deterrent saw the new bomber as a more capable and survivable replacement for the aging B-52 Stratofortress in our strategic triad of ICBMs, submarine-launched ballistic missiles (SLBM), and long-range bombers. Others in the defense community thought penetrating manned bombers were becoming obsolete and preferred to invest money in new weapons, such as standoff cruise missiles. Speaking for the sea service and many in the defense reform community, Secretary of the Navy John Chafee expressed the view that maintaining the "nuclear triad" was becoming too expensive, and he obviously did not mean to do away with SLBMs. Then there were the nuclear disarmament zealots who, for ideological or religious reasons, were against existing strategic weapons, let alone any new ones.

Controlling the cost of the future bomber in a period of inflation became our biggest concern, leading to a major review of
the new bomber's design called "Project Focus" (a nickname
also used for a similar effort with the F-15). Soon after my arrival in 1969, I became an advocate of controlling its cost by
eliminating some of the features asked for in the original AMSA
concept and being more flexible in relaxing contract specifications. Exercising my oversight of electronic programs, in June
1971 I strongly recommended postponing much of the B-1's
expensive avionics package, especially its multimode radar and
electronic countermeasures. I did not think we should yet be
spending money on a rapidly evolving technology that would
soon be rendered obsolete. As time went on and the design
matured, the choices became harder. We even had to eliminate
two of five planned prototypes and stretch out the test program
from 12 to 24 months.

By the time I became secretary, Congress was getting restless
about the B-1. Indeed, George Brown and I were grilled about
its delays and increased costs during our confirmation hearing by the Senate Armed Services Committee. In August 1973 I
requested an independent review of the program by a panel of
more than 30 experts, headed by Raymond Bisplinghoff, deputy
director of the National Science Foundation.[20] Its report, submitted in November 1973, identified many technical risks and
basically concluded that the B-1 development program was too
ambitious for its funding and needed more preproduction development and testing. It also pointed out serious morale and
continuity problems faced by Rockwell's workforce because of
layoffs between delivery of the third prototype and a production
decision. Although we scrubbed requirements and tightened up
contract management, only Congress could ease the budgetary
constraints that were the root cause of many problems. In February 1974 I asked Senator Stennis to help us fund a fourth prototype in fiscal year 1975 and possibly a fifth in 1976.

Some cost-saving proposals we rejected as too severely com-
promising the B-1's mission, such as replacing its variable
sweep wings with fixed wings. The plane's exceptionally complex crew escape capsule, which tests had found unstable at
speeds over 300 knots, did offer potential savings. In October
1974, after Rockwell told us they wanted another $300 million

above the original estimate to fix the capsule, I called my old friend, Gen Russell Dougherty (then SAC commander), basically saying, "Russ, we have two choices: you can either get the capsule or get the B-1." As expected, he told me to "forget the capsule." Other economy measures included eliminating variable engine inlet ramps (thus lowering its top speed from Mach 2.2 to just over the speed of sound), reducing titanium content, and simplifying avionics.

Some of the B-1's critics could not be satisfied no matter how hard we tried to correct problems and control costs. An anecdote related to the completion of the first B-1 serves to illustrate the mind-set we were dealing with at the time. In September 1974 we sent routine letters over my signature inviting some key members of Congress to witness its rollout ceremony at the Rockwell Plant in Palmdale, California. Two members, both outspoken critics of defense spending, declined our invitation. Representative Otis Pike of New York explained politely that he "would not want to add anything more to the cost of the program." Representative Patricia Schroeder of Colorado was more critical, lambasting an "egregiously inappropriate" ceremony that constituted "improper lobbying by an executive agency for a program requiring Congressional approval." In her letter of regrets, she further wrote, "I find it inconceivable that the Department of Defense intends to celebrate such a monumental expenditure."

Even so, there was still a large crowd in attendance when Rockwell rolled out its first B-1 on 26 October 1974. Flight-testing began two months later, and soon we were able to order a fourth prototype incorporating many design changes. In December 1976 the DSARC approved full-scale production; however, Congress passed an amendment deferring any final decision until after the presidential election. By June 1977 inflation and a reduction in the planned size of the fleet had raised the projected per unit cost over the politically sensitive $100 million threshold, and President Carter chose to cancel future B-1 production. Secretary of Defense Harold Brown and Gen David Jones, Air Force chief of staff, supported this decision. Many airpower advocates thought Jones should have resigned in protest; instead, he became chairman of the JCS.

Meanwhile, a series of tests and studies involving the four existing B-1As kept the program on life support until 1981, when President Reagan announced his decision to rapidly procure 100 B-1Bs so that the Air Force would have a modern penetrating bomber until an advanced technology stealth bomber might become available. I supported the idea of resurrecting the B-1 but not the hurried production without adequate testing that ensued.[21] As for the stealth bomber, when the Cold War ended I was very skeptical about whether its continued procurement would be worth the astronomical cost per plane. I still worry about the reaction if and when the first B-2 is lost, whether in combat or an accident.

Birth of the Stealth Fighter

Regarding future stealth aircraft, I departed as secretary before the Air Force got actively involved. I was of course well aware of how reduced radar signatures added to the survivability of the SR-71 Blackbird and various unmanned vehicles. As I learned later, the Defense Advanced Research Projects Agency (DARPA) under George Heilmeier initiated some exploratory studies in 1974 on extremely low radar observability, largely at the behest of Mal Currie. Although Jim Schlesinger told me many years later that he got Dave Jones to advance DARPA some Air Force money for a stealth competition between Lockheed and Northrop, I don't remember being informed at the time. On 1 November 1975, a few weeks before I left the Air Force, those two companies were awarded $1.5 million contracts for the first phase of an experimental test-bed program.[22] After further development under the code name Have Blue, the stealth program really took off under the active stewardship of William J. Perry, who became head of Defense Research and Engineering in January 1977, leading to the secret deployment of the F-117A Nighthawk.

Although I can't claim any credit for the stealth aircraft program, I take a lot of pride in helping advance the distinguished career of Bill Perry. Not only did I hire him for his first job in 1951 (see chap. 1); I also encouraged him to go to the Pentagon in 1977. When Harold Brown offered him the DDR&E job, he was reluctant to leave his company (ESL Inc.) and sell its stock. I told him I knew what he was going through and sent him an

133

editorial I had written for *Aviation Week and Space Technology* concluding that, if I hadn't gone into government service, "I'd probably regret not having done so for the rest of my life."[23] Bill later told me this made it easier for him to make up his mind. After serving as undersecretary of defense for research and engineering in the Carter administration, he went on to become the deputy secretary and then secretary of defense in the William J. Clinton administration. I believe he was the most competent and respected member of Clinton's national security team and one of our best secretaries of defense.

Transforming Air Warfare with AWACS

The major program with which I was involved the longest was the Airborne Warning and Control System. I still consider AWACS a technological *tour de force.* It evolved from research done by the Electronic Systems Division with support from MITRE on Over-land Radar Technology (ORT). New Pulse Doppler radars and computer software allowed accurate tracking of aerial targets over land by factoring out surface echoes (called ground clutter) and factoring in speeds of moving targets. In 1967 Boeing and McDon-nell began working on proposals for an airframe to carry the revolutionary radar and associated data processing equipment.

From almost the day I arrived in the Pentagon, however, I expressed concern about the projected costs of the AWACS program, which at the time had an ambitious development and production schedule to deploy a full fleet of aircraft. Every time the radar contractors made an improvement, I was duly impressed with their ingenuity, but I was still cautious about trying to push the technology too fast. On 14 May 1969 I submitted a memo to Secretary Seamans with the following idea to restructure the program.

> The fleet of 40 or more will cost several billion dollars. If we were really trying to save money, we would first check out the key technical item—in this case, the radar. We could conduct a radar fly off. We could select a prime AWACS contractor, but ask him to hold down the non-radar portion of the system until the radar works. After it works, we could fund him to build a couple of austere AWACS aircraft, put them in the field and get some feedback. Then, and only then, do the full system configuration and then build in quantity.

My scenario is essentially what unfolded. Although design-ing the distinctive rotating radar dome ("rotodome") took some time, the airframe itself was not our primary concern. In July 1970 the Air Force awarded Boeing a contract for two prototype aircraft. Mainly to save money, we decided to convert Boeing's plentiful and dependable 707-320B airliner into the AWACS platform, starting with the two test-bed models (given the mili-tary designation EC-137D). In addition to providing the air-frames, we also selected Boeing as the prime contractor for integration of the various components. The main focus of the AWACS program was its radar, command and control, commu-nications, data processing, and other electronic subsystems. I generally fought attempts to reduce these capabilities too dras-tically in pursuit of an "austere" AWACS as penny-wise and pound-foolish. For example, some OSD officials including Dave Packard and members of DDR&E suggested we limit AWACS to air defense only. Fortunately, the AWACS concept of opera-tions evolved from air defense of North America to theater air operations overseas, and the Tactical Air Command replaced the Aerospace Defense Command as primary sponsor. This was significant. Funding for the air defense of North America was rapidly declining, but battle management for the tactical air forces was of growing importance.

As regards the all-important radars, my suggestion to Air Force colleagues and DDR&E that we slow down overall system development to conduct a radar "fly off" competition became a key element in the acquisition strategy. ESD used the first two EC-137Ds to test the Westinghouse and Hughes radars in early "brass board" configurations. Employing a variety of combat air-craft as targets, a combined USAF-Boeing test force in Seattle evaluated the two radars in several geographical environments from April to September 1972. The following month, we selected Westinghouse as the winner. In January 1973 the Air Force be-gan full-scale development. The more Air Force tacticians and OSD analysts learned about AWACS capabilities, the more they began to see its most important mission as orchestrating a high-intensity air campaign, especially in Central Europe.

For budgetary reasons, we had to cut back the number of preproduction aircraft, now designated officially as the E-3A, from six to four (to include reconfiguring the EC-137s). On the whole, however, the program did not encounter major budgetary problems, largely because Maj Gen Kenneth Russell at ESD made realistic cost estimates and met them. I considered Ken an ideal program manager. Activities included system integration along with a series of developmental and initial operational tests and evaluation phases involving many other aircraft, including a free-style exercise against a formidable aggressor force added to allay doubts about its survivability. In both 1973 and 1975 we deployed AWACS aircraft to Europe. The first demonstration greatly impressed NATO officials as well as Dave Jones, then commander of United States Air Forces in Europe, with its potential value, and the second deployment helped test its capabilities in the European and Mediterranean environments.

When Dave became Air Force chief of staff, he led a proactive campaign to convince skeptics on the revolutionary potential of AWACS. In addition to members of the fighter pilot community who had never liked being told what to do by air controllers, we worked hard to win over the other services, OSD, Congress, the General Accounting Office, and the defense press. I wonder how many AWACS critics ever admitted later how wrong they had been. Based on the success of initial tests, the DSARC authorized acquisition of the first block of production aircraft in April 1975. By then we had already begun development of a more advanced model, starting a series of evolutionary improvements that has been going on ever since. We also continued to push our North Atlantic allies—except the United Kingdom (UK), which insisted on using its own Nimrod—to fund the E-3 for NATO's planned fleet of multinational AEW aircraft (a decision finally made in 1978).

I consider the AWACS along with the F-16 as the greatest acquisition success stories of my Air Force years. Almost as soon as they became operational at Tinker AFB, Oklahoma, in the late 1970s, E-3s began deploying overseas as veritable "electronic gunboats" to support American diplomacy in various international crises. Aptly named the Sentry, the E-3 helped to deter hostile actions without the provocation of deploying combat forces. Although not originally designed as a "spy plane," it also has provided increasingly valuable intelligence. The full military value of AWACS was demonstrated to the world during the 1991 Gulf War, when continuously orbiting E-3s masterfully controlled one of the most intense and successful aerial operations in history. Today's Air Force pilots probably find it hard to imagine fighting a major air battle without E-3 Sentries watching over them and the enemy.

Remotely Piloted Vehicles—Technology Ahead of Its Time

With much less visibility than the manned aircraft programs, I also oversaw the development of what today are called unmanned or uninhabited aerial vehicles (UAV). They used to be called pilotless aircraft, drones, or remotely piloted vehicles. Drones had long been used as practice targets for training purposes, as had obsolete fighters flown by remote control devices. I first became acquainted with adapting drones for reconnaissance at HRB in the mid-1950s with the Army's camera-carrying SD-1. In more secrecy, I later became involved with the development of drones for reconnaissance (photo and SIGINT), electronic counter-measures, decoys, weapons delivery, and even dropping propaganda leaflets. Some projects were developed through the NRO, but most responded to tactical requirements, especially in Southeast Asia. Security restrictions kept this story from the public and even from much of the Air Force.

Teledyne Ryan Aeronautical was the company most involved in building drones. Starting in 1960 it began work on the first of many reconnaissance drones, many of them speedily adapted from its Q-2 Firebee target drone, which was first flown in 1951. (Q is the military designation for drones.) With almost 7,000 having been built, it is still in use more than 50 years later. The basic Firebee—also referred to by Ryan's "147" designation—evolved from a simple airframe guided by a nearby operator to longer-range vehicles controlled by airborne controllers from a considerable distance or flying autonomously, using computer navigation programs.

The Air Force first flew reconnaissance drone missions over North Vietnam and China in 1964. With SAC as operating command, the Air Force used modified C-130 Hercules transports (DC-130s) for air-launched versions, and the Navy had some ship-launched models. Ryan eventually built more specialized designs, and use of the term *RPV* became more common. One obvious incentive for relying on RPVs was to avoid losing crew members over hostile territory. Because of their relatively low cost and stringent security classifications, the Air Force managed these programs with a strictly compartmented quick-reaction structure for streamlined procurement called "Big Safari."

One of my earliest experiences with an RPV project occurred in the wake of North Korea's provocative shoot down of a US Navy EC-121 in April 1969, with the loss of 31 crew members. Only three weeks later, some people from Ryan and the Air Force Logistics Command SPO briefed me on their concept for modifying the 147T photographic reconnaissance model to a 147TE signals intelligence model (code-named Combat Dawn). With concurrence from the National Security Agency and Dave Packard, I approved the project. In February 1970 the new RPV began operations from Osan Air Base, South Korea, monitoring the North Korean and Chinese coasts. The Air Staff officer who coordinated this project later remarked on our streamlined acquisition process. "Ryan had a basic system that could readily be adapted to meet changing requirements, and the Air Force had a management technique that cut through the red tape and got the job done on a compressed time schedule. The Under Secretary of the Air Force and the Deputy Secretary of Defense saw to that."[24]

To provide more sophisticated capabilities, the NRO and Ryan also developed the highly classified Compass Arrow in the late 1960s to fly deep into China. With a wingspan of 48 feet and range of 2,000 miles, it was designed to cruise at nearly 15 miles altitude while gathering electronic intelligence or taking photos, showing details as small as one foot across. It could navigate automatically or be "flown" manually by an operator in the DC-130 launch aircraft. Compass Arrow was ready to deploy by late 1971, but friendlier relations with China kept it from being used. To present a small radar image and avoid surface-to-air missiles, Compass Arrow's vertical surfaces canted inward (as with the SR-71); its smoothly curved body used radar-absorbing materials, and its engine was mounted on top to reduce the heat signature. Although never operational, lessons learned from Compass Arrow contributed to later stealth aircraft and UAVs.

About that same time, we increased efforts to acquire more types of RPVs. Program responsibility was transferred from Air Force Logistics Command, which managed modifications of existing equipment, to Air Force Systems Command, which developed new equipment. In early 1971, as a signal that RPVs were at last being recognized as organic military assets, the Air Force devised an official military designation scheme. For example, Compass Arrow became the AQM-91A, and Ryan's 147 series of drones were designated as AQM-34s or BQM-34s, depending respectively on whether they were exclusively air-launched or could be both air- and surface-launched.[25] We also sponsored development of an interim long–range, propeller-driven RPV with a program named Compass Dwell. Although three companies built prototypes, budget priorities kept Compass Dwell from entering production. If nothing else, the program motivated Ryan to find ways of reducing flying-hour costs for some of its models.

In 1972 the tempo of RPV operations over North Vietnam reached its peak. The RPV's ability to take clear photos under low cloud cover in conditions too dangerous for manned aircraft proved especially valuable. In some cases, our prisoners of war (POW) were heartened to see drones flying right over their prison camps—much to the consternation of the guards. On 31 May 1972 in the midst of Linebacker operations, I partly lifted the veil of secrecy by presenting a paper on remotely piloted vehicles to a meeting of the Electronic Industries Association. After discreetly mentioning the recent use of drones "for certain reconnaissance functions," I predicted, "Today we are on the brink of realizing some operational breakthroughs from our past research." In addition to their potential cost savings over manned systems, I emphasized how RPVs could help prevent aircrews from becoming casualties or prisoners. "With RPVs," I explained, "survival is not the driving factor." In addition to reconnaissance, I predicted we were on the threshold of using them for selected strike missions, such as air defense suppression, and as radio relay platforms. We were even studying their potential for air superiority missions in the more distant future. Reliability would be a challenge. RPVs carrying expensive communications, intelligence, and navigation equipment would need to approach the reliability of manned aircraft to be cost-effective. To perform such a variety of missions at a reasonable cost, we looked for a basic airframe and engine with modular avionics, the same practice used with Ryan's versatile model 147 in the past.[26]

Despite occasional publicity about downed RPVs recovered by North Vietnam and China, it was not until end of February 1973 that the US government officially confirmed the use of RPVs for reconnaissance in Southeast Asia. The general public finally got to see some of their pictures in the 23 April 1973 issue of *Aviation Week and Space Technology*. These photos belatedly countered widely believed communist propaganda and claims by antiwar activists that the B-52s had "carpet bombed" Hanoi and Haiphong, causing wide-scale destruction in civilian neighborhoods. It was too bad these photos could not have been made public earlier, as I would have liked. In all, 77 drone sorties during the 11-day operation obtained coverage of more than 600 objectives, while 12 RF-4C missions were able to do the same for only 49 objectives.

In April 1974 I released additional information on the use of RPVs in Southeast Asia to a forum at Wright-Patterson AFB, revealing that the Air Force had flown more than 2,500 sorties, thereby proving their utility in combat. Although not exactly cheap, remotely piloted vehicles were a small line item in the defense budget. While speaking to a room of about 500 people, I joked that the amount of money we have in our budget for RPVs was roughly enough to pay the travel expenses of all the people who came to the meeting. With too much optimism, I predicted R&D funding for RPVs would increase rapidly in the coming years because of their potential to reduce casualties in high-threat environments, to perform missions discreetly in sensitive areas, and to reduce costs. I once again warned, however, "If RPVs are to be a truly viable element of our combat forces, they must have functional dependability approaching comparable manned systems."[27] I believe that still remains a real challenge for large UAVs today.

By my final years with the Air Force, Aeronautical Systems Division had about 15 RPV programs in progress involving such companies as Teledyne Ryan, Boeing, Lear Siegler, and Fairchild. Among the more promising was the BGM-34, which had three interchangeable nose sections for real-time reconnaissance, electronic countermeasures, or strike missions, such as designating targets with lasers, dropping 500-pound bombs, or launching TV-guided Maverick missiles. With Hughes, Sperry Corporation, and RCA as contractors, ASD was also developing a drone control and data-retrieval system with the ability to simultaneously operate large numbers of RPVs in a hostile environment.

Another of our more interesting programs was the Compass Cope fly off competition for long-endurance, high-altitude electronic surveillance and photoreconnaissance. Ryan and Boeing independently developed narrow-winged, twin-tailed prototypes similar in appearance. Boeing's YQM-94A Gull (also designated Compass Cope-B) made its maiden flight at Edwards AFB in July 1973. With a 90-foot wingspan, it was the largest RPV not originally a manned bomber ever to fly. Unfortunately, it crashed on its next flight. A second YQM-94A, delivered more than a year later, was more successful, completing one mission of more than 17 hours. Ryan's entry, a pair of YQM-98A Terns (Compass Cope-R) began test flights in August 1974. On 4 November 1974 one of them set an endurance record for unmanned, unrefueled flight of more than 28 hours. That record would stand until being broken by an RQ-4A Global Hawk in April 2000. Despite its promise, the Compass Cope program was cancelled shortly after its record flight, based on a belief that manned spy planes could meet most requirements.[28]

For the next two decades, remotely piloted vehicles did not fare well in the Air Force. Shortly after I left as secretary, their development slowed almost to a standstill. Indeed, by the early 1980s the only such vehicle still operational in the US military was Ryan's humble old Firebee target drone.[29] Finally, in 1995, many years after the other services fielded small UAVs, the Air Force once again had an operational unmanned air vehicle when it activated its first RQ-1A Predator unit. Outfitted with Hellfire antitank missiles, some Predators have even been used to strike enemy targets—a capability the BGM-34 had demonstrated with both PGMs and Mavrick missiles back in the early 1970s.

Although there were valid budgetary and technical reasons for the Air Force's long interlude in UAV development, I believe the service's aviation culture and doctrine played a role. The long-standing use of the term "drone" hints at the disrespect long given to aircraft without pilots. For blue suiters, advocating pilotless aircraft never tended to be very career enhancing. "The real issue," as I was quoted saying in 1974, "is to get people to accept the RPVs. It is only natural for the Air Force to be biased toward the manned system."[30] Deep down, I think rated officers until quite recently saw unmanned vehicles as a job threat. After Vietnam, operation of RPVs came under the purview of TAC, dominated by fighter pilots, and their contributions in Southeast Asia were largely forgotten. The lack of interest persisted until new technical innovations made their potential obvious to the Air Force at large.

Precision Weapons—Unlocking the Potential of Airpower

In addition to the RPV, another innovation that proved itself during the Vietnam War was the PGM. These types of weapons, which became popularly known as "smart bombs," consisted mainly of Navy-developed Walleye electro-optical (EO, i.e., television-guided) bombs and the Air Force–developed Paveway series of mostly laser-guided bombs (LGB).[31] The Walleye was first used in 1967, scoring a few impressive hits on targets in North Vietnam. The more powerful laser-guided bombs were not available until after the suspension of Rolling Thunder in May 1968. During the next few years, Seventh Air Force dropped many thousands of LGBs, mostly against the Ho Chi Minh Trail. The main limitation for both laser- and the more expensive television-guided bombs was their need for clear visibility. Furthermore, guiding bombs with early model laser designators required aircrews to orbit over the target, exposed to enemy fire. We also sorely needed night capabilities, but technical challenges, such as keeping sensors supercooled, slowed the development of infrared guidance.

During the three years that high-risk missions over North Vietnam were suspended, Systems Command continued to work on improvements to LGBs. I had concluded that we had wasted a lot of bombs as well as lives trying to hit small targets in the past, and I was quite enthusiastic about the potential of guided bombs to make air operations much more efficient. Many in the Air Force, however, were not anxious to spend too much on them. On a trip to Southeast Asia in February and March 1971, I visited the 8th Tactical Fighter Wing at Udorn Air Base in Thailand. Commanded by Col Larry Killpack, this was Seventh Air Force's lead unit for developing tactics with precision-guided munitions. In a memo on 11 March 1971, I reported to Seamans: "There is no doubt in my mind that if the Air Force were filled with people like we have here, not only would our effectiveness in Southeast Asia have been considerably greater, but also the demand for these weapons would have increased." Killpack later told me that, as far as he was concerned, we should never drop unmodified ordnance again. (Although then speaking in hyperbole, this almost became the case for American aircraft over Kosovo and Serbia 25 years later.) I also recall walking down the hall with Dave Packard, discussing the results achieved with guided bombs. Referring to Rolling Thunder, he lamented that, if we could only turn the clock back, we could have avoided 90 percent of our losses.

The North Vietnamese offensive in 1972 offered lucrative targets for our improved laser-guided bombs. By the time it started, Seventh Air Force had received six brand new "Pave Knife" laser designator pods that could swivel independently of the aircraft's motion. They proved highly effective in destroying hardened and heavily defended targets, such as bridges, with unprecedented accuracy and far fewer downed aircraft. Their performance hinted at the quantum leap in precision to be revealed two decades later over Iraq and fully exploited several years later against Serbia. The development of PGMs has made modern air operations not only far more effective, but by reducing collateral civilian casualties, more humane as well.

With longer range and higher speed than PGMs, rocket-propelled air-to-ground missiles (AGM) had been pursued ever since World War II. In general, the Navy outpaced the Air Force in developing air-launched tactical missiles. In late 1972, however, the Air Force introduced the AGM-65 Maverick electro-optically guided missile to combat in Vietnam. I had been very impressed with the Maverick's potential, especially since visiting the Hughes plant in 1970. It featured a true "launch and leave" capability, which greatly reduced exposure to enemy air defenses. In the years immediately thereafter, I pushed hard for developing an infrared version of the Maverick that would finally give the United States a reliable night-attack capability in the mid-1980s. Unlike RPVs, which seemed to compete with manned aircraft, PGMs and AGMs enhanced their capability. Perhaps that is one reason why work on them continued unabated after the Vietnam War. The oft-misused cliché, "surgical strike," was finally approaching reality.

143

Nuclear Delivery Systems—Maintaining the Strategic Balance

As for bigger missiles, specifically intercontinental ballistic missiles, the size of the Air Force ICBM force had stabilized at 1,054 missiles in 1967, a number later frozen in place under terms of the Strategic Arms Limitation Talks (SALT) interim agreement in May 1972. The major modernization of our missile force during my Air Force years was deployment of the Minuteman III, which began in 1970 but had been set in motion during Harold Brown's tenure. The Minuteman III was equipped with three multiple independently targetable reentry vehicles (MIRV). By mid-1975, our ICBM modernization program was virtually complete, with 450 Minuteman IIs, 550 Minuteman IIIs, and 54 Titan IIs on alert with SAC.

Meanwhile, the buildup and modernization of the Soviets' ICBM force, under way since their humiliation in the Cuban missile crisis, disturbed many Americans. I was a strong advocate of maintaining a credible nuclear force, but I thought there was some fear-mongering going on, both within and outside the government. Even President Nixon and Secretary Laird were making speeches to the effect that we had leveled off on improving our strategic capability since 1965, whereas the Russians had continued a rapid buildup. They were referring to total megatons and the fact that, with the notable exception of the B-1, the United States had not started any major new nuclear delivery systems. To put the situation in a more balanced perspective, I wrote a hypothetical speech on Western nuclear capabilities as seen from the vantage point of the commander of Soviet Strategic Rocket Forces. I summarized this perspective for Mel Laird on 17 May 1971. In 1965, I explained, the US Minuteman force was not fully deployed, and the missiles that were deployed were not reliable. Now they not only had a reliability rate above 95 percent, but we were "MIRVing" many of our Minutemen, thereby tripling their target coverage. Further, the Navy has done the same with Poseidon, and we had begun production of the AGM-69 Short Range Attack Missile (SRAM), which would significantly increase the lethality of our B-52s. I said that if I were the Soviet commander, I might view this large-scale American nuclear weapons buildup with alarm, especially with B-1s in the pipeline. Mel acknowledged that my perspective made some good points. I think he tried to strike a reasonable balance between preparedness and overreaction in dealing with nuclear weapons issues, which were almost always contentious.

The initial SALT agreement in 1972 did not alleviate the fears of many that our ICBMs might be vulnerable to a Soviet first strike. MIRV technology soon became a two-edged sword as the Russians began to put even more multiple warheads on their larger missiles. Because the scenarios of a nuclear war were so theoretical, opinions about the strategic balance were an important ingredient in national policy. In a memo that I wrote on 20 August 1974, after visiting the Lawrence Livermore Laboratory in California and being briefed on their scientists' latest ideas, I explored some of the problems we faced.

> There is the . . . question of perceptions. If [the Soviets] build many hundreds of larger missiles (as they have done) and then MIRV a large fraction of them, they will have sufficient warheads to knock out a major fraction of the Minuteman force. We don't claim to be able to decapitate a major fraction of their missile force, so they could be perceived to be ahead of us, at least psychologically. We must decide if they will threaten the Minuteman fields, and if so, be ready to deploy a less vulnerable system.

One possible solution was to develop an airmobile ICBM that I dubbed the "Minuteman IV," two or more of which could be launched from a modified C-5 or 747. This proposal was studied in 1975 but never caught on. Survivability of land-based missiles remained a major issue as we embarked on designing the 10-warhead Missile-Experimental (MX), which would finally be deployed as the Peacekeeper ICBM in 1986. While I was with the Air Force and for many years thereafter, the chief issue with the MX program was not developing the missile itself but deciding how to base these high-value targets so that they could survive a theoretical first strike. By the end of the decade, the Air Force had studied no less than 40 different basing schemes, both hardened and mobile.

Even though we planned to modernize the Air Force's two legs of the nuclear triad with the B-1 and MX programs, we also pursued various ways to enhance the capabilities of our aging fleet of B-52G and H Stratofortress bombers. As mentioned above, the B-52 already carried the supersonic AGM-69 SRAM, which it could launch with nuclear warheads at targets up to 70 miles away, primarily to suppress Soviet air defenses. One of the more visible and controversial of our new programs was the AGM-86A Subsonic Cruise Armed Decoy (SCAD), which was intended to replace the obsolete ADM-20 Quail decoy developed in the 1950s. The 14-foot-long turbofan-powered SCAD was designed to closely emulate the big bomber in its flight profile, radar signature, and electronic emissions—and thereby help overtax Soviet air defenses. In 1971 the Senate Armed Services Committee directed the Air Force to develop a SCAD that could also carry an armed warhead. Adding a nuclear weapon on the SCAD proved impractical for a variety of technical and cost reasons. Some B-1 critics, such as Senator Bill Proxmire, thought we were reluctant to weaponize the SCAD because this would tend to undermine justification for a new penetrating bomber.

The Air Force began full-scale engineering development of the SCAD in mid-1972, with ASD using an innovative management structure in which its SPO exercised direct control over five associate contractors who were to build the various components. Despite all attempts to control costs, the projected price for even the unarmed version escalated sharply from about $500,000 to $1 million per copy. In the face of tight budgets, this led to program cancellation in June 1973, much to the consternation of certain congressmen who charged the Air Force with sabotaging the program to protect the B-1. Not all of the work on the SCAD went to waste. Some of its airframe and propulsion technologies were used for the new generation of cruise missiles.

One long-standing dilemma involving B-52s that I took by the horns after becoming undersecretary was positive control over their nuclear weapons. ICBMs were, by nature of their basing and supervision, very secure and tightly controlled. Mobile nuclear delivery systems, however, posed more potential problems. Despite safeguards to prevent the imaginary scenario featured in the apocalyptic motion picture, *Dr. Strangelove*, I became concerned that the Air Force still had some gaps in command and control over certain nuclear weapons.

In the early 1960s Sandia National Laboratory had developed special electromechanical devices to prevent unauthorized arming or launch of nuclear weapons. Known as permissive action links (PAL), they were initially used to secure the American nuclear bombs carried by NATO aircraft. The Air Force later had PALs installed on most of its nuclear weapons—except those carried on SAC's B-52s. One day I asked the vice CSAF, Gen John Meyer, "why not B-52s?" He said it was for the credibility of our strategic bomber force. When I observed hypothetically that one captain (the aircraft commander) could decide to start World War III, he explained that the whole crew would have to agree to take the needed steps. But in theory, I countered, a persuasive aircraft commander might convince the other crew members to go along. To make a long story short, it took some time for the people at SAC to accept the idea, but eventually PALs were also installed on the nuclear weapons carried by B-52s.

For the bigger picture of nuclear policy, Jim Schlesinger took the lead while I was secretary, building on some ideas advanced earlier by Johnny Foster to update the Single Integrated Operational Plan (SIOP) for nuclear warfare. Dr. Schlesinger's main goal was to maintain "essential equivalency" between American and Soviet strategic systems. His major achievements included introduction of more flexible targeting options to supplement the existing doctrine of mutually assured destruction (MAD) and set the stage for development of counterforce capabilities that could threaten Soviet ICBMs and command centers if their capabilities became a threat to ours.

Looking back at the Cold War, and knowing as much as I did about the awesome power in our nuclear arsenal and that of the USSR to destroy civilization, I thank God that deterrence proved successful on both sides. Having seen the destruction wrought at Nagasaki by even a small atomic bomb, I pray that no such weapon is ever used again.

Notes

1. John L. McLucas, "Achieving an Effective Air Force in Peacetime," *Air Force Policy Letter for Commanders*, December 1973, 10 (from a transcript of a speech given to the Channel City Club, Santa Barbara, 6 September 1973).

2. Puryear, *Destined for Stars*, 198.

3. By the early 1970s, the traditional term *procurement* (which McLucas continued to prefer) was gradually being supplanted by the somewhat broader term *acquisition* in referring to the range of activities involved in obtaining weapon systems and other equipment. For an overview, see Lawrence

R. Benson, *Acquisition Management in the United States Air Force and Its Predecessors* (Washington, D.C.: Air Force History and Museums Program, 1996).

4. John L. McLucas, "The New Look in R&D Management," part 1, *Armed Forces Management*, December 1969, 13.

5. Ibid., part 2, January 1970, 13.

6. In 1987 the Air Force abolished AF/RD and combined most of its staff with that of the newly created assistant secretary of the Air Force for acquisition (SAF/AQ). This reorganization occurred largely in response to the Goldwater-Nichols DOD Reorganization Act and the report of David Packard's Commission on Defense Management, both in 1986.

7. McLucas, "New Look in R&D Management," part 2, 14.

8. In 1957 the Air Force abolished the Air Proving Ground Command, its independent operational test agency since 1941. Lawrence R. Benson, *History of Air Force Operational Test and Evaluation* (Kirtland AFB, N.M.: AF Operational Test and Evaluation Center, 1992), 5–8.

9. McLucas, "New Look in R&D Management," part 2, 13.

10. In addition to SPOs, system offices (SYSTO) at Headquarters AFSC and program element monitors (PEM) on the Air Staff also helped administer acquisition programs.

11. In the mid-1990s, the undersecretary of defense for acquisition and technology announced a "new" policy to consider the cost of a system as an "independent variable" rather than an outcome of its acquisition process. This would seem very similar to "design to cost."

12. Data based on total obligating authority (TOA) for Major Force Program VI (R&D) as presented in the "Air Force Almanac," *Air Force Magazine*, May 1976, 139.

13. The term *FCRC* has been replaced in recent years by Federally Funded Research and Development Center (FFRDC).

14. In describing this difficult process, Killian later acknowledged, "The divesture of the laboratory was not accomplished without great cost to the Institute [MIT] and hurt to Dr. Draper." James R. Killian, *The Education of a College President: A Memoir* (Cambridge, Mass.: MIT, 1985), 276.

15. Quoted in *Air Force Magazine*, November 1975, 42.

16. The first set of hearings by Senator John McClellan's Investigations Subcommittee in 1963 had focused on alleged improprieties in the contract award. McClellan's hearings in 1970 primarily looked into performance problems and rising costs blamed on mismanagement.

17. Haldeman, *Diary*, 394.

18. Quoted by Edgar Ulsamer, "A Hard Look at the US Technological Posture," *Air Force/Space Digest*, May 1969, 76.

19. For his perspective on the case and criticisms of the military-industrial complex, see A. Ernest Fitzgerald, *The High Priests of Waste* (New York: Norton, 1972) and *The Pentagonists: An Insider's View of Waste, Mismanagement, and Fraud in Defense Spending* (Boston: Houghton Mifflin, 1989). For Bob Seamans's side of the C-5 controversy, see the privately published edition of *Aiming at Targets* (Beverly, Mass.: Memoirs Unlimited, 1994), 195–201

(copy on file at Air University Library), which contains information on the Fitzgerald case not included in the later NASA edition because of legal concerns.

20. In 1969 Bob Seamans had appointed Bisplinghoff, a former MIT professor and NASA official who was a member of the Air Force Scientific Advisory Board, to head a similar technical review of the C-5 program.

21. For a president to, in effect, make a production decision in advance of development and testing was highly unusual. It was based on an optimistic assumption that the B-1B represented a relatively simple upgrade of the already tested B-1A; however, the heavier B model incorporated numerous changes, especially in its avionics, which caused many years of embarrassment and expense before it finally matured.

22. Alison J. Crickmore and Paul F. Crickmore, *F-117 Nighthawk* (Osceola, Wisc.: MBI Publishing Co., 1999), 12–14.

23. For a longer extract from this editorial, which was printed in the 25 April 1977 issue of *Aviation Week and Space Technology* (hereinafter cited as *AW&ST*), see chapter 7.

24. William Wagner, *Lightning Bugs and other Reconnaissance Drones* (Fallbrook, Calif.: Aero Publishers, 1982), 168. This book presents an insider's history of Ryan's RPVs.

25. Several Firebee derivatives and a Compass Arrow later went on display at the USAF Museum in Dayton.

26. John L. McLucas, "An Outline of Air Force Views on RPV Potentials," *Teledyne Ryan Aeronautical Reporter*, Summer 1972, 9–13.

27. John L. McLucas, "The Role of RPVs in the Air Force," *Commanders Digest* (DOD), 16 January 1975, 12–16 (transcript of speech given in Dayton, Ohio, 30 April 1974).

28. One of the Ryan YQM-98s ended up at the Pima Air Museum in Tucson, Arizona, while the surviving Boeing YQM-94A went on display at the Air Force Museum in Dayton.

29. The "ancient" DQM-34 was even used to spoof Iraqi air defenses in both the 1991 and 2003 wars. See David Fulgrum, "Targets Become UAVs," *AW&ST*, 28 July 2003, 54–55.

30. Edgar Ulsamer, "Secretary McLucas Looks at Pressing Air Force Needs," *Air Force Magazine*, January 1974, 44.

31. Paveway was originally called PAVE Way, with the first word an acronym for "precision avionics vectoring equipment." For a recent study of the improvements in both technology and training that began during this period, see C. R. Anderegg, *Sierra Hotel: Flying Air Force Fighters in the Decade after Vietnam* (Washington, D.C.: AF History and Museums Program, 2001).

THIS PAGE INTENTIONALLY LEFT BLANK

Reflections
of a Technocrat

PHOTO SECTION

THIS PAGE INTENTIONALLY LEFT BLANK

John and his sister Jean, ca. 1924

Ensign McLucas in 1944

Japanese fighter diving at the USS *St. George* on 6 May 1945, during the Battle for Okinawa. Note the antiaircraft artillery blast right in front of the aircraft, which hit the ship seconds later.

John McLucas as president of HRB in 1958

Harold Brown, director of Defense Research and Engineering

Pioneering infrared image of Manhattan, taken on the night of 9 January 1958, by HRB's "Reconofax" scanner

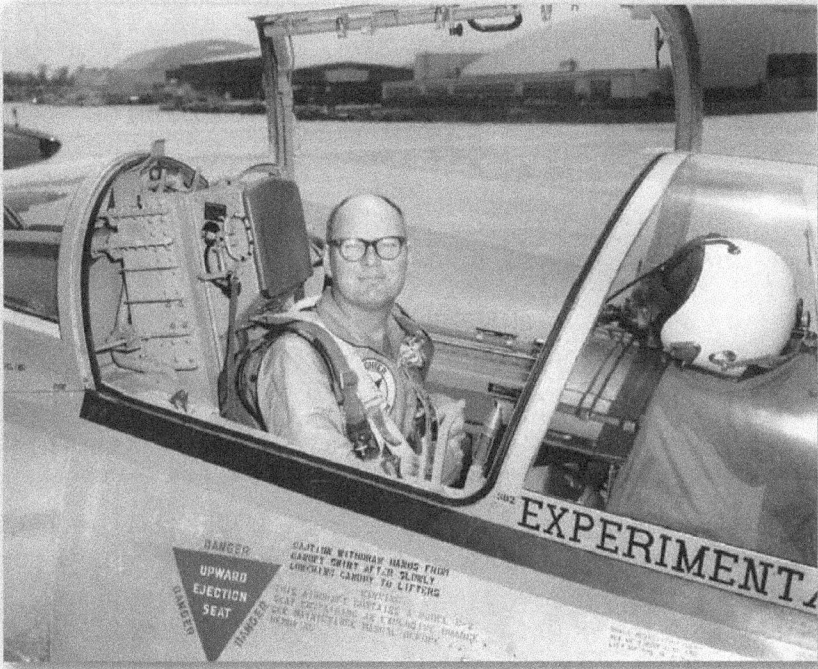

McLucas in backseat of an F-104B Starfighter, early 1960s

An early F-111A fighter-bomber during flight testing, mid-1960s

Task Force Alpha infiltration surveillance center at Nakhon Phanom Royal Thai AFB, reputedly the largest building in Southeast Asia, late 1960s (photo courtesy of the late Corey Loney)

ASID III seismic sensor

The McLucas family in 1969. *Left to right*: John Knapp (Pat's brother), Pam, John Jr., Pat, John, Susan, and Rod.

Personal Secretary Kathy Rizzardi

Walking a Pentagon corridor behind Gen Jack Ryan, Secretary of Defense Mel Laird, and Secretary of the Air Force Bob Seamans, early 1970s

Deputy Secretary of Defense David Packard

Presenting an award to Bob Naka, 1972

The Pentagon Leadership in 1972. *Seated clockwise from left to right*: Gen Jack Ryan (USAF), John McLucas (representing Bob Seamans), Gen Creighton Abrams (USA), Robert Froehkle, Kenneth Rush, Melvin Laird, John Warner, Adm Elmo Zumwalt, Adm Thomas Moorer, and Gen Robert Cushman (USMC). Other ranking members of the OSD and Joint Staff are standing in back.

McLucas visiting a Social Actions race relations class in 1972 with Brig Gen Lucius Theus (*upper right corner*)

With Lt Gen Daniel "Chappie" James, who became America's first black four-star general in September 1975

Secretary McLucas with generals David Jones and George Brown, ca. 1974

With Undersecretary James Plummer, General Counsel Jack Stempler, and Military Assistant Brig Gen William Usher in July 1975

EC-137D developmental AWACS aircraft in Boeing's Seattle Plant, early 1970s

Air Force DC-130 carrying four Ryan 147 remotely piloted vehicles

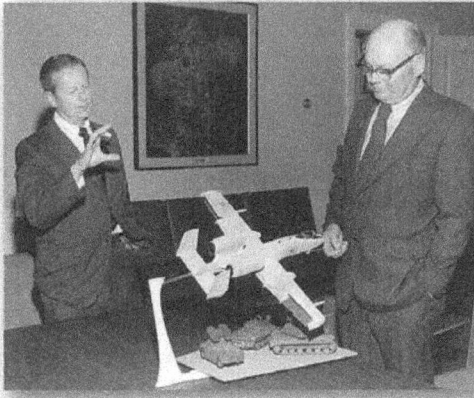

Examining an A-10 model with Ed Uhl, president of Fairchild Industries

With T. Wilson, Boeing president

Looking at a model of the YC-15 with Sanford "Sandy" McDonnell of McDonnell Douglas (in foreground)

With GD's YF-16 test pilot Neil Anderson at the Paris Air Show, June 1975

Thor booster launching the final Corona Satellite, 25 May 1972

Exhibit of the Corona KH-4 camera and return capsule at the National Air and Space Museum

McLucas visiting a reconnaissance film storage area, probably at the Defense Intelligence Agency

With James Schlesinger, Gen David Jones, and Joseph Laitin in January 1976

Being sworn in as FAA administrator with Secretary of Transportation William Coleman, President Ford, and Pat McLucas in November 1975

Meeting with President Ford and White House chief of staff Richard Cheney in the Oval Office

Checking out the cockpit of a Boeing 720, 1976

Large antennas at a Comsat ground station, early 1980s

With top Soviet space official Vladimir Kotelnikov, 1984

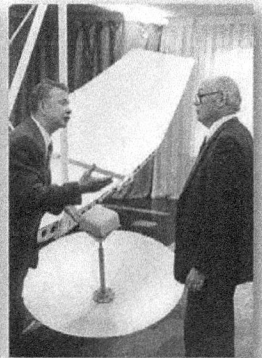

With Alexei Bogmolov at the USSR's Bear Lake Ground Station, 1984

Relaxing with Arthur C. Clarke in Sri Lanka, 1987

McLucas promoting the "Mission to Planet Earth" initiative

McLucas with the three previous directors of the NRO: Joseph Charyk (*seated at left*), Alexander Flax (*seated at right*), and Brockway McMillan (*upper right*), ca. 1990

Release of Pegasus from a B-52 on its first space mission in May 1990

Pegasus launch vehicles under construction (photo courtesy of Orbital Sciences Corp.)

John and Harriet during a visit to Scotland and Ireland, 1995

THIS PAGE INTENTIONALLY LEFT BLANK

Chapter 5

What Can Now Be Told

The National Reconnaissance and Air Force Space Programs

If the Vietnam War and domestic turmoil were the most depressing aspects of America in the 1960s, our progress in space exploration was to me the most uplifting. I can still remember my excitement in July 1969 when watching on television the blurred images of *Apollo 11*'s lunar landing. Yet I believe the most impressive result of the Apollo Program may have been the new awareness we gained of our own planet from viewing pictures of a beautiful blue and white Earth hanging in the pitch-black sky above a bleak moonscape. Three decades later, the early sense of wonder many of us felt about going into space has long since faded into memory, but gratitude for this precious planet has not diminished. Even so, many have asked, why should we spend so much to send instruments and people into space? At a space history symposium in 1995, I listed some basic reasons for us to do so (not necessarily in any order of their importance).

- Enhance national security and advance foreign policy

- Exercise world leadership

- Expand scientific knowledge and advance education

- Broaden our vision of life and inspire our youth

- Maintain technological and competitive advantages

- Improve terrestrial services (e.g., communications, navigation, weather forecasts)

- Compare the utility of people and robotic devices

- Promote new commercial opportunities

- Explore the solar system and universe[1]

Although the public's attention in the 1960s and 1970s focused on the accomplishments of the National Aeronautics and Space Administration, the United States actually had four major space efforts under way. As with President Kennedy's challenge for NASA to beat the Soviets to the Moon, our other space efforts were also to some degree motivated by the Cold War. The Air Force, as the DOD's de facto executive agent for space acquisition, pursued a variety of space projects, ranging from communications and early warning satellites to a manned orbiting laboratory, and it was responsible for launching other defense payloads. Starting in 1965, the Comsat Corporation began the era of commercial space enterprise with the first Early Bird communications satellite. Meanwhile, hidden from view, the National Reconnaissance Office (NRO) was developing and operating the sophisticated satellites that allowed privileged members of the US government to see and hear what was going on in the world's most closed and secretive nations. Until 1992 the NRO's very existence and name were officially classified at a level higher than secret—with its innovative work buried even deeper in the so-called black world of special access programs, each protected by specific code words and strict "need to know" procedures. I am grateful to have had the opportunity to oversee both the Air Force and NRO space programs when each scored some major achievements and prepared for future breakthroughs and subsequently to help manage Comsat during some of its best years.

Directing the National Reconnaissance Office in Its Golden Age

I had been interested in remote imaging since my early days at HRB, but I first became knowledgeable about the NRO and some of its products in 1967 as a member of the Advisory Committee for the Defense Intelligence Agency (chaired at the time by Gene Fubini). In this capacity, I had the opportunity to review satellite photos of various Soviet systems, so I already knew the importance of the NRO's work when the prospect of returning to the Pentagon presented itself. I had also known the current director, Dr. Alexander H. "Al" Flax, since he came to the Pentagon in 1963 as assistant secretary of the Air Force for R&D. As one of the last presidential appointees from the former administration remaining with the Air Force, Al was anxious to see me sworn in so he could move on. He didn't presume to tell me how to run the office, but he was very helpful in explaining the background behind current programs and his opinions of them.

As the NRO's fourth director, I had the privilege of running the organization during what might be considered its golden age. By this, I mean after its turf wars had been mostly settled but before it became subject to increased oversight from Congress and micromanagement by the OSD. Before and just after the secret establishment of the National Reconnaissance Program (NRP) and the NRO in 1961, the USAF undersecretary, Dr. Joseph V. Charyk, and the CIA's deputy director for plans, Richard Bissell, served as codirectors of the satellite reconnaissance effort. Soon, however, personality conflicts and mutual suspicions between the CIA's Directorate of Science and Technology and the Air Force–dominated headquarters of the NRO led to virtual internecine warfare. Starting in 1965, the situation gradually improved with the departure of the brilliant but combative Alfred D. "Bud" Wheelon from the CIA and the appointment of Al Flax as NRO director. That year also saw chartering of a small NRP Executive Committee (ExCom), with the deputy SecDef, CIA director, and president's science advisor as voting members. The ExCom's roles included setting spending priorities and allocating responsibilities among the NRO's DOD elements and the CIA.[2]

I think Al Flax deserved a lot of credit for restoring much of the cooperative atmosphere between the Air Force, the OSD, and the CIA that had prevailed during the early days of the Corona program. Even so, the NRO was caught in a sort of cross fire between the CIA on one side and the regular Air Force on the other. Some in the CIA considered the NRO to be a Trojan horse, intent on taking over the development of all satellite reconnaissance. For many Air Force people, however, the NRO represented a hijacking of what they considered key elements of the USAF's inherent air and space mission.[3]

For all its importance, the NRO was still a rather small organization, with only about 300 personnel. It consisted of a headquarters staff in the Pentagon, manned largely by Air Force people assigned to the SecAF's Office of Space Systems, and four major program elements, identified by the letters A through D. In those days the veil of secrecy sheltered the NRO from uninformed criticism, and unlike most of the DOD, funding was not the main limiting factor. For a technocrat like me, being in charge of an operation that could push technology—rather than be pulled by requirements or stifled by lack of resources—was very rewarding.

At that time, the NRO had a small but very appreciative customer base. Lyndon Johnson, in one of his less guarded moments, had confided in 1967 that information from satellites "justified spending ten times what the nation had already spent on space."[4] Because our systems had become so crucial to determining Soviet military capabilities, Richard Nixon and Henry Kissinger continued to make sure the NRO received relatively generous funding. It is unlikely they would have entered into the Strategic Arms Limitation Talks without the ability of American reconnaissance satellites to keep track of Soviet missiles and related facilities. In fact, the SALT I Treaty specified that its terms would be monitored by "national technical means," a euphemism mainly for reconnaissance satellites. Photographic collection priorities were set by an interagency group called the Committee on Imagery Requirements and Exploitation (COMIREX), while another body called the SIGINT Overhead Reconnaissance Subcommittee (SORS) of the US Intelligence Board (USIB) provided guidance on signals intelligence. At the White House level, select members of the National Security Council (NSC), known as the "40 Committee" under Nixon, oversaw national intelligence policies and issues.

As for my immediate superiors, I've already described the excellent supervision received from Secretary of Defense Mel Laird and Deputy Secretary Dave Packard when I was Air Force undersecretary. Likewise, I don't believe my relationship with Richard Helms could have been any better. I mainly dealt with him in his role as a member of the ExCom, but occasionally I'd go over to see him at Langley. The NRO deputy director, being assigned to a CIA position, closely interacted with the agency, especially its scientific and engineering community. In both my DOD and USAF jobs, I tried hard to improve the NRO's image with the Air Force's military leadership. Yet much to my disappointment, lots of blue suiters—including some of the very best of the Air Force's senior officers—still resented the existence of the NRO.

I was more than satisfied with the way the Executive Committee worked during my first few years at the NRO and considered its three voting members as my board of directors. Dave Packard, by virtue of his position as deputy SecDef, chaired ExCom meetings except for a few occasions when Mel Laird got involved. Quite often when I went down to Dave's office with a question involving the NRO, he would call Dick Helms on the phone, and the two of them would settle the issue without a formal meeting. When the ExCom convened, usually about once per quarter, the presidential science advisor—Dr. Lee A. DuBridge from Caltech and later Dr. Edward E. David from Bell Labs—added additional expertise and perspective. My NRO deputy would go over the proposed agenda and list of outside attendees with Packard in advance and serve as secretary. Only a few other outsiders, such as the DOD and CIA comptrollers and the DDR&E, would regularly attend in advisory capacities. During my first few years, the ExCom was a very compatible group, well informed on matters of both policy and technology, but small enough to make timely decisions. My role was to present project proposals, budget adjustments, schedule changes, and other agenda items. Most decisions were reached by consensus. With such a streamlined bureaucracy, I thought we had about the best-run operation in Washington.

Unfortunately, the role of the ExCom started to fall apart during my last year with the NRO. In March 1971 Jim Schlesinger, then at OMB, completed a review of the intelligence community, commissioned by President Nixon. His recommendations resulted in Nixon issuing a memorandum in November 1971 that increased the responsibility and authority of the director of central intelligence (DCI) over the intelligence budgets of DOD departments and agencies, to include formation of an Intelligence Resources Advisory Board. A few months later, a revised NSC directive called for the DCI to chair and staff all intelligence committees and advisory boards, set intelligence requirements and priorities, and submit a consolidated intelligence program to OMB. When the president decided to send Helms to Iran, Jim Schlesinger agreed to take over as DCI in February 1973, after receiving a pledge that he would chair all the intelligence committees. When Jim became ExCom chairman, the new deputy SecDef, in the person of Bill Clements, withdrew from the committee because he outranked Schlesinger at the time. As a result the assistant SecDef for intelligence, Albert C. Hall, became the OSD member. Among other ramifications, this meant the Ex-

173

Com could no longer approve the NRO's budget but only recommend funding.

I had known Al Hall ever since he was a fellow deputy in DDR&E and got along well with him, but his presence on the ExCom opened the door to getting more OSD functions involved in the NRO's business. I remember a call from Dr. Gardiner Tucker, the assistant secretary of defense for program analysis, saying he would like to have some of his people cleared so they could see our programs. I said that didn't sound too unreasonable—until he sent me a list with 42 names on it from just one office! I thought to myself, "the bureaucracy is going to start taking over," which indeed is what steadily happened thereafter. Meanwhile, President Nixon dismissed Ed David and dissolved the PSAC, which had been a valuable source of counsel on space and defense programs since the Eisenhower administration.[5] These changes lowered the status of the NRO in the Pentagon and weakened its link to the White House, relegating its business to the lower working levels of the NSC.

Somewhat ironically, Jim Schlesinger was barely beginning to exercise the DCI's new authorities when he became secretary of defense in July 1973. Moving to the Pentagon probably gave him a different perspective on power sharing between the CIA and the DOD. The changes strengthening the role of the DCI would be there for William E. Colby, who took over the CIA in September 1973. By then the NRP ExCom retained only a shadow of its former importance.

During my tenure very few outside agencies dealt directly with the NRO. The two most immediate customers for information garnered by our satellites were the CIA's National Photographic Interpretation Center (NPIC) and the National Security Agency (NSA). The NPIC, established at the final days of Eisenhower's term, served as a clearinghouse between the NRO and authorized users of our imagery, while signals intelligence went directly to the NSA, which was many times larger than the NRO but still almost as secret.[6] At the executive level, I got along well with the NSA's directors, who included Vice Adm Noel Gayler from August 1969 through July 1972, followed by Air Force lieutenant generals Samuel C. Phillips in August 1972 and Lew Allen Jr. one year later—the latter two already having been close associates of mine. At lower levels, however, there was often friction over how much prerogative NSA people had to task NRO-operated signals intelligence assets. Both agencies' security classification rules often complicated the sharing of information and the flow of communications. Dr. F. Robert Naka (my longest-serving deputy) and I tried to break down these barriers through administrative means.

Congress probably exercised less control over the NRO than any other federal organization of its significance. Only select members of the relevant subcommittees of the House and Senate Appropriations and Armed Services Committees provided rather relaxed oversight. When they wanted to be briefed, I would take two or three of our experts with some beautiful charts and slides to a secure conference room on the Hill. Only the chairman and a couple of his most trusted compatriots would be with us in the room. With most technical details of satellite operations being over their heads in more ways than one, their general attitude could be simply stated as "Gee, you guys are spending a lot of money, but we have to admit you are doing a lot of wonderful things, and the information you are collecting is really important. We're sorry it costs so much, but I guess we have no choice but to pay for it."

Although details on past intelligence budgets remain classified, I can say that the NRO—because of the expense of building and launching satellites—received a large share of the intelligence community's total funding. Much of this money was allocated with relatively few strings attached. Nevertheless, for each fiscal year from 1966 through at least 1973, the National Reconnaissance Program stayed within its overall appropriations and often volunteered or accepted significant reductions. Because of the compartmented nature of NRO programs and our flexible contracting practices, we at the headquarters did not centrally manipulate budgets for work being done in the field. As attested to by a later NRO deputy director who served as its comptroller in the mid-1970s, "technical feasibility [was] the only factor that limited American space-based reconnaissance efforts during the 1960s and early 1970s—not funding, acquisition, or congressional cooperation, which usually followed in a most rapid and efficacious manner."[7] Although perhaps a bit overstated, his observation reflected the funding priorities enjoyed by the NRO during my tour.

Not until 1975 did Congress begin to dig deeply into the affairs of the CIA and other elements of the intelligence community amidst revelations about covert activities overseas and spying on

175

American citizens at home. I always felt good that the NRO had a "clean" mission, compared to the sometimes unsavory work performed by other intelligence agencies. Even with the more extensive congressional oversight of intelligence that followed the hearings of 1975, the NRO continued to receive relatively mild scrutiny until the mid-1990s.

The less structured procedures of the past allowed the NRO to form close partnerships with select members of the air and space, electronics, film, and optics industries. Harking back to the U-2, the NRO's contracts focused on meeting performance goals rather than complying with detailed technical specifications. The DOD-CIA connection also allowed the NRO (using the DCI's special statutory procurement authorities in section 413J, Title 50, of the *US Code*) to avoid normal regulations and payment rules when necessary. Not being slowed by numerous reviews or bogged down with paperwork, vouchers, and audits, we could immediately apply lessons learned from satellites in orbit to payloads being built in the factory. As one of the true Air Force–NRO space pioneers later explained, "This tight loop allowed us to go through a generation of design every year or so."[8] I was generally impressed with the contractors who worked on launch vehicles and satellites for the Air Force and NRO. They included Lockheed, TRW, Martin-Marietta, McDonnell Douglas, Hughes, Boeing, General Electric, Itek, Perkin-Elmer, and Eastman Kodak.

Key NRO People and Organizations

While I was director, most of the NRO's headquarters staff of about 30 people was located in a less than luxurious office suite behind the doors of Pentagon Room 4C1000 or in a vault deep in the basement. When I arrived, the top military officer in the building was an old colleague from DDR&E, Col (soon Brig Gen) Lew Allen Jr., who was shown on unclassified personnel listings as the secretary of the Air Force's director of space systems. In reality, Lew served as staff director of the NRO.[9]

During my first few months, the NRO's deputy director was James Q. Reber, a long-time CIA employee. We thought it was time for a change, and I wanted to replace him with the chief scientist from MITRE, Bob Naka. (Like most Japanese-Americans, he and his family had been interned during World War II.) Dick Helms already knew Naka and readily agreed to the move. Despite his highly classified reconnaissance work in the past, Bob admitted in a late-1990s interview that he "hadn't the foggiest notion" about the existence of the NRO until I offered him the job.[10] Upon accepting, he asked if he was now supposed to come to the Pentagon and just disappear. So we decided to give his position an unclassified cover title: deputy undersecretary of the Air Force for space systems. This identity afforded him more standing within the Air Force than his more covert predecessor. Because of my undersecretary duties, I delegated Naka authority for most of the NRO's day-to-day operations, but he kept me fully informed. Bob returned to the private sector in September 1972 after staying one year longer than initially agreed. At my suggestion, he would come back to the Pentagon for a tour as chief scientist of the Air Force from 1975 to 1978.

My next deputy director was Robert D. Singel, a longtime CIA officer and an expert in the field of electronic intelligence. He returned to CIA headquarters in July 1974, to be replaced by Dr. Charles W. Cook. I had known Charlie Cook ever since coming to the Pentagon in the early 1960s and thought highly of him. Although a CIA veteran, he came to us after working in DDR&E. He continued to serve as deputy director until 1979.

Program A: The Air Force Element

Out on the West Coast, collocated with AFSC's Space Systems Division at Los Angeles Air Force Station in El Segundo, California, the SecAF's Office of Special Projects (aka NRO Program A) performed a variety of support functions as well as developing some innovative systems of its own. Perhaps most notable was a family of close-look satellites that supplemented Corona.[11] The original system and its immediate successor would serve us well for many years. Program A had come a long way in the late 1960s, thanks in large part to the innovative management of Brig Gen John L. Martin Jr. To impose more discipline on the acquisition process without adding the red tape that often stifled normal programs, Martin devised a specialized incentive structure for satellite contracts that greatly improved their performance and reliability. Based on his stellar performance, Martin was promoted to major general in March 1968 and moved in July 1969 to Headquarters AFSC. (I would later work closely with him at Comsat Corporation.)

When John left Program A, I was glad we had Brig Gen William G. "Bill" King, another outstanding space expert, ready and waiting to replace him. Bill had been in the reconnaissance satellite business ever since serving as project officer for the WS-117L Advanced Reconnaissance System in the 1950s. As he approached his time for mandatory retirement, we sent Lew Allen to El Segundo in September 1970 to serve as his assistant until taking over Program A in April 1971. With a doctorate in physics, Lew was technically brilliant. After leaving the NRO in January 1973, he served as deputy director of the CIA and director of the NSA before becoming the first (and so far, only) PhD to become chief of staff of the Air Force. After retirement, Dr. Allen became director of the Jet Propulsion Laboratory in Pasadena, California, and later chairman of the board of the Draper Laboratory in Cambridge, Massachusetts.

In January 1973 I replaced Lew with Brig Gen David Bradburn, who was promoted to major general in 1974. Dave had been working for me as director of the NRO staff since April 1971. Like Bill King, he was an old-timer in the satellite reconnaissance business, having started with the Air Force's WS-117L program in 1957. Among his accomplishments at Program A, he helped contribute to the successful repair of NASA's Skylab orbiting laboratory, some key parts of which had been damaged during its unmanned ascent on 14 May 1973.[12] On 21 June I was proud to give Dave a commendation for his leadership in performing "an unparalleled engineering feat" that further stated, "General Bradburn's forceful and decisive actions directly saved the Skylab program from severe curtailment." The last director of Program A during my Air Force years was Brig Gen John E. Kulpa, who had directed the very valuable Defense Meteorological Satellite Program in the late 1960s. As with all the others at Program A, he had previously been the NRO's staff director, holding that job from January 1973 until September 1974. He served as Program A director from August 1975 until January 1983, his long tenure there indicating how valuable he was to the National Reconnaissance Program.

Program B: The CIA Element

In contrast to the changes of leadership at Program A, only one man headed the CIA's Program B during my entire Air Force tour. He was Carl E. Duckett, who served as deputy director of the CIA for science and technology from April 1967 until April 1976.[13] Carl was a consummate politician and forceful advocate for Program B's projects. I respected Carl, and I think he reciprocated, but he was very loyal to what he conceived as the CIA's best interests. The main component of Program B was the CIA's Office of Special Projects (OSP), which was headed by John J. Crowley from its formation in 1965 until November 1970. He was replaced by Harold L. Brownman, a key contributor to the revolutionary imaging system mentioned later. Brownman led the office until March 1973, when it was abolished in one of Duckett's reorganizations. The OSP was replaced by the new Office of Development and Engineering (OD&E), with Leslie C. Dirks in charge. I thought highly of Dirks, who had long been one of the CIA's top technical experts.

Bob Naka took an early interest in getting Program B to work more closely with Program A. For example, he took a trip in the fall of 1969 with John Crowley to visit one of Program B's contractors, whose project also involved Program A. Bob was surprised that Bill King was not in attendance, especially after he encountered Bill in the same building on another errand. Naka asked Crowley why General King was not invited to their meeting, and John said his office never did that. Bob told him that from now on, the Program A director would always be invited to meetings that directly involved his mission. Bob remembers saying, "either he comes, or no one attends." Crowley agreed, and he and Bob developed a fine relationship.[14]

Military personnel often worked at the CIA. In 1971, for example, we transferred Maj Jimmie D. Hill from Program A to Program B. While at Program B, Jimmie also tried to encourage the CIA to share more of its technology with Program A, although he encountered some resistance because of residual anti–Air Force attitudes. After retirement from the military side of the Air Force in February 1974, he returned to the NRO as its comptroller. Jimmie went on to become director of the NRO staff in 1978 and deputy director of the NRO from 1982 to 1996, twice serving 10 months as acting director. He was a major force behind reorganizing the NRO along functional lines in the mid-1990s. Jimmie thought the competition between the program elements had been beneficial in the early years, but he later concluded that this competition could be destructive when more technical approaches became available than there was money to pursue them. As Jimmie told me after his retirement, "We needed to get competition out of the government but keep it in industry."

Program C: The Navy Element

Compared with the scope of Programs A and B, the Navy's Program C was relatively simple and straightforward, focusing mainly on electronic intelligence. Program C had two directors during my time with the NRO. Rear Adm Frederick J. Harlfinger II, who directed Program C from August 1968 until January 1971, continued like his predecessors to have primary duties as assistant CNO for intelligence and commander of the Naval Intelligence Command. In view of this, Harlfinger and his predecessors were largely figureheads to give Program C an officer with flag rank.

He was succeeded by Capt Robert K. "Bob" Geiger. What Bob lacked in rank, he made up for in space-related experience. He had been assigned to Program A in El Segundo and then to our Pentagon staff from 1966 through 1970. Bob's unclassified position was project manager of the newly established Navy Space Projects Office. I was so impressed with Bob that I went personally to Admiral Zumwalt and told him Captain Geiger really deserved a promotion. Much to my delight, he became a rear admiral in July 1974. His time with the NRO ended in July 1975 when he was appointed chief of naval research, a really prestigious assignment.

Program D: Aircraft Operations

Program D was an interesting amalgamation of Air Force and CIA resources. When the NRO was formed, it inherited oversight of the CIA's U-2 and new A-12 Oxcart supersonic reconnaissance aircraft. It also became involved in remotely piloted vehicles. Oxcart operations ended in 1968, replaced by somewhat less classified Air Force SR-71s operated by the Strategic Air Command

(SAC). The CIA continued to conduct U-2 operations and even received six new U-2Rs—a larger more capable redesign of the original.[15] In addition, the NRO occasionally tasked SR–71 missions through Program D. I once suggested to Mel Laird that perhaps our new satellites were making it unnecessary to fly regular SR–71 missions over Cuba. Mel replied, however, that he wanted to keep sending a message to Fidel Castro.

Although our space-based reconnaissance was steadily improving, airborne platforms retained several advantages. Except for the USSR and certain other high-threat areas, aircraft could fly when and where needed, return frequently to the same targets, and linger over an area of interest for extended periods. As for visual spectrum imagery, U-2 and SR-71 photographs were still generally sharper than those taken from space and available in a more timely fashion. As it had from the beginning of the U-2 program, the CIA relied heavily on Air Force personnel (both "civilianized" and active duty) and logistical support in performing its aerial reconnaissance mission. Air Force officers supervised Program D. Col Frank W. Hartley was its director when I came to the NRO. The next (and last) director was Col Bernard "Buzz" Bailey, who replaced Hartley in July 1972. Even before Buzz took over, Program D was living on borrowed time.

Soon after arriving at the NRO, I began to wonder why we (and by extension, the CIA) were still operating our own little fleet of aircraft, especially when we had so much else on our plate with major satellite programs. Making SAC the single manager for all U-2s would be more efficient in terms of both operational control and support costs. In December 1969 I raised this idea with Dave Packard, who got Richard Helms to consider it as a future action. Within a few weeks, however, President Nixon decided to keep the CIA's U-2 capabilities intact until 1971, at least partly to keep providing covert support to Nationalist Chinese U-2 flights.[16] Then, in early August 1970, the 40 Committee recommended continuing the CIA's U-2 program through 1972. A few days later Henry Kissinger requested satellite coverage of the Suez Canal area, where he was trying to broker a cease-fire between Israel and Egypt. Unfortunately, the available Corona could not provide the level of detail needed, so Kissinger asked for U-2 coverage. To his chagrin, SAC responded that it would be unable to deploy an element of its U-2s from Laughlin AFB at Del Rio, Texas, for several weeks. Helms, however, said the CIA could send some of its U-2Rs from its Detachment G at Edwards AFB to the eastern Mediterranean in less than a week. In an operation called Even Steven, the first of these was taking photos of the cease-fire zone on 9 August 1970. The quick response of the CIA's U-2Rs helped keep Program D alive for another few years.[17]

Because of my well-known desire to shut down Program D, Buzz Bailey was in a somewhat awkward position during his tour, but he seemed to take the situation in stride. In June 1973 CIA director Schlesinger told the 40 Committee that the agency's operation of U-2s was no longer necessary, and on 30 August, the committee approved termination of the CIA's U-2 program in one year. Nixon's opening of relations with Communist China led to the phaseout of coastal surveillance flights from Taiwan in May 1974. In October 1974, with concurrence of the new director of Central Intelligence (DCI), William Colby, the NRO finally transferred its aerial reconnaissance mission to the JCS, with SAC taking over all U-2 operations, finally implementing my proposal of five years earlier.

As an aside, just a month earlier, the general public witnessed some of the SR-71's fantastic capabilities when a SAC aircrew flew from New York to London, a distance of 3,490 nautical miles, in just under 1 hour and 55 minutes—an average speed of more than 1,800 mph. After being a major attraction at the international Farnborough Air Show, this aircraft's return flight on 13 September took it from London to Los Angeles in less than 3 hours and 48 minutes. Both of these records still stand. As secretary of the Air Force, I initiated this well-publicized demonstration. When I announced my desire to show off the SR-71, some interested officials protested that this would break security classification rules. I informed them that I already had President Ford's approval.

Passing the Torch

After being appointed as the new secretary of the Air Force in July 1973, I knew my responsibilities would not allow me to devote enough time to the NRO. It took longer than anticipated, but in December 1973 I finally turned over the reins of the NRO to James Plummer (see chap. 3).

Ever since serving as Lockheed's program manager for Corona, Jim had been a key contributor to the National Reconnaissance Program. I believe him to be one of the best-qualified individuals ever to become director. I told Jim he had total authority for the NRO, and from then I focused full time on being secretary of the Air Force.

Selected Reconnaissance Programs

I was at the NRO as it began to transition to the second generation of satellite reconnaissance technology. The best example of the first generation was Corona, essentially an optical search system that returned photographic negatives from orbit to be developed and studied days or weeks after the film had been exposed.

The main hallmark of the next generation of imagery satellites would be near-real-time return of images and other data from orbit through use of new technologies and communications links. Although most details about other imagery and SIGINT satellites from my time with the NRO still remain officially classified, enough information has appeared in public to give the interested reader a rough idea about some of these programs. Because of security rules governing my former government service, this memoir cannot discuss operational activities or corroborate the information on specific programs (except Corona) that has appeared in open sources.

The Last Years of Corona

By the time I arrived on the scene, Corona satellites with their Agena upper-stage launchers were a fully mature and smoothly functioning system.[18] The major issue was whether to continue improvements and extend production beyond 1972—a concern heightened by problems experienced with six of seven missions between September 1968 and July 1969. I was quite nervous about having enough Corona satellites left to maintain required coverage. We therefore took steps to keep as many experienced government and contractor personnel as possible working on the Corona program, to refurbish remaining components, and to procure adequate spare parts. Meanwhile, a study group during the second half of 1969 concluded that no additional Coronas should be procured beyond those already under contract. In February 1970 I submitted this committee's report to DCI Richard Helms with my concurrence. Helms and Lee DuBridge, the president's science advisor, agreed with our recommendation, as had Dave Packard.

This decision to end Corona was not quite final. Even in those days of strict secrecy, there was some consideration of using Corona's technology for civilian applications. In 1969 NASA approached the NRO with a tentative proposal to adapt Corona for use as an earth resources survey satellite. We were a bit intrigued with the idea, which would preserve our contractors' capabilities to remanufacture the satellites in case of an emergency, but NASA soon dropped the proposal.[19] Then, in the fall of 1970, the State Department's Bureau of Intelligence and Research, with the concurrence of the DIA, requested that we consider procuring a reserve of Corona satellites for quick response to world crises, since their larger and more expensive successor would not be available in enough numbers for such use. In early 1971, after additional study of these and other concerns as weighed against the projected costs of reopening the production line, I polled the ExCom on whether to order more Coronas. The members' response confirmed the previous year's decision to terminate the program.

Ten of the 11 Corona missions from September 1969 until May 1972 were successful. These later missions could remain in orbit up to 19 days and return two packages of 16,000 feet of film covering more than nine million square miles. In addition to continuing to monitor the Soviet Union and China, Corona was looking down at other areas of interest, such as the Middle East. We launched the 145th and final Corona mission on 25 May 1972 and recovered photos from its KH-4B camera system a week later. Aware of the historical significance of this mission, Bob Naka had brought the launch directive to me to sign. In view of his closer involvement with the program, I told him to do so instead. During Corona's 12-year lifetime, about 120 successfully orbited satellites returned 2.1 million feet of film.[20]

Naka also realized that we had no satellites left over for posterity, so he had the contractors put together a display model using the last return capsule, a developmental version of the camera, and other spare parts. Soon thereafter, we set up a classified Corona exhibit at NPIC headquarters. This display would eventually be transferred to the National Air and Space Museum. The Corona program was declassified by President Clinton in February 1995, and its approximately 860,000 images were transferred to the National Archives and Records Administration for scientific and historical purposes. At the time of this writing, program information about other retired systems from my years with the NRO has yet to be declassified, although the National Imagery and Mapping Agency (NPIC's successor) did release some additional film from the 1960s and 1970s to the National Archives in October 2002.

Toward the Next Generation

Although we focused much of our attention on evolutionary improvements to photographic capabilities during my tenure, we were also looking toward a true revolution in overhead reconnaissance. While photos returned from orbit were great for finding and monitoring fixed targets, the time it took to retrieve film and analyze the results did not help much in tracking fast-moving crises. What we still needed was a more timely reconnaissance system, as envisioned during the WS-117L program of the 1950s, but with much sharper acuity than possible then.

When I came to the NRO, cutting-edge imaging technology was finally catching up with the requirement to deliver images from orbit in near real time. The proposed satellite system would be very costly, and thus it raised frequent debates over budget priorities. First, we had to consider a less ambitious capability being developed by Program A that would be cheaper and available sooner at lower risk. In view of financial constraints, this evolutionary approach appealed to Mel Laird and also to Jim Schlesinger at OMB. The CIA's Directorate of Science and Technology, led by Carl Duckett, advocated the more revolutionary electro-optical system that OSP specialists such as Leslie Dirks had been exploring for years. It would employ light-sensitive diodes and, later, charge-coupled devices (CCD) to capture images. Such silicon-based arrays of pixels would generate small electrical charges well suited for digital processing and transmission.

I ate one of my most memorable lunches with Dick Helms, Dave Packard, Ed David, and the venerable Edwin "Din" Land to discuss the options for the next generation satellite. Land, just as farsighted then as when he revolutionized photoreconnaissance in the 1950s, made a strong pitch for moving ahead with Program B's near-real-time system—making immediate converts of all of us in the room. Brainstorming all the applications this breakthrough would mean for our national defense and foreign policy, we all grew increasingly enthusiastic.

Apparently not content with the ExCom's support, Duckett slipped over to Capitol Hill and lobbied Senator Allen Ellender, chairman of the Appropriations Committee, on the advantages of Program B's proposal. Although I did not appreciate Carl's maneuver when I learned of it, in this case he did it for a good cause. Key Air Force and DOD people considered Program A's project as being more timely and cost-effective, but as it became obvious that there would not be enough money to continue developing both, I came to consider Program A's system as no more than a possible backup.

In June 1971, after many months of debate at lower levels, the President's Foreign Intelligence Advisory Board (PFIAB) considered the new system. Dr. Land was a member, so his opinion carried a lot of weight. I recall hearing from someone who sat in on a White House briefing that Din Land really impressed Nixon when he explained that the new system (unlike the complicated film mechanisms used to date) had no moving parts. Not only would it be much more capable, it promised to be even more dependable. Much like Eisenhower many years earlier with Corona, Nixon trusted Land's judgment and, in effect, made the production decision for the new system. We were given a timetable: to have it operational before Nixon left office, that is, January 1977. By then, of course, Ford had long since replaced Nixon. The first president to see the fantastic new satellite's imagery would be Jimmy Carter—on the day after his inauguration.[21]

Overseeing development of this complex and revolutionary new system and a related data transmission network were among the top priorities during my time with the NRO. Technical challenges involving the satellite itself and the program's digital display technology were formidable. Program B managed the satellite program, but it also wanted to manage the data relay system. I did not agree. The Air Force was already doing a lot with communications satellites, and because of other programs winding down, Program A would have people available to do this work. The system would also have less classified applications for other types of satellites. So I made sure Program A and the Air Force's Space and Missile Systems Organization got this project. At my going away party as NRO director, some of the attendees presented a caricature of me on a stepladder painting a facsimile of the relay satellite in Air Force blue.

As with many pioneering applications of technology, the cost of the new imaging program escalated, and its schedule was hard to keep. Some skeptics continued to consider this program too risky and expensive, draining funds from others of more immediate value. I wish I could say more about this debate, but I think most of the opponents would later agree that the new satellite was well worth the cost.

Another more publicized example of American air and space technology being developed about this time was the Space Transportation System (STS), better known later as the space shuttle. To make it more economical, NASA wanted the STS to support as many DOD space missions as possible. Johnny Foster at DDR&E was the Pentagon's point man on the shuttle, and he consulted with me on its potential. We agreed that it would only make sense for the DOD to use the shuttle if it could carry our largest payloads. After considering future military spacecraft requirements, we told NASA that the shuttle would need a cargo bay 60 feet long by 15 feet in diameter. NASA's leaders agreed, so in effect we determined the ultimate size of the shuttle, which originally had been planned with a cargo bay about 40 feet long by 12 feet in diameter. Although in some respects NASA might have been better off with a somewhat smaller STS that would not put so much stress on its engines, it then could not have accommodated some of the space agency's largest payloads, most recently the 43-foot-long Chandra X-Ray Astrophysics Telescope launched in 1999.

NASA was interested in obtaining Air Force money for R&D work on the shuttle as well as for a West Coast launch facility at Vandenberg AFB, California. I was willing to provide some relatively modest funding but reluctant to become a major financial supporter. Although agreeable to adding the shuttle to our inventory of launch vehicles, I never intended that it become the sole means of getting all major satellites into orbit, as later became the policy. Fortunately, Edward "Pete" Aldridge, after he became undersecretary of the Air Force and director of the NRO in 1981, insisted on having some expendable launch vehicles as a backup. Following the *Challenger* disaster, I once told him, "Pete, every day I wake up and thank you for not leaving us dependent on the Space Shuttle." I do think that the basic STS design was a good compromise between requirements, funds available, and the technology of the 1970s. In one aspect of its design, however, NASA may have missed an opportunity. Since the early planning stages, I thought some of its external tanks should have been modified to go into orbit for use as cheap space stations rather than just falling back into the ocean.[22]

Some Thoughts on Secrecy and Bureaucracy

I had been taught at an early age not to "bear false witness." The realities of the secular world have often required governments to deceive their adversaries—and sometimes their own citizens. Being responsible for a covert activity such as the NRO posed something of a moral dilemma, since keeping its operations secret required the invention of cover stories, which are basically just plausible-sounding lies. I consoled myself that our lack of candor was necessary for the sake of national security during the Cold War. Over time I broke off pieces from our secret domain that I believed no longer required special concealment, such as U-2 operations and meteorological data. During 1972 I raised the question of declassifying the existence of the NRO with Dick Helms, who had overall responsibility for its security policy. Concerns about the "slippery slope" scenario of one disclosure leading to others deterred us from making changes at that time. As regards our classified programs, I believed when possible we should say nothing at all rather than put out false information. For example, on a visit to the Air Force Museum in Dayton, Ohio, I saw an exhibit on the aerial recovery system for capsules from orbit claiming they were for scientific experiments. I said the exhibit should be taken down until we could release information about their true purpose.

Within a few years after leaving the NRO, I found it most distressing to learn how some of the crown jewels of its technology were betrayed to the Soviets. Although occasional leaks to the press and clever observations by amateur space watchers had begun to open some peepholes through the walls that concealed our satellite operations, Eisenhower's original goal of almost total secrecy survived virtually intact for 20 years.[23] Based largely on an uncensored congressional report, the first extensive article about the NRO appeared in the *Washington Post* on 9 December 1973. The NRO did a damage assessment, primarily for the benefit of any concerned congressmen.[24] Although some may have initially panicked at the disclosure, I don't recall being very upset by the article, and the US government's classification policy for the National Reconnaissance Program remained largely unchanged for another 18 years.

The Soviet government no doubt knew more about US satellites than the American public, but we were confident that many details about our advanced technologies remained a mystery even to the KGB. I was therefore shocked and saddened about the material sold to the Soviets by a young TRW employee named Christopher Boyce, whose case went to trial in 1977, and the following year's revelations about the sale of other valuable satellite data by a junior CIA analyst named William Kampiles.[25] Although neither of these trials publicly highlighted the role of the NRO, the cases called into question our national technical means of verifying Soviet compliance with arms control agreements, particularly the proposed SALT II treaty. To someone who was all-too-well aware of the effort and expense that went into protecting information about reconnaissance satellites, it was very frustrating to see how easily security had been breached.

As time went by, the number of "uncleared" people within government and industry who knew about the existence of the NRO multiplied. On the outside, more and more books, articles, and news reports began mentioning the NRO by name and attempting to describe its satellite programs in increasing detail.[26] As a result I began to conclude that the government was keeping its head in the sand by maintaining all the old secrecy rules implemented during the Eisenhower and Kennedy administrations. Martin Faga, NRO director at the time, later joked that I was in his office every month saying, "You've got to get this place declassified."[27] On 31 October 1991 I sent a long letter to Secretary of Defense Richard "Dick" Cheney on this issue, with copies also going to DCI Robert Gates and National Security Advisor Brent Scowcroft. As a member of the National Academy of Sciences' Committee on Science, Engineering, and Public Policy, I was involved in an eight-part television documentary on space as one of my projects in support of the International Space Year (see chap. 8). Bob Seamans's son Joe was one of its cinematographers. This project ran into a stone wall when we sought comments by government officials on the value of satellite imagery. My correspondence to Dick Cheney made a strong case for relaxing the outdated classification rules governing satellite reconnaissance. (See appendix B.)

I received a cursory reply to this letter from Duane Andrews, assistant secretary of defense (C^3I), saying that they were working on the issue. I also called Brent Scowcroft at the NSC more than once to get his opinion, but he never answered directly. Martin Faga, however, formally raised the declassification issue in December 1991, and in August 1992 an outside task force submitted a report to the DCI that recommended admitting the NRO's existence. As might be expected, I was generally pleased by the announcement distributed without fanfare on 18 September 1992 declassifying the existence of the NRO and NRP.[28] Other relaxations of security followed in the next few years, most notably the decision in 1994 to allow commercial development of high-resolution (one meter) electro-optical satellites and declassification of the pioneering Corona program in 1995. I regret, however, that the veil of secrecy covering several other old programs from the 1960s and 1970s remains in place at the time of this writing.[29]

Inevitably, today's larger, centralized, and more visible National Reconnaissance Office is a much different organization from the one I knew, which operated in a simpler, more benign environment. For better or worse, the trend that began during my last year as director toward making the NRO a more regular part of the national security bureaucracy persisted. Today's NRO supplies more timely and sophisticated information to many more customers than we did. I do not think, however, that the speed and scope of the secret breakthroughs made by the NRO during the pioneering era of the 1960s and 1970s can be re-created in the future. Knowledge about "spy satellites" has become too pervasive, and the various technologies used are now too mature or expensive to permit such revolutionary progress. I am confident, however, that continued, if incremental, improvements in multispectral sensors, networking, and data processing can yield great advances in the state of the art for satellite reconnaissance if properly managed.

The Air Force in Space

At the same time the Air Force was giving essential support to the NRO, it also conducted a less classified but very active space mission of its own. The Air Force's space efforts began in the mid-1950s with multiple satellites being planned under the umbrella of the WS-117L program. In the 1960s it pursued a variety of satellite-based programs to include watching for Soviet rocket launches with the Missile Defense Alarm System (MIDAS), detecting nuclear detonations with Vela Hotel satellites, and transmitting radio signals with orbiting communication systems, starting with the Initial Defense Communications Satellite Program (IDCSP). At least as important for the future, the Air Force also built an extensive space infrastructure of launch facilities at Cape Canaveral Air Force Station, Florida, and Vandenberg Air Force Base, California, as well as a satellite control facility at Sunnyvale, California, linked to tracking and control sites scattered around the world. Also on a global scale, the Air Force developed the Space Detection and Tracking System (SPADATS), which used telescopic cameras and other devices to identify the growing number of satellites and space debris monitored at the North American Defense Command's Combat Operations Center in Colorado. Finally, the Air Force and its contractors adapted the Douglas Thor, Convair Atlas, and Martin Titan ballistic missiles to serve as the DOD's primary boosters for launching satellites. Various models of Lockheed's Agena, first developed for the WS-117L program, served as the main upper-stage rocket as well as a satellite if placed in orbit. By the end of the 1960s, all of these launch systems were achieving an impressive record of reliability.[30]

Much of the credit for the Air Force's growing space capabilities could be attributed to farsighted leaders like Bennie

Schriever, who culminated his military career as the first commander of Air Force Systems Command from 1961 to 1966. One year after his retirement, AFSC's Space Systems and Ballistic Missile Divisions, which were both descended from the Western Development Division founded by Schriever in 1954, were reunited into the Space and Missile Systems Organization (SAMSO). Headquartered at Los Angeles Air Force Station, SAMSO performed "cradle to grave" management of Air Force launchers, many satellites, and related systems.[31] In those days, space systems were so highly technical and few in number that they were considered to be in a continual state of R&D. Leading SAMSO during most of my time as undersecretary was Lt Gen Samuel C. "Sam" Phillips. He assumed this position in September 1969 after having spent the previous five years on loan to NASA running the Apollo lunar landing program. Sam was one of my favorite Air Force generals—highly intelligent, personable, and technically competent. In August 1972 duty again called him away from the Air Force when he was made director of the NSA and replaced by Lt Gen Kenneth W. Shultz.

Aerospace Corporation succeeded the company Simon "Si" Ramo founded in the mid-1950s to support the predecessors of SAMSO.[32] Since its formation in 1960, Aerospace's president had been Dr. Ivan A. Getting, a highly respected scientist who had pioneered what became known as system engineering during World War II. Having just run a similar corporation (MITRE), I was of two minds about Aerospace. It was very proficient in the work it performed, but it seemed to have become somewhat slow and expensive. In my first year on the job, I came to agree that the contractors building our launch vehicles and satellites had gained enough experience that neither they nor SAMSO still needed Aerospace to be so heavily involved. In an "eyes only" memo to Secretary Seamans on 11 August 1970, I mentioned that Lt Generals John O'Neill and Sam Phillips, past and present SAMSO commanders, had complained of having "to dig through several layers at Aerospace . . . to get down to someone who can discuss a detailed technical problem with them." Because its "superstructure" had grown so large, I wrote, "what we really ought to do is have a complete housecleaning of the management at Aerospace." This proved easier said than done. My continued frustration was evident when I wrote in a memo to Walter LaBerge (my assistant secretary for R&D) on 25 November 1974, "Someone's got to cut the overhead at Aerospace." With the influential Ivan Getting continuing as president until 1977, I don't think we ever made much progress in changing Aerospace's management structure.[33]

Sam Phillips was none too happy with SAMSO's location in Los Angeles, where the cost of living made it difficult to attract and retain military and civil service personnel. He proposed moving the organization up the coast to Vandenberg AFB. Getting, however, polled his people and said many would seek other jobs in the Los Angeles area rather than move. In view of the importance of Aerospace's workforce and the political implications of such a move (which would require new construction at Vandenberg), I disappointed Sam by not actively supporting his proposal. The Air Force's later acquisition of nearby Fort McArthur provided some of the family housing badly needed for blue suiters assigned to Los Angeles.

Unlike NRO Program A, SAMSO usually had to follow standard acquisition procedures. In this regard, Headquarters AFSC performed oversight, much as it did to other product divisions. Although not a space expert, George Brown provided good leadership in this area. His replacement in August 1973 was none other than Sam Phillips, who—with the possible exception of Schriever—is probably the most celebrated space pioneer ever to wear four stars. At the next echelon, the Air Staff and the Secretariat both had offices to oversee space programs. While Bob Seamans was secretary, he relied heavily on Grant Hansen, and I later counted on Walter LaBerge. Walt established a separate deputy for space programs in 1974. On the Air Staff, the deputy chiefs of staff for R&D had long had a separate space directorate, headed by colonels or brigadier generals, to serve as their focal point.

Despite the Air Force's accomplishments since becoming the DOD's executive agent for space procurement in 1961, the other services' growing awareness of the value of satellites led Melvin Laird to modify the existing arrangement. On 8 September 1970 DOD Directive 5160.32, *Development of Space Systems*, stipulated that responsibilities for space systems would henceforth be assigned much like other weapon systems. On 12 February 1971 reflecting the Air Force's unhappiness with the new policy, I sent a memo to Dave Packard on the implications of the DOD directive and our misgivings about the Navy's growing ambition to build its own space systems despite "the heavy investment we have in facilities and people at SAMSO and the ranges." I further argued, "Until SAMSO shows its inability to be responsive, I don't think we are justified in duplicating its capability in another Service."[34] After getting my memo, Laird modified the directive by requiring proposed space programs be coordinated with the Air Force before development. Nevertheless, I was still unhappy with the situation. At a secretaries' lunch on 11 March, I again raised our concerns about the need for a consistent space policy. Packard soon showed me a memorandum wherein he asked DDR&E to sponsor a plan using service and JCS inputs to lay out consolidated space mission requirements. I thought this was a step in the right direction. Although space efforts became more fragmented among the services, the challenge of competition may have helped motivate the Air Force not to neglect its space mission any worse than it did in the face of tight budgets during the rest of the decade.

Manned Space Systems

Some in the Air Force had never been content with operating only unmanned space systems, while NASA put all American astronauts into orbit. After McNamara canceled the Air Force's one-man space plane called the X-20 Dyna-Soar in 1962, he authorized the Air Force to develop the Manned Orbiting Laboratory (MOL). This was to be a combination of NASA's two-man Gemini space capsule and a 41-foot-long canister. Harold Brown, who had been the mastermind of the MOL concept when he was DDR&E, became its overseer as secretary of the Air Force. During his term, MOL's mission evolved from experimental projects to overhead reconnaissance, ocean surveillance, and satellite inspection. Human eyesight, it was believed, could enhance the use of other optical systems.

When I came to the Air Force, the Nixon administration was looking closely at the MOL program, with its estimated cost having doubled from $1.5 billion in 1965 to $3 billion in 1969. Bob Seamans, who worked closely with the Air Force on the MOL when he was with NASA, strongly believed the Air Force should have manned space missions in its future. Because of

MOL's reconnaissance role, I became involved in helping determine the program's future. Aware of the rapidly improving capabilities of unmanned systems at the NRO, I was unconvinced that the MOL would be worth its cost. When I shared my opinions with Bob, he thought they were sort of heretical. Dave Packard agreed with my judgment, but Mel Laird and the JCS supported continuation of the program. In late May Johnny Foster testified on the Hill that the MOL was still a valid requirement, but President Nixon had already decided to terminate the program. Mel Laird gave Bob Seamans a last chance to save the MOL, arranging for him and the program manager, Maj Gen James Stewart, to have an audience with Nixon and Kissinger in the Oval Office. They made the best case possible, but to no avail. Dave Packard announced termination of the MOL on 10 June 1969. Various contractors and the Aerospace Corporation lost many jobs as a result.

Although there was little tangible to show for the $1.4 billion spent on the MOL, some of its R&D would later help NASA in developing the three-man Skylab space station and the STS. Canceling the MOL saved at least $1.5 billion over the next three years, and the NRO received some of these funds to complete various subsystems and fold them into unmanned satellites. Future American space programs also benefited from the human capital invested in MOL. Seven of the 17 crew members in training became NASA astronauts, three of whom later became flag officers.

At about the time the MOL was cancelled, NASA began the space shuttle program, which became the means by which the DOD was expected to conduct manned operations in space. In January 1973 the OSD formed the Defense Department Shuttle User Committee to identify potential military applications and coordinate them with NASA. As a member I became intrigued with its potential for on-orbit servicing or recovery of valuable satellites. Since many DOD missions used polar orbits, they would need to be launched at Vandenberg. This meant we might recoup some of the construction costs that had gone into Space Launch Complex 6 "Slick 6," built in preparation for the MOL. Unfortunately, the substantial funds spent there over 20 years ultimately went to waste when the *Challenger* disaster led the Air Force to shut down the facility.

Planning for DOD shuttle missions raised the issue of how to command and control military space operations. As a result, no fewer than four of the Air Force's major commands volunteered during 1974 for this new mission: Air Force Systems Command, which operated communications satellites; Aerospace Defense Command, which operated early warning satellites; Strategic Air Command, which operated meteorological satellites; and the Military Airlift Command (MAC), which considered the shuttle another "transportation system." This intraservice rivalry illustrated the fragmented nature of the Air Force's space activities. The question of operational responsibilities for space remained largely unresolved during my final year as secretary, but I would have an opportunity to revisit this issue several years later.

Watching the Weather from Above

The Defense Meteorological Satellite Program (DMSP), developed primarily to support the NRO and SAC, is a real success story.[35] When I arrived, DMSP was still a special access program with the numerical designation 417. I thought its continued classification was becoming outdated and believed its data could be valuable for civil and scientific purposes. In its Block 5 configuration, DMSP could show both visible light and infrared images as well as record temperature and moisture readings, all of which could be quickly disseminated to weather stations around the world. On 7 December 1971 I told Dave Packard it was a mistake to keep the program so highly classified in view of its already widespread use for tactical applications, and he let me pursue lifting its special access status. I remember convincing Al Hall, the assistant secretary of defense for intelligence, that it was unfair to the taxpayer to keep its data so secret. In late 1972 we began to furnish data routinely to the National Oceanic and Atmospheric Administration (NOAA) and its National Weather Service.

On 16 March 1973 I held a news conference to announce declassification of the DMSP under a more general name, the Data Acquisition and Processing Program.[36] The photographs we released received widespread publicity. Some of the composite views of North America at night showing the lights of metropolitan areas soon became popular for wall posters. In addition to improving civilian weather forecasting and climatic data, the DMSP provided some fringe benefits.[37] The infrared sensors also detected such fires as the routine burning of fields in Vietnam or forest fires in North America. During October 1973 I showed some visiting defense reporters DMSP photos of Egyptian oil facilities in the Sinai knocked out of commission by Israel to illustrate the kind of information that could be obtained from even low-resolution imagery. The day before I had checked with Al Hall before declassifying them, and he had no objections. I promised the reporters some copies but had to renege on this offer when Jerry Friedheim, the assistant secretary of defense for public affairs, said he didn't think that would be a good idea.[38]

The Air Force continued to make incremental improvements to the DMSP with Block 5A, B, and C satellites, launched between 1970 and 1976. In view of DMSP's civilian applications, in late 1972 the OMB asked the Departments of Defense and Commerce to consider a consolidated civilian-military weather satellite program. An interagency committee determined that having the Air Force operate DMSP satellites for both military and civil use would be the most economical option. Henry Kissinger, however, rejected this proposal for political and diplomatic reasons. In the mid-1990s I found it interesting that NOAA was finally given responsibility for controlling DMSP along with civilian weather satellites.

Early Warning and Surveillance

Although the MIDAS program had achieved some successes, a better and more reliable system was needed to provide full-time surveillance of ballistic missile launches in the Soviet Union and detect intermediate-range and submarine-launched missiles elsewhere around the globe. In June 1969 this system was given the innocuous name of Defense Support Program (DSP). The ambiguity of this name masked another great success story for the Air Force in space. On 6 November 1970 a Titan IIIC at Cape Canaveral launched the first DSP satellite. It failed to reach a geosynchronous orbit as planned, but its highly elliptical orbit still allowed it to function well enough to provide valuable test data. With a length of 23 feet and weight of 2,000 pounds, the heart of the DSP satellite

197

was a 12-foot infrared telescope, which TRW designed to continuously scan the surface for the heat signatures of missile exhaust plumes.

Because a satellite in a stationary orbit covering the Eurasian landmass would be on the opposite side of the world from the United States, we needed an overseas ground station to receive its signals. It soon became obvious that the most secure location for this downlink would be deep in the Australian Outback. In March 1969 a team led by Mike Yarymovych (at the time Grant Hansen's deputy for requirements) selected a site near Woomera, a missile test range in South Australia. In April 1969 Australian prime minister John Gorton announced the new facility, which was given the unclassified name Joint Defence Space Communications Station Nurrungar (which means "to hear" in an Aboriginal language). I visited the partially completed Nurrungar station on my trip to Australia in December 1969 (see the F-111 section in chap. 4). In his history of the DSP, Jeffrey Richelson quotes me as saying Nurrungar was "at the end of the world" and a place I only wanted to visit once.[39] Located in a shallow depression, the area seemed more desolate than even our most remote sites in Nevada.

In May 1971 the just-completed ground station took over control of a new DSP satellite launched into a geostationary orbit 22,300 miles over the Indian Ocean and began downlinking its data on Soviet and Chinese missile launches. Although the station at Nurrungar was valuable at that time, I considered it a potential liability in the longer term. My papers record that I told Secretary Laird in March 1971 that growing political opposition in Australia made me feel even more strongly that we should not remain permanently dependent on that country's ground stations for our space activities. I was also concerned about the long-term expense of housing a large base population at Nurrungar and recommended exploring such other alternatives as using a satellite relay system. For the time being, however, the need for speedy analysis and distribution of the data forced us to station a large contingent at Nurrungar, which by year's end had a total population of more than 800 airmen, civilian employees, and dependents. The election of a Labour Party government in 1973 confirmed some of my fears and confronted us with a very delicate episode in our mutual defense relations. This ended only when the British governor-general fired the Labour prime minister (an action some Australians thought was a CIA plot). The immediate threat to our Australian facilities was averted, but little could I have guessed then that the Air Force would continue to operate Nurrungar until 1998.

The deployment of Soviet ballistic missile submarines in the late 1960s led to the need for DSP satellites to also watch over the oceans closer to North America. To downlink from these satellites, I strongly supported establishment of a ground station at Buckley Air National Guard Base near Denver. Although I thought Buckley had special advantages, there was some opposition to this choice within both the Air Force and the OSD. After I got Dave Packard's approval, Buckley was officially chosen as the DSP's new ground station in June 1970. We also established a DSP training facility at nearby Lowry AFB, Colorado, which could serve as a backup station. By early 1973 the Aerospace Defense Command's three DSP satellites were fully operational.

In addition to spotting ballistic missile launches, a special study showed the DSP was also sensitive enough to detect much smaller, shorter-burning rockets, such as SA-2s fired at our bombers over North Vietnam in late 1972 and missiles fired against Israeli aircraft in October 1973. Almost 20 years later, the capabilities of the DSP would become famous when used to warn coalition forces in the Gulf War of 1991 about Iraqi Scud missile launches. Although the DSP always had a theoretical role in helping fight a nuclear war, the way it boosted confidence in our early warning capabilities and made our strategic forces less susceptible to false alarms no doubt relaxed the threat of nuclear confrontation for the rest of the Cold War.

With help from the Atomic Energy Commission and its laboratories, the Air Force also developed Vela Hotel satellites, TRW-built systems that carried X-ray, gamma ray, and neutron sensors into 60,000-mile-high orbits to watch for nuclear tests. We launched the last advanced Vela satellites in 1969 and 1970, keeping up to eight satellites in operation at one time. After that, the Air Force gradually supplanted these long-lived satellites by installing compact nuclear detectors on DSP and global positioning system satellites, forming a network called the Integrated Operational Nuclear Detonation Detection System (IONDDS). This reflected our expanding practice of "piggy backing" multiple missions on major satellites to save money.

Worldwide Communications

For the US military, the age of satellite communications began with completion in 1967 and 1968 of what became known as the Defense Satellite Communications System (DSCS) I. Having numerous satellites circling in low orbits meant that ground stations had to keep tracking one after another with multiple moving antennas. Nevertheless, the satellites themselves were reliable, and a few of them remained operational until the mid-1970s. DSCS I also served as the basis for both NATO II and British Skynet communications satellites launched in 1969 and 1970.

In view of the almost insatiable demand of US forces worldwide for dependable, high-volume communications, these early systems represented only a start. In March 1969 the Air Force selected TRW as the prime contractor for DSCS II, the first military communications satellite designed for a geosynchronous orbit. Unfortunately, we experienced a lot of quality control problems. The first pair of DSCS satellites, launched in November 1971, suffered numerous glitches before prematurely going off the air. As a result, we had to redesign the satellite and were unable to launch the second pair of DSCS II satellites until December 1973. These worked reasonably well, with one of them operating until 1993—four times its designed life span. Then our third pair in May 1975 failed to reach orbit. In view of the pressing need for DSCS II services, these disruptions were frustrating and somewhat embarrassing. Not until 1978 did the Air Force finally achieve an operational four-satellite network for global communications.

Reflecting the Navy's overriding requirement for communicating with its ships on the seven seas, it sponsored and funded the Fleet Satellite Communications System (FLTSATCOM). Although this was the first program under the new DOD space acquisition directive for which the Air Force was not in charge, I persuaded Admiral Zumwalt to let SAMSO manage most of the procurement of the satellites. The Navy also agreed to some add-ons to accommodate nuclear command and control channels for the Air Force Satellite Communications (AFSATCOM) System. TRW won this contract in 1971. As with DSCS II, development of the FLTSATCOM was slow and frustrating. The satellite had to be partially redesigned because of problems like cross-modulation among its multiple communications channels, and the first satellite was not launched into geosynchronous orbit until February 1978. I don't think the Air Force did any worse than the Navy would have done in managing the program, but perhaps no better either. Ironically, I was at Comsat Corporation in 1977 when the company began leasing capacity to the Navy on its Marisat satellites as an interim solution for the delay of FLTSATCOM.[40]

About that time or shortly thereafter, we were visited by a delegation of top Air Force procurement officers wanting to learn why the Air Force's communications satellites were so much more expensive and troublesome than Comsat's commercial satellites. Some of the Comsat people bragged, in effect, that "it's because we're smarter than you are." "Wait a minute," I explained, "I've seen it from both sides, and I think the people involved are equally smart, but I'll tell you one difference. In the Air Force you're always trying to push the envelope; at Comsat we want to get something that's guaranteed to work." The Air Force also had to consider such stringent military requirements as hardening against the electromagnetic pulse of nuclear blasts, encrypting against enemy intercepts, and protecting against jamming.

Navigation and Position Finding

Over the centuries, the desire of mariners and explorers to know their location has given birth to many new instruments. The US Navy—supported by Johns Hopkins' Applied Physics Lab—pioneered the development of satellites for aiding ship navigation with its Transit program, which began operation in 1964. In the late 1960s, both the Navy and the Air Force began work on improved three-dimensional navigation systems that would indicate altitude as well as latitude and longitude and work fast enough for use by aircraft. The Navy program was called Time Navigation (Timation for short), while the Air Force program had the less catchy designation 621B, the title of which was Satellite System for Precise Navigation. Ivan Getting at Aerospace was a staunch supporter of this and earlier concepts.

In 1969 I found the 621B concept had matured into a plan for 20 satellites. Meanwhile the Naval Research Laboratory already had some experimental Timation satellites in orbit. With the Army also proposing its own land navigation system, the OSD formed an interservice committee to correlate the three projects. These efforts continued on parallel tracks until Col Bradford W. Parkinson took over our 621B program late in 1972. An Annapolis graduate, Brad began working with the Navy's Timation team on ways to combine the two programs. When Mike Yarymovych became chief scientist of the Air Force in early 1973, he immediately became a strong advocate of a joint program.[41] After many meetings to iron out details, the DSARC approved this new Defense Navigation Satellite System as a joint program in December 1973.

The new system incorporated the 621B's innovative signal format and Timation's higher orbits and atomic clocks. North American Rockwell became prime contractor for the satellites. I gave the program my maximum support. In May 1974 we officially renamed it the global positioning system, a name devised by Yarymovych and Parkinson to more clearly indicate its utility to skeptical military leaders. The DOD later tried to redesignate the program as Navstar, but the GPS name also stuck.[42] As quoted in *Air Force Magazine* at the time, I predicted the potential of satellite navigation and position finding "is virtually unlimited and largely untapped" and "offers revolutionary potential for blind weapon delivery, standoff systems, and—to a degree—the elimination of weather and visibility as major factors in military operations."[43] Not everyone was as enthusiastic about GPS, and the program suffered by not having an operational command to act as its sponsor. I remember having to use my authority as SecAF to restore GPS funding that the Air Staff had deleted because of SAC's reluctance to rely on any navigation system not based on inertial guidance. Some within the Air Force were also unhappy about having to pay most of the bills for a support system that would benefit other services and the private sector. After I left, Mal Currie at DDR&E was an essential advocate for keeping the GPS alive for the remainder of the Ford administration.

From my vantage point at Comsat, I was delighted when the Air Force launched the first developmental GPS satellites in 1978. Because of limited appropriations, expanding the constellation took a painfully long time. After its successful use in the Gulf War, however, everyone finally seemed to want GPS receivers. With a large enough constellation finally in place, the GPS achieved its full initial operational capability in 1993. I felt gratified during that same year when the FAA administrator asked me to head a committee on how commercial aircraft could use a global navigation satellite system. When I gave the committee's final briefing in 1994, I held up my customized Virginia license plate—"GPS-NOW"—and said let's get on with it! I also prodded the Air Force and other services to adopt GPS as a navigation aid on their passenger aircraft. After an Air Force transport carrying Secretary of Commerce Ron Brown to Croatia crashed into a Balkan mountain in 1996, I publicly expressed my amazement that the DOD had spent billions to put the GPS constellation in orbit but had not spent the miniscule amount needed to install receiver terminals in its passenger aircraft.[44]

Antisatellite Weapons

The perceived need to intercept Soviet satellites, including threatened orbital weapons, led to a preliminary Air Force system known as Program 437, which could carry either nuclear warheads or inspection cameras atop Thor missiles launched from Johnston Island in the Pacific. Its intercept capability was successfully exercised several times in 1968. That same year the United States and Soviet Union signed a treaty banning weapons of mass destruction in space, which removed much of the threat for which Program 437 had been designed. The Air Force placed it on standby status in 1970. In view of the diminished likelihood it would ever be used and the detrimental effects if it was, as well as its residual cost, I strongly recommended closing the program down. Finally, in April 1975, the OSD did just that.

The Soviets, however, had by then developed a low-orbiting antisatellite (ASAT) system that relied on close maneuvering and conventional explosives. Despite a moratorium on testing this satellite, its existence provoked concerns both within and outside the US government. To provide a US capability, President Ford in 1975 approved development of a homing antisatellite missile to be launched from F-15s, a program that was eventually cancelled by Congress in 1988. I supported beginning work on this ASAT missile so that we would have this capability if needed as a deterrent, even if it remained mostly an R&D program.

The NRO and Air Force in Retrospect

Despite progress made on some key systems, the Air Force's policies for space basically remained in a holding pattern during the 1970s and early 1980s. Much of its senior leadership tended to consider space-related activities as either R&D endeavors or DOD support functions not directly related to that era's unofficial Air Force mission statement, "to fly and fight." No doubt the privileged status enjoyed by the National Reconnaissance Office contributed to the Air Force's ambiguity about its roles in space.

In the mid-1990s I contacted six retired Air Force four-star generals (five of them former commanders of AFSC), two former NRO directors, and some other knowledgeable people to get their views about the NRO and its impact on the Air Force's space mission. By and large, these men expressed confidence in the NRO's management record, but their opinions varied significantly on whether we really needed to resort to something as unique as the NRO. To match its streamlined procurement capability, some said that the Lockheed Skunk Works model could have been extrapolated to reconnaissance satellite programs under Air Force management. It would have sufficed, they thought, to have the CIA state requirements, get the Air Force to build and operate the systems, and then let the CIA handle the products. Summing up, I isolated the following points as the most commonly agreed upon positive and negative features of the NRO during its first two decades.

Compared with normal Air Force acquisition, the NRO's advantages included (1) an ability to move swiftly, exploiting the state of the art in technology; (2) well-managed programs with stable budgets and, in general, relatively modest cost overruns; (3) limited visibility to naysayers, including those in Congress (and government micromanagers); (4) the quality and continuity of its personnel, which fostered esprit de corps and corporate memory; (5) multiservice and interagency staffing, which could draw on expertise in the Navy, Army, CIA, and NSA; and (6) its total focus on the space mission, which the Air Force often shortchanged in favor of more immediate concerns.

On the other hand, the existence of the NRO (1) split the defense space program when a unified effort would have been better, at least conceptually; (2) allowed the CIA to dominate collection activities to include building hardware; (3) overclassified its work, keeping information from many who could have benefited; (4) hamstrung the Air Force and other services in learning how to apply valuable space assets to tactical needs; (5) bred jealousy and negative attitudes toward space among some key Air Force personnel; and (6) fostered perceptions of extravagance because of its secrecy and easy access to funds.

My own opinion is that the pros significantly outweighed the cons. Unfortunately, as Al Flax said on this issue, history does not provide us with the alternatives. We only know what happened with the NRO in existence, not what other arrangements might have worked better or worse. Yet it cannot be denied that the Air Force's leadership seemed reluctant to incorporate the growing potential of space technologies into its operations and organization. In July 1980 I had the privilege of chairing the Air Force Scientific Advisory Board's Summer Study on Space. My group included Bennie Schriever and a dozen other distinguished military and civilian members. We concluded, "the Air Force is inadequately organized for operational exploitation of space and has placed insufficient emphasis on inclusion of space systems in an integrated force study."[45] We also urged building a mixed fleet of launch vehicles rather than relying only on the shuttle. To give the study report more visibility, Schriever sought to schedule my briefing at the next Corona conference of the top Air Force leaders in October 1980.* To our disappointment, Bennie was told that the agenda was too crowded to include our short briefing. Even so, this study apparently had some influence on subsequent improvements in Air Force space policy.[46]

---

*The nickname of these conferences had no connection to the satellite reconnaissance program of the same name. The Air Force leadership normally conducted three Corona meetings a year, with Corona West held in the fall at the Air Force Academy.

With the advent of the Reagan administration, growing defense budgets encouraged bolder approaches for the Air Force's space mission. As case in point, Undersecretary and NRO director Pete Aldridge was able to begin acquiring expendable launch vehicles prior to the *Challenger* disaster.[47] On the organizational front, pressure from Aldridge, uniformed Air Force space advocates, the GAO, and influential congressmen convinced Lew Allen to approve creation of an Air Force space command in 1982. This began the normalization of space activities from the realm of R&D to that of operations. One year later, Reagan's Strategic Defense Initiative began to shower funds on new projects designed to stop ballistic missiles. Bennie Schriever and Si Ramo tried to enlist me on a team of outside experts supporting acceleration of the SDI, but I preferred to remain on the sidelines, favoring instead a robust R&D program.[48]

After the Gulf War of 1991 clearly demonstrated the military value of assets in orbit, the Air Force began to emphasize more strongly its space roles and missions. In January 2001 a congressionally chartered commission on space organization and management, convinced that military operations in space are inevitable and warning of a potential "Space Pearl Harbor" (too alarmist in my opinion), recommended various actions to elevate the organizational status of space within the US government, especially the DOD.[49] After Donald Rumsfeld, who chaired this commission, became secretary of defense, many of its recommendations were implemented. Some of these changes reminded me of our practices 30 years earlier. I was glad to see the Air Force being made DOD's executive agent for space, with considerably broader authority than it had enjoyed from 1961 to 1971. I also liked having directorship of the NRO returned to the Air Force undersecretary. Another change with echoes of the past was the consolidation of space R&D, procurement, and operations under Air Force Space Command. Although on a much larger scale, this reminds me of System Command's "cradle to grave" management of some space systems in the 1960s and 1970s. As any reader of this chapter might guess, I like the idea of having the SecDef and DCI cochair a committee on space intelligence matters reminiscent of the NRP's ExCom. I also agree with elevating space issues within the National Security Council, as was the case when I ran the NRO.

I am not in favor of certain other proposals by the commission. Unlike many space enthusiasts, both in and out of uniform, I do not consider waging war with space weapons to be necessarily inevitable or desirable, and I like the idea even less now that the Russians are no longer so much of a technological competitor. As regards postulated threats to our satellites, even if some unfriendly nation could secretly develop an ASAT system, the sheer proliferation of low-orbiting satellites and the difficulty of reaching those in higher orbits would tend to negate its effectiveness. Furthermore, our early warning and intelligence resources could quickly identify the launch site, enabling us to retaliate against whoever might be foolish enough to try such an attack. In addition to building ballistic missile defenses before adequate testing, there are also hopes to launch orbital weapons with which to attack future enemies. I think there are better ways to spend defense dollars. I still agree with Eisenhower's original philosophy to keep space as a sanctuary from offensive weapons as long as possible.[50] If pursued, I hope such projects will be limited to research unless or until there are threats real enough to justify deployment.

Notes

1. John L. McLucas, "The U.S. Space Program since 1961: A Personal Assessment," in *The U.S. Air Force in Space: 1945 to the 21st Century*, ed. R. Cargill Hall and Jacob Neufeld, Proceedings of Air Force Historical Foundation Symposium, Andrews AFB, Md., 21–22 September 1995 (Washington, D.C.: Air Force History and Museums Program, 1998), 76–101. For more on the author's ideas about space, focusing on civilian programs (with a foreword by Arthur C. Clarke), see John L. McLucas, *Space Commerce* (Cambridge, Mass.: Harvard, 1991).

2. Robert L. Perry wrote a detailed history of the formative years of the NRO, *Management of the National Reconnaissance Program, 1960–1965* (1969; [R. Cargill Hall, as NRO historian, edited a redacted version of this monograph] Chantilly, Va.: NRO History Office, 2001). The first account of the CIA–Air Force rivalry available to the general public appeared in William E. Burroughs, *Deep Black: Space Espionage and National Security* (New York: Random House, 1986), 130–31, 192–200. Jeffrey T. Richelson's *The Wizards of Langley* (Boulder, Colo.: Westview, 2001), 22–130, described this contentious period from the perspective of former CIA officials. Alfred Wheelon presented some of his memories with "Corona: A Triumph of American Technology" in *Eye in the Sky: The Story of the Corona Spy Satellites*, ed. Dwayne A. Day, John Logsdon, and Brian Latell (Washington, D.C.: Smith-

sonian Institution, 1998), 29–47, while Gerald Haines, an NRO historian, briefly summarized the rivalry in "The National Reconnaissance Office: Its Origins, Creation, and Early Years," ibid., 143–56.

3. See R. Cargill Hall, "Civil-Military Relations in America's Early Space Program," in *The US Air Force in Space*, 19–31.

4. Evert Clark, "Satellite Spying Cited by Johnson," *New York Times*, 17 March 1967, 13.

5. Nixon later delegated the science advisor role to Guy Stever, who was director at the National Science Foundation. Influential people such as Jim Killian were dismayed by Nixon's action and helped remedy it by pushing for the congressionally chartered creation of the Office of Science and Technology Policy by President Ford in May 1976. See Guy Stever, *In War and Peace: My Life in Science and Technology* (Washington, D.C.: Joseph Henry Press, 2002), 202–26.

6. See James Bamford, *Body of Secrets: Anatomy of the Ultra-Secret National Security Agency* (New York: Doubleday, 2001), for some purported information on SIGINT satellites. An earlier book by Bamford, *The Puzzle Palace: A Report on NSA, America's Most Secret Agency* (Boston: Houghton Mifflin, 1982), first revealed some details about the agency.

7. Clayton D. Laurie, *Congress and the National Reconnaissance Office* (Chantilly, Va.: NRO History Office, 2001), 17, quotes from a presentation by Jimmie D. Hill to the Senate Defense Appropriations Subcommittee on 26 April 1994.

8. Maj Gen David D. Bradburn, USAF, retired, "Evolution of Military Space Systems," in *The U.S. Air Force in Space*, 63.

9. To supplement Dr. McLucas's memories of people discussed in this chapter, biographies were consulted on the NRO's 40th anniversary CD-ROM, *Freedom's Sentinel in Space* (Chantilly, Va., NRO, 2000), and its Web site http://www.nro.gov; the USAF Web site's biography database http://www.af .mil/lib/bio/; the Air Force Space and Missile Pioneer Award page on the Air Force Space Command Web site http://www.spacecom.af.mil/hqafspc/ history/pioneers.htm; as well as some published sources cited in other notes, especially Richelson's *Wizards of Langley*.

10. Dwayne A. Day, *Lightning Rod: A History of the Air Force Chief Scientist's Office* (Washington, D.C.: Headquarters USAF, 2000), 198.

11. For some details on this system (the KH-7), which Dr. McLucas was unable to describe here prior to its declassification, see Jeffrey T. Richelson, "A 'Rifle' in Space," *Air Force Magazine*, June 2003, 72–75.

12. See McLucas, *Space Commerce*, 159–61, for more details on this pioneering achievement.

13. For Duckett's expansion of the S&T directorate, see Richelson, *The Wizards of Langley*, 131–94.

14. Bob Naka made a special trip to Medford, Oregon, in the mid-1980s to see a terminally ill John Crowley and reminisce fondly about their time together with the NRO.

15. For a good source on these aircraft, see Curtis Peebles, *Dark Eagles: A History of Top Secret Aircraft* (Novato, Calif.: Presidio Press, 1995).

16. For a firsthand account of the Republic of China's U-2 operations against the People's Republic, which included overflights from 1962 to 1968, see Hsichun Mike Hua, "The Black Cat Squadron," *Air Power History*, Spring 2002, 4–19.

17. The CIA U-2 unit at Edwards was identified for unclassified purposes as the Air Force's 1130th Aerospace Technical Development and Training Group. Among the extensive literature published about the U-2, the most authoritative on the years through 1974 is Gregory W. Pedlow and Donald E. Welzenbach, *The CIA and the U-2 Program* (Langley, Va.: CIA Center for the Study of Intelligence, 1998), 247–57 (a redacted version of an earlier classified monograph, *The U-2 and OXCART Programs, 1954–1974*, printed by the CIA History Staff in 1992). For additional documentation, see Jeffrey Richelson, ed., "The U-2, OXCART, and the SR-71: U.S. Aerial Espionage in the Cold War and Beyond," National Security Archive Electronic Briefing Book, no. 74, 16 October 2002, http://www.gwu.edu/~nsarchiv/NSAEBB/NSAEBB74/.

18. There are now a number of books and articles about Corona, for example, Curtis Peebles, *The Corona Project: America's First Spy Satellites* (Annapolis: US Naval Institute, 1997). For a compendium of essays by scholars and reminiscences by participants, see Day, Logsdon, and Latell, *Eye in the Sky*. Jonathan A. Lewis examined one of the contractors in *Spy Capitalism: ITEK and the CIA* (New Haven, Conn.: Yale, 2002). Frederic C. E. Oder, James Fitzpatrick, and Paul Worthman prepared a detailed firsthand account, *The Corona Story* (1987; repr., Chantilly, Va.: NRO, 2001), CD-ROM. A heavily redacted version was included on the compact disk, *Freedom's Sentinel in Space*, prepared in commemoration of the NRO 40th anniversary in 2000. A recent book about both the U-2 and Corona is Philip Taubman's *Secret Empire: Eisenhower, the CIA, and the Hidden Story of America's Space Espionage* (New York: Simon & Schuster, 2003).

19. NASA instead put its available funds into the specialized Earth Resources Technology Satellite later renamed Landsat-1. To limit uses of its imagery, the NSC's 40 Committee specified that Landsat not have a resolution better than 20 meters.

20. For more on program accomplishments, see Oder, Fitzpatrick, and Worthman, *Corona Story*, and Dwayne Day, "The Development and Improvement of the Corona Satellite," in *Eye in the Sky*, 48–85. For data on the cameras and a list of launches, see appendices A and B, 231–45.

21. Taubman, *Secret Empire*, 351; and Thomas C. Reed, *At the Abyss: An Insider's History of the Cold War* (New York: Ballantine Books, 2004), 184–85.

22. For more on efforts to promote use of the external tanks, see McLucas, *Space Commerce*, 171–76, and chapter 8 of this memoir.

23. Apparently, the first mention of the NRO in the mainstream press occurred in two articles by Benjamin Welles: "Foreign Policy: Disquiet over Intelligence Setup," *New York Times*, 22 January 1971, 1, 8, and "H-L-S of the C.I A.," *New York Times Magazine*, 18 April 1971, 34–54.

24. Senate, *Questions Related to Secret and Confidential Documents*, 93d Cong., 1st sess., 12 October 1973, S. Rept. 93-456; and Laurence Stern, "A

1.5 Billion Secret in the Sky," *Washington Post*, 9 December 1973, A1, 9; NRO Report, "Analysis of 'A 1.5 Billion Secret in the Sky'," December 1973 (redacted copy at www.gwu.edu/nsarchiv/NSAEBB35/11-06htm).

25. The Boyce case became the subject of a best-selling book by Robert Lindsey, *The Falcon and the Snowman: A True Story of Friendship and Espionage* (New York: Simon & Schuster, 1979, rev. 1980), and several years later, a movie of the same name. Although less dramatic, the Kampiles incident was also extensively reported at the time, for example, "Former CIA Officer Arrested in Secret Satellite Manual Sale: The Case of William P. Kampiles," *AW&ST*, 28 August 1978, 22–23; George C. Wilson, "Soviets Learned of Spy Satellite from U.S. Manual . . .," *Washington Post*, 23 November 1978, 1, 12.

26. The first book about the NRO appeared in 1986 with publication of William Burroughs' *Deep Black*, followed by Jeffrey T. Richelson, *America's Secret Eyes in Space: The U.S. Keyhole Spy Satellite Program* (New York: Harper Business, 1990).

27. Transcript of remarks by Martin C. Faga at John McLucas's 80th birthday party, Belle Haven Country Club, 26 August 2000.

28. Martin Faga formally recommended its declassification to the SecDef and DCI on 30 July 1992, subject: "Changing the National Reconnaissance Office to an Overt Organization," which was prompted by a threat from the Senate Select Committee on Intelligence to do so unless the president certified "grave damage to the nation would result." See National Security Archives, "The NRO Declassified," http://www.gwu.edu/nsarch1/NSAEBB35; and Jeffrey Richelson, "Out of the Black: The Disclosure and Declassification of the National Reconnaissance Office," *International Journal of Intelligence and Counter Intelligence*, Spring 1998, 1–25.

29. Shortly after this chapter was submitted for security review, the National Imagery and Mapping Agency (NIMA) released declassified imagery from the KH-7 system and the KH-9's mapping camera to the National Archives and the US Geological Survey: NIMA News Release PA 03-01, "NIMA Sponsors Historical Imagery Declassification Conference . . .," 3 October 2002, http://www.nima.mil/general/2002/presrel/3oct02.html; James Asker, "U.S. Declassifies More Cold War Recce Satellite Imagery," *AW&ST*, 4 November 2002, 68–70. In November 2003 NIMA was renamed the National Geospatial-Intelligence Agency (NGA).

30. The most comprehensive history of the Air Force's space mission is David Spires, *Beyond Horizons: A Half Century of Air Force Space Leadership* (Maxwell AFB, Ala.: Air University Press, rev. 1998), used here extensively for background information. For a shorter history, see Curtis Peebles, *High Frontier: The United States Air Force and the Military Space Program* (Washington, D.C.: Air Force History and Museums Program, 1997), republished commercially by Diane Publishing, Collingdale, Penn. For a reference on the missiles that became launch systems, see John C. Lonquest and David F. Winkler, *To Defend and Deter: The Legacy of the United States Cold War Missile Program* (Washington, D.C.: DOD Legacy Project, 1996). For the early period, see Neufeld, *Ballistic Missiles*.

31. Timothy C. Hanley and Harry N. Waldron, *Historical Overview: Space and Missile Systems Center 1954–1995*, November 1997, http://www.los angeles.af.mil/SMC/HO/Smchov.htm.

32. Ramo-Wooldridge Corporation had begun this function in 1954, but after it merged with Thompson Products in 1958 to form TRW, Space Technology Laboratories (STL) had been spun off from TRW to do system engineering. As a profit-making company, STL was distrusted by other defense contractors, and Congress recommended establishment of a nonprofit entity, resulting in the establishment of Aerospace Corp.

33. See Ivan A. Getting, *All in a Lifetime: Science in the Defense of Democracy* (New York: Vantage Press, 1989), for his perspectives. On pages 439–43 he defended Aerospace's salaries. Stephen B. Johnson, *The United States Air Force and the Culture of Innovation* (Washington, D.C.: Air Force History and Museums Program, 2002), analyzes the role of Aerospace Corp. and its predecessors through 1965.

34. This memo in Dr. McLucas's personal papers was recently published in David N. Spires, *Orbital Futures: Selected Documents in Air Force Space History*, vol. 1 (Peterson AFB, Colo.: AF Space Command, 2004), 45–46.

35. See R. Cargill Hall, *A History of the Military Polar Orbiting Meteorological Satellite Program* (Chantilly, Va.: NRO, September 2001), for a technical history of the DMSP.

36. "Air Force Declassifies Weather Satellite Data for NOAA," *Aerospace Daily*, 7 March 1973.

37. John L. McLucas, "A New Look from USAF's Weather Satellites," *Air Force Magazine*, June 1973, 64–67.

38. John McLucas, "USAF's Role in Space," *AW&ST*, 18 February 1974, 21.

39. Jeffrey T. Richelson, *America's Space Sentinels: DSP Satellites and National Security* (Lawrence, Kans.: University Press of Kansas, 1999), 52.

40. See McLucas, *Space Commerce*, 46–53, and the Comsat section in chapter 7 of this book.

41. Day, *Lightning Rod*, 190–94.

42. See Bradford Parkinson and James Spilker Jr., eds., *Global Positioning System: Theory and Applications*, vol. 1 (Washington, D.C.: AIAA, 1996), 6–7.

43. Edgar Ulsamer, "Secretary McLucas Looks at Pressing Air Force Needs," *Air Force Magazine*, January 1974, 41–42.

44. John Mintz and Don Phillips, "Military Ordered to Update Air Safety Equipment on All Passenger Planes," *Washington Post*, 17 April 1996, A7.

45. Letter, Dr. John L. McLucas to Dr. Raymond Bisplinghoff, chairman of the Air Force Scientific Advisory Board, 13 August 1980, cited by Spires, *Beyond Horizons*, 327.

46. Ibid., 198–200.

47. Edward C. Aldridge, "The Air Force Civil-Industrial Partnership," in *The U.S. Air Force in Space*, 144–49. According to him, "the cost of that transition from the Space Shuttle back to an ELV [expendable launch vehicle] capability . . . amounted to about $16 billion."

48. While president of the American Institute for Aeronautics and Astronautics (AIAA), McLucas expressed this viewpoint with an editorial in *Aerospace America*, December 1985, 6.

49. *Report of the Commission to Assess United States National Security Space Management and Organization, Pursuant to Public Law 106-65*, 11 January 2001, http://www.defenselink.mil/pubs/space 20010111.html.

50. McLucas consistently supported this position over the years. For example, see his editorial, "Space: Sanctuary or Menace," *Aerospace America*, May 1985, 6. The consolidated supervision of NRO and all other less classified DOD space programs under the Air Force undersecretary lasted only until the summer of 2005, when Secretary Rumsfeld nominated different men to become undersecretary and to serve as NRO director.

Chapter 6

Facing Other Issues—Overseas and at Home

When I came on board as undersecretary, dealing with the war in Vietnam was at the top of the agenda. Indeed, one of the reasons I returned to Washington was to help extricate our forces from what I had come to consider a "black hole" in Southeast Asia—one that the previous administration, regardless of its original intentions, had stumbled into. Ironically, my immediate predecessor, Townsend Hoopes, helped set the stage for Lyndon Johnson's decision to stop escalating the war and not seek reelection. A month after the enemy's Tet offensive of January 1968, Gen William Westmoreland, commander of US Forces in South Vietnam, through Gen Earle Wheeler, chairman of the JCS, submitted a request for more than 200,000 additional soldiers. When Hoopes feared the White House might be receptive to this buildup, he leaked word of the troop request to the *New York Times*. The ensuing publicity added to the swelling chorus of antiwar sentiment that led up to Johnson's surprise announcement on 31 March 1968.[1]

Although not such an antiwar activist as Hoopes, I returned to the Pentagon more of a dove than a hawk (in the overused parlance of that time). My sincere desire was to help end the war with as little harm to our national interests as possible. As explained in a statement I prepared in late February 1969 for the Senate Armed Services Committee, "It was the hope and belief that Mr. Nixon wanted to resolve and could resolve the situation in Vietnam which made me willing to come to Washington as a part of this administration." Like Nixon during his campaign, I didn't have a specific plan, but I was hopeful one could be devised.

Trying for Peace with Honor in Vietnam

Soon after reporting for duty in the Pentagon, I learned that Melvin Laird had already done a lot of the necessary thinking about a new policy for Vietnam. After an extensive visit to South-

east Asia during the first part of March, the recently appointed
secretary of defense submitted a long report to the White House.
Based on the apparent progress of "pacification" efforts in many
parts of South Vietnam's countryside and the "de-American-
ization" program being pushed by Gen Creighton W. Abrams
Jr. (who had replaced Westmoreland in Saigon), Laird recommended an accelerated upgrading of South Vietnamese forces
so that they could do more and more of the fighting themselves.
As the corollary to this, Laird proposed withdrawing at least
50,000 American troops in 1969 and many more in the future.
The president quickly approved the basic thrust of this program
and the name that Laird preferred: *Vietnamization.*[2]

To offer tangible evidence that Vietnamization was more than
just a clever slogan, Laird soon set in motion a steady program
of phased American troop withdrawals. Although not on an in-
flexible long-term schedule, these personnel reductions soon
took on an inexorable momentum and became embedded in
DOD programming and budgeting as well as in congressional
and public expectations. As time went on, Nixon and Kissinger
discovered to their consternation that they were unable to stop
the drawdown no matter what happened in Southeast Asia. This
was exactly what Laird intended. More than any other leader
in the Nixon administration, Mel was attuned to the growing
sentiment against the war and the grave threat it posed on the
home front. Although anxious to prevent a communist victory,
he considered this domestic crisis an even greater danger to
our national interest than the fate of the Republic of Vietnam or
Kissinger's concerns about America's international credibility.

In the Pentagon, Mel Laird also tried to change the impression that the war in Southeast Asia was always the main order
of business. For example, he directed that staff meetings and
press briefings not constantly start with the latest reports from
Vietnam. Nevertheless, the war remained a festering sore that
we could never forget for long. Almost every morning, Bob Seamans and I were briefed on the latest intelligence and operations in Southeast Asia. As time went on, I learned more details
about what was happening over there, to include our improving
technology and tactics. However, this did not give me cause to
change my overall opinion that we needed to bring our ground
forces home with all deliberate speed. I continued to feel the

war was a distortion of national priorities that detracted from our ability to maintain readiness across the board. But getting out of Vietnam was far more difficult than getting in, and with a diplomatic settlement proving elusive, it took longer than I (and many others) initially envisioned.

USAF Operations in Southeast Asia

In many respects, the Air Force fought four somewhat different air wars in Southeast Asia. These were in South Vietnam, North Vietnam, northern Laos, and southern Laos (to later include Cambodia). Each area of operation had its own rules and complications. In addition to various bases in South Vietnam, a large portion of our participating Air Force units was located in Thailand, with B-52 bombers operating from as far away as Guam.[3]

Most of the US Air Force of that era had been designed and trained primarily to fight a high-intensity war with Warsaw Pact nations in Europe or to strike deep into the Soviet Union—not to loiter over the jungles and rice paddies of a third-world nation trying to find targets worth hitting. South Vietnam was already the most heavily bombed nation in history. Laos would have the dubious distinction of becoming the second most. I hate to think of all the ordnance we wasted splintering trees and making craters in the Indo-Chinese countryside. For this—and many other reasons of politics, strategy, and tactics—results from the first several years of air operations had been disappointing. In 1968 I had chaired an Air Force Scientific Advisory Board summer study on engaging targets in Vietnam. Based on this, MITRE's work, and other sources, I concluded that we did not have very good capabilities for finding targets in that environment, especially at night and in bad weather, or of hitting fleeting targets when we did. I think this study may have helped a bit in pushing future development of better night-vision equipment, fire-control devices, smarter munitions, improved gunships, and other equipment described in a previous chapter.

While serving as undersecretary, I was no doubt considered something of a skeptic, if not a downright pessimist, by many of my military colleagues. Despite their disappointments on the course of the war, most of them still tended to have the admirable "can-do" attitude expected of those in uniform. Unfortunately, this sometimes impeded realistic analyses of the true challenges we faced in Southeast Asia. For example, I would get optimistic briefings on how effective our interdiction missions were in slowing traffic on the Ho Chi Minh Trail. When they cited results, such as how many choke points we had cut or enemy trucks we had "killed," I would consider how difficult it was to make accurate damage assessments. But even if their numbers had been correct, I was never convinced that we were stopping enough supplies to cripple the communist war effort. The enemy forces were very economical in their logistical needs as well as resilient in repairing damage and hiding their equipment and supplies.

Despite my engineering background, I thought our military had become too infatuated with raw statistics—body counts by the Army and Marines, sorties flown and bombs dropped by the Air Force and the Navy—the kind of data that could be fed into computers and readily portrayed on charts. Some of the blame for this mind-set went back to Bob McNamara and his statistical control philosophy, but those in the chain of command who reported optimistic or inflated data made the situation even worse. Inputs from the South Vietnamese were especially suspect. As time went on, we refined some air capabilities that proved effective against the increasingly conventional warfare being waged by North Vietnam. By then, however, our leaders were no longer seeking a purely military solution to the war, even if one had still been possible.

Regardless of my opinions, overseeing air operations in Vietnam was never part of my job description. In accordance with the Defense Reorganization Act of 1958 and the evolving Unified Command Plan that laid out war-fighting responsibilities, the operational chain of command ran from the president to the secretary of defense (advised by the Joint Chiefs of Staff and their chairman) to the responsible military commands. In the case of Southeast Asia, the commander in chief of the Pacific Command (CINCPAC), headquartered in Hawaii, had regional responsibility for fighting the war, with the exception of B-52 missions conducted by the Strategic Air Command (SAC). For operations within South Vietnam, however, CINCPAC had to delegate responsibility to a huge subcommand, the Military Assistance Command Vietnam (MACV), headquartered at Saigon. The Air Force component of the US Pacific Command (PACOM) was Pacific Air Forces (PACAF), also headquartered in Hawaii. Under PACAF were the Thirteenth Air Force, headquartered in the Philippines, and the Seventh Air Force, headquartered in South Vietnam but with many units based in Thailand. Partly because of Thai sensitivities about these units reporting to a headquarters in South Vietnam, Seventh Air Force was ostensibly subordinate to Thirteenth Air Force for administrative matters (and therefore referred to as 7/13 Air Force). The Seventh Air Force commander was "dual hatted" as MACV's deputy for air operations.

Because service secretaries were not in this operational chain of command, neither Bob Seamans nor I were directly involved in making or executing decisions on the conduct of the war. In fact, we were not even informed that Air Force planes, mostly B-52s, were covertly bombing Cambodia, starting in March 1969. Nixon and Kissinger wanted this bombing done with the utmost secrecy, and so it was. Seamans and I would see reconnaissance photos of bombing results in briefings and SecDef staff meetings, but no one ever used the word *Cambodia* in connection with any of these images. Without knowing the exact geographical coordinates of a photograph, I assumed the bombs were falling in Vietnam or Laos. Mel Laird (whose initial reluctance to conduct this bombing in secret had been overruled by the White House) later said he thought I knew about it, but truthfully I had no idea. So, without the knowledge of the Air Force's civilian leadership, the Cambodia bombing went on for 14 months, eventually totaling 3,875 sorties.

About the time I became secretary, a disgruntled SAC officer finally revealed the secret bombing campaign. It was a great embarrassment to all concerned. At his staff meeting on 14 July 1973, Jim Schlesinger tried to console me by noting that the bombing took place at direct orders of the National Security Council, with the JCS and Air Force people involved in the subterfuge just doing as they were told. During a subsequent probe of the matter by the Senate Armed Services Committee, Senator Symington called former secretary Seamans as a witness. At the hearing on 25 July, General Ryan confirmed that Seamans had not been informed, which Bob said made him "damn mad," especially since he signed a report to Congress stating that any bombing in Cambodia was limited and accidental.[4] Although I was a practitioner of compartmenting information on a "need to know" basis in my NRO role, I felt strongly that, in the case of such a significant activity, we should have been informed. In view of confidentiality rules governing operational matters within the joint staff, I don't think the Air Force generals involved can be faulted for not volunteering this information to their civilian bosses. Even so, the revelation that not even the Air Force secretary knew where SAC planes were bombing discredited the Defense Department and put another nail in the coffin of Nixon's Vietnam policy.

Conventional bombing by B-52s could be very devastating on enemy formations and morale, but only if intelligence allowed them to actually hit valid targets. Bob Seamans was skeptical about the effectiveness of routine and often indiscriminate B-52 missions over South Vietnam, which significantly increased under the Nixon administration.[5] At Mel Laird's request, he had discussed their value with General Abrams and Amb. Ellsworth Bunker, both of whom considered the B-52s indispensable. After making a lengthy visit of my own to Southeast Asia, I reported to Seamans on 11 March 1971, "Everyone in the theater admits that BDA [bomb damage assessment] on B-52s is basically nonexistent." Mel Laird, never enthusiastic about large-scale bombing, agreed with this assessment. For example, at a DOD staff meeting in May 1971, he questioned the worth of much of the B-52 bombing effort and cautioned that we shouldn't keep flying as many sorties as allocated without better assessing their effectiveness.

Laird's cautious attitude did not carry much weight in the White House. As revealed later in some of the many books about the Nixon administration, the president and Henry Kissinger frequently kept Mel "out of the loop" on Vietnam operations. They went behind his back to Admiral Moorer, the JCS chairman, or one of the chiefs in violation of the chain of command. Nixon advisor John Ehrlichman later wrote, "Laird actually refused to carry out some of Nixon's instructions concerning the conduct of the war, particularly some of the Air Force operations. So Henry Kissinger and the president cultivated the Joint Chiefs of Staff. . . . Laird knew what was happening, of course, but he didn't object, since Henry's gambit left Laird free to disown the operations."[6] Kissinger's assistant, Alexander Haig, went even further in condemning Laird, saying Congress's failure to support South Vietnam "happened because Mel Laird sold the country down the river."[7]

Although soon overshadowed by news of the administration's secret Cambodia bombing, a previous case involving the falsification of bombing reports became something of a scandal within the Air Force. This was known as the "Lavelle affair." It was partly a consequence of the often restrictive and arbitrary rules of engagement for attacking North Vietnamese targets, a controversial aspect of the war that has been criticized by veterans, defense analysts, and airpower historians ever since.[8] In essence, Gen John D. "Jack" Lavelle, commander of Seventh Air Force, bent if not broke the existing rules of engagement allowing aircrews to make "protective reaction" strikes in self-defense against North Vietnamese threats. Believing he had guidance from his superiors (including Admiral Moorer and General Abrams) to be more proactive, Lavelle instructed some of his subordinate commanders to hit targets that merely supported the enemy air defense network or had the potential of becoming threats in the future. Unfortunately, he also told them not to report the absence of overt enemy activity in these cases. One of his wing commanders went even further and interpreted this as permitting the insertion of false entries about enemy opposition in after-action reports.

A concerned sergeant eventually brought this phony reporting practice to the attention of Senator Harold Hughes of Iowa, a prominent critic of the war. The Air Force inspector general soon launched an investigation. The incidents in question involved relatively few missions. Yet in view of the falsification of official records, General Ryan and Secretary Seamans, after conferring with Mel Laird, quietly retired Lavelle as a lieutenant general on 7 April 1972. By early May some congressmen began to seek more information, and General Ryan disclosed that Lavelle had been relieved for irregularities. Although his firing was considered unfair treatment by many of Lavelle's friends, this action did not satisfy some in the press or Senator Hughes. Hughes did his best to make a major issue of this incident and bombing practices in general. Both the House and Senate Armed Services Committees held hearings on the subject, but they did not support Hughes' demands that the Air Force court-martial Lavelle and take disciplinary measures against other officers involved. Jack Stempler, our general counsel, helped greatly in the resolution of this matter. The Senate committee did, however, reduce Lavelle's retirement rank to major general. Although Hughes kept all the others involved in the secret bombing (mostly colonels) off promotion lists during his last two years in the Senate, three of Lavelle's implicated subordinates eventually went on to earn four stars: Alton D. Slay, Jerome O'Malley, and Charles Gabriel.

The affair was painful for me on a personal basis. From December 1967 until September 1970, Jack Lavelle had served as deputy director and then director of the Defense Communications Planning Group, the joint organization formed to develop the electronic barrier against North Vietnamese infiltration (described in chap. 2). Because of MITRE's involvement in this project, I became close friends with Jack, who I found to be a very personable individual. I continued working occasionally with him after becoming undersecretary. In March 1971, when he was vice-commander of Pacific Air Forces, I visited him at Hickam AFB, Hawaii. He and his wife treated my entourage to a wonderful luau at their house. Like many others who knew and liked Jack, I was very sad to watch his career end so ignominiously, but as I explained to Senator Hughes when he grilled me about Lavelle, "I was on the sidelines" of this case.[9]

After Jack's forced retirement, I attended a party in his honor with about 20 other people involved with the DCPG. Dr. John Foster, the DDR&E, had overseen its work. Seated next to me at the head table, Johnny arrived late and asked if we were supposed to make any speeches. I said yes and showed him my 3-by-5-inch cards, with five points on Lavelle's contributions. When Johnny was called upon as the first speaker, he said the things I was planning to say. So I had to improvise about what a great guy Jack Lavelle had been to work with. I never saw him again before his death several years later.

As fate would have it, Lavelle's forced retirement occurred on 7 April 1972, just as North Vietnam's full-scale "Easter Invasion" of South Vietnam was about to provoke the most effective US air operations of the war. To beef up forces in Southeast Asia, the Air Force and Navy had launched one of the largest long-range deployments of combat squadrons and aircraft carriers prior to Desert Shield in 1990. Ironically, the rules of engagement that had so bedeviled Lavelle were already being rewritten.

When monsoon weather during the first two weeks of the North Vietnamese offensive kept the Air Force from responding "with maximum aggressiveness" as ordered by President Nixon, he and Kissinger sarcastically denigrated the Air Force for its ineffectiveness and timidity. They personally selected Gen John W. Vogt, director of the Joint Staff, to be Lavelle's replacement. Henry Kissinger respected Vogt, who had been a student of his at Harvard, and Nixon told Vogt to do whatever was necessary to stop the invasion.[10] The White House kept a direct line to the new Seventh Air Force commander, whom they upgraded to also be the deputy commander of MACV. With American ground forces now mostly withdrawn, Vogt was much more relevant to military operations than General Abrams. When MACV closed down in March 1973, Seventh Air Force headquarters moved to Nakhon Phanom in Thailand as a new joint command, and Vogt became the senior American officer in Southeast Asia. In my experience, he always seemed to speak with authority, and I considered him the rare combination of an effective combat leader who had truly mastered the political arena.

Although some South Vietnamese army units (with American advisors playing key roles) performed reasonably well against the North Vietnamese offensive, others did not. In the end American airpower determined the outcome.[11] After raising the stakes by having the Navy mine Haiphong harbor on 9 May, President Nixon renewed air strikes in the north, code-named Linebacker. These interdiction missions and the intensified air support in the south marked the first time in military history that airpower was so influential in defeating such a large-scale ground offensive. Despite using some troubling new antiaircraft weapons, the North Vietnamese tanks, artillery, and massed forces played into the strengths of American airpower. Linebacker featured extensive employment of our recently perfected laser-guided bombs, improved LORAN-guided navigation and radio-directed bombing, better electronic countermeasures, night and all weather strikes by F-111s, and heavy bombing by B-52s in the north. By the summer of 1972, the North Vietnamese had suffered huge losses in men and materiel. Unable to adequately resupply their front lines, they were forced to withdraw from populated areas. More than 100,000 North Vietnamese soldiers, however, continued to occupy large tracts south of the old demilitarized zone and along the Cambodian border.

Within the Pentagon, early enthusiasm for Linebacker was generally more prevalent among the uniformed leadership than their civilian counterparts. Mel Laird initially considered the spring offensive as a good test of Vietnamization and opposed large-scale bombing. Inasmuch as Bob Seamans was involved, he supported Mel's position. My own opinion was that, just when we almost had both feet out of the Vietnam quagmire, we were putting one foot back in. In the short term, I was too pessimistic. As commander in chief, President Nixon of course had the final say. He worked directly with Tom Moorer on many of the military actions. Even General Abrams was among those surprised by the mining at Haiphong. Laird, however, was generally able to mitigate Linebacker by virtue of his day-to-day authority to select or reject specific targets, often in harmony with Kissinger's attempts to orchestrate the bombing for diplomatic reasons.

Although successful militarily, the air campaign resulted in collateral damage to the Air Force's budgets for fiscal years 1972 and 1973. Congress had imposed a tight ceiling on Vietnam War expenditures, which suddenly shot through the roof. We had to deal with unprogrammed costs for such items as munitions, spare parts, lost equipment, fuel, and combat pay. Although Mel Laird was able to get some supplemental appropriations out of Congress, only by a heroic effort by Spence Shedler and his staff in Financial Management were we in the Air Force able to balance our books. In the end we also had to cut force levels more than planned, reduce readiness, and postpone some modernization initiatives.

As for results of Linebacker air operations, I was gratified by the performance of many of our new or improved weapon systems. These included Maverick air-to-ground missiles, Paveway laser-guided bombs, improved reconnaissance drones, and something called the Target Identification System Electro-Optical (TISEO) that allowed aircrews to identify enemy aircraft at a greater distance. I was also satisfied with some of the NRO's contributions to combat operations, such as the use of certain electronic intelligence sensors to fix enemy positions.

By the fall of 1972, attention turned to the long-running peace talks with the North Vietnamese in Paris. Based on Kissinger's advice, Nixon ended bombing in the northern half of North Vietnam on 23 October. General Vogt and some other military leaders thought this was premature, although the coming of the monsoon season would have hampered continued attacks there in any case. Despite an optimistic declaration by Kissinger that "peace is at hand" just before the November presidential election, negotiations quickly bogged down. Demands by Pres. Nguyen Van Thieu of South Vietnam for a complete withdrawal of North Vietnamese forces combined with the renewed intransigence of the North Vietnamese delegation caused a stalemate. This set the stage for Linebacker II, President Nixon's audacious attempt to break the deadlock by proving to Thieu that the United States was willing and able to take decisive military action, while at the same time forcefully persuading North Vietnam to negotiate more sincerely.

From 18 December through 29 December (except for 36 hours over Christmas), the full force of US airpower remaining in-theater was finally brought to bear around Hanoi and Haiphong. Despite the loss of 15 B-52s, the intensive bombing campaign seemed to shake North Vietnamese morale and self-confidence. In return for a halt to the bombing, they went back to the Paris peace talks and signed a cease-fire agreement on 27 January 1973. Later that same day, in one of his last acts as secretary, Mel Laird suspended the military draft five months ahead of schedule. For most American forces, the long war was finally over.

Although apprehensive when the bombing started—as was Bob Seamans and most of my civilian colleagues—I was relieved when Linebacker II went as well as it did. Memoirs and studies since the war confirm my recollection that Mel Laird, who had only a few weeks remaining as secretary of defense, was also reluctant to unleash the B-52s near Hanoi and Haiphong, as was Henry Kissinger. On the other side, Nixon and Al Haig were among the most hawkish about using the big bombers.[12] Considering the circumstances, collateral damage and civilian casualties were remarkably light. North Vietnam later reported 1,318 killed in Hanoi and 305 in Haiphong. Nevertheless, initial communist propaganda and exaggerated claims by visiting American and European antiwar activists helped raise a firestorm of criticism about inhumane "carpet bombing," both in the United States and around the world. Not until April did the belated publication of reconnaissance photos taken by our remotely piloted vehicles help prove otherwise (see chap. 4).

Return of our prisoners from their captivity was probably the most satisfying aspect of ending the war. Air Force C-141s began flying to Hanoi on 14 February to pick them up. During the past four years, we had been learning more and more about how badly our prisoners were treated, especially during the early years of the war. As a small compensation for their sacrifices, the services did their best to welcome the POWs back with Operation Homecoming. Under this carefully planned project, we tried to treat their physical ailments and aid their transition back to normal lives as much as possible.

The signing of the cease-fire treaty in Paris coincided with the seating of the 93d Congress, which soon made more American military action almost impossible. Reflecting war weariness and growing distrust of the Nixon administration, the new Congress took measures to curtail the war effort, including a ban on any further American combat operations in Vietnam, Laos, and Cambodia after 15 August 1973. So, less than a month after I became secretary of the Air Force, the administration's four-year-old policy of building up South Vietnamese forces was about to be put to the ultimate test.

Vietnamization and Withdrawal

During my earlier tour in the Pentagon (1962–64), the Air Force's mission in Vietnam had evolved from advisory and training roles to combat support and active (if largely covert) counterinsurgency operations. I left for Paris just days before the naval incidents in the Gulf of Tonkin led to overt US combat operations. When I returned to the Pentagon almost five years later, the Air Force stood at its peak strength of 61,400 personnel in Southeast Asia (out of a total American troop deployment of 543,000). Our objective under Vietnamization was steadily to reduce this number and return to an advisory and training role, albeit on a much grander scale than in the early years. In May 1969 Secretary Laird informed the JCS that "Vietnamizing" the war was the DOD's highest priority, and in August he instructed the joint chiefs and the services that our goal was to make the Republic of Vietnam's military capable of coping with a combined attack by Vietcong guerillas and the North Vietnamese Army.

224

For the first several months of Vietnamization, which officially began on 1 July 1969, we were not entirely sure of its timetable or impact on the Air Force—especially in view of the complexity of our weapons systems compared to most of those that the Army and Navy would transfer to South Vietnam. At one meeting with Laird and Packard in mid-October 1969, I asked whether the Air Force should be counting on an early withdrawal from Vietnam and therefore de-emphasize work on weapons systems optimized for use in Southeast Asia in favor of our other commitments. On the contrary, they told me, we needed to plan more seriously to equip the South's air force, and they complained that the Air Staff was already too interested in pushing weapons like the F-15 and B-1 and not enough about those that would support Vietnamization. Even so, I expressed concern about Vietnamese airmen being able to take over the kind of equipment needed to replace USAF capabilities. Upon returning from a trip to South Vietnam, Curtis Tarr (assistant SecAF for manpower and reserve affairs) predicted it would take several years for them just to do routine maintenance of our aircraft, and "they had no hope to overhaul them for a decade."[13] On 28 October 1969, I put my thoughts on this issue into a memorandum that Bob Seamans forwarded to Mel Laird. Some of my concerns about our Vietnamization plan are evident in the following extracts.

> This plan will approximately double the size of the VNAF [South Vietnamese Air Force]. But even after this equipment is turned over, the VNAF will not have the capability to conduct air-to-air or large-scale interdiction campaigns; it is aimed at the counterinsurgency threat. A Vietnamization program to speed the withdrawal of U.S. forces while leaving the South Vietnamese able to resist not only the VC [Vietcong] but the NVA [North Vietnamese Army] as well is a different situation. Assuming NVN [North Vietnam] continues to pose an air threat of the present magnitude, before we withdraw completely we should give the VNAF some more advanced air force capabilities. . . . [But] should we assume that the Vietnamese can use efficiently some of the sophisticated systems we have developed for ourselves? How can we train enough technicians to maintain these systems, or conversely, how can we simplify the systems we turn over to them? . . . Even with this advanced capability in the hands of the VNAF, some U.S. Air Force units may be needed nearby to squelch attempts by the NVN to overrun the country.

Notwithstanding these all-too-real concerns, I was a sincere advocate of Vietnamization. I believe the Air Force as a whole also began to take the new policy very seriously. At the headquarters level, most of the detailed planning and coordination was done on the Air Staff side, but within the Secretariat, Phil Whittaker in Installations and Logistics, Curtis Tarr in Manpower and Reserve Affairs, and their people were deeply involved. Curtis, who had been a college president and had a strong background in education, took a special interest in training issues. The scope of Vietnamization made impractical the earlier practice of bringing most Vietnamese airmen to the United States for technical and flight training. Curtis made numerous trips to South Vietnam and did a lot of work on expanding the South Vietnamese Air Force's in-country training capacity, teaching Vietnamese airmen English, sending Air Force and contractor training detachments to Vietnam, having South Vietnamese airmen work side-by-side with their American counterparts for on-the-job training, and writing special manuals for the Vietnamese Air Force. Phil Whittaker took a close interest in the difficult challenge of building up South Vietnamese supply and maintenance capabilities. Phil Hilbert, my deputy for international affairs, was heavily involved in transferring aircraft and equipment to South Vietnam. Unfortunately, we didn't have enough of the simple hardware that would have been compatible with South Vietnamese maintenance capabilities yet still be operationally effective.

When Vietnamization began, our first priority was to train enough South Vietnamese airmen to handle the aircraft inventory they would be getting in later years. During 1969 the South Vietnamese Air Force almost doubled in size, from 17,500 to 36,000 personnel, while its aircraft inventory grew only from 400 to 450 airplanes and helicopters (most of the latter provided by the US Army). At the same time, Vietnamese pilots began to fly more sorties, including combat missions. During the next two years we significantly expanded the South Vietnamese aircraft fleet. By the end of 1971, the South's air arm had grown to over 700 fixed-wing aircraft and 500 helicopters, most of them housed in US-built facilities being turned over to the South Vietnamese Air Force. The USAF also trained the South Vietnamese to take over most of the tactical air control network that tasked missions and guided aircraft to their targets.

To American observers during those early years of Vietnamization, morale appeared good and, considering the time available, the
South Vietnamese seemed to be doing pretty well in most areas. In
its first real test, the South Vietnamese Air Force flew 20,000 combat sorties in helping US airpower turn back North Vietnam's 1972
invasion, as well as airlifting many men and supplies. The aircraft
we provided to South Vietnam were simpler to operate and maintain than most of the USAF's latest generation, but many were old
and tired. They included small transports (mostly the C-7 Caribou
and C-123 Provider, which could use short and primitive airstrips),
older gunships such as the AC-47 Spooky and AC-119 Shadow,
light forward air control aircraft such as the O-1 Birddog and O-2
Skymaster, and ground attack aircraft such as the A-1 Skyraider
and A-37 Dragonfly (the latter adapted from the T-37 jet trainer).
To improve air defense capability against North Vietnamese MiGs,
we also added to the South Vietnamese inventory of supersonic
(but short-range) F-5 Freedom Fighters.

One crucial mission capability not provided was interdiction,
which we considered far too complex for the South Vietnamese
at the time. In addition to large fighter-bombers like the F-4 Phantom supported by aerial tankers, effective interdiction required
sophisticated intelligence, navigation, electronic countermeasures,
defense suppression, and accurate munitions "delivery." The
South Vietnamese Air Force also lacked rescue units to pick
up downed pilots (a capability very important to the morale
of aircrews). The administration's assumption was that major
North Vietnamese movements of men and materiel into South
Vietnam would be a violation of a future truce agreement and
thereby justify renewed American bombing. Ironically, it was
an honest statement about this commitment that got Bob Seamans into hot water with the White House.[14]

As the expected truce drew closer late in 1972, we sped up
delivery of military equipment that would presumably be curtailed by a future agreement. Project Enhance Plus was the
name of this massive effort. By year's end, South Vietnam
ranked as the fourth largest air force in the world, with 61,000
personnel and 2,000 aircraft of more than 20 types. Operating
and maintaining this force was expensive, about as much as
the cost of the million-man South Vietnamese Army. In addition to prohibiting USAF and Navy airpower from again coming

227

to South Vietnam's rescue, Congress in 1973 and 1974 denied much of the requested funds for supplying their forces, which had become accustomed to American logistical abundance. Escalating oil prices and worldwide inflation played havoc with South Vietnam's economy, and to tighten its fiscal belt, the government cut essential funding from the Air Force.

Meanwhile, increasingly formidable North Vietnamese antiaircraft weapons shot down some of South Vietnam's irreplaceable aircraft and pilots, discouraging commanders from risking others. Much of their existing equipment also sat unused for lack of enough skilled maintenance specialists. Furthermore, most South Vietnamese Air Force units were parceled out to regional Army commanders for political reasons, sacrificing centralized command and control. By mid-1974, after most USAF personnel had been withdrawn, the General Accounting Office issued a report critical of the performance and discipline of South Vietnamese airmen being trained in maintenance and support skills by American contractors. The Vietnamese were also suffering major breakdowns in managing supplies and spare parts. During these years I was routinely briefed on the progress and problems of Vietnamization. Although hoping for the best, I feared that it would take a full generation for the South Vietnamese to effectively operate a modern air force. North Vietnam, however, would not give them the luxury of time.

When the second North Vietnamese invasion came in March 1975, South Vietnam's Air Force proved woefully unable to help stem the tide. The collapse of leadership, starting with President Thieu and extending downward through the chain of command, precipitated a disastrous retreat by the South's ground forces and the rapid communist victory. South Vietnamese Air Force leaders begged us in vain for help. As Saigon fell on 30 April, some of their flyable aircraft evacuated airmen and their families, but most became spoils of war for the victors. As we watched events unfold, our morale at the Pentagon was in the pits. Everyone knew that without US airpower, the result was inevitable, but we had no choice but to accept fate. After the fall of Saigon, I remember Jim Schlesinger trying to console the depressed attendees at his staff meeting that all had not been in vain—that in the long run, the history of Southeast Asia may have been changed for the better by our apparently futile intervention in

Vietnam. In other words, no more dominoes beyond what had once been French Indochina would fall to communism.

Looking back with the advantage of hindsight, I believe the Air Force and airpower in general should have been more influential on the military course of the war than they were. Although we eventually achieved some success with new high-tech weapons, much of our equipment and tactics was incompatible with the situation. In the final analysis, however, the political challenges of intervening in a civil war on the mainland of Asia were probably too intractable for any reasonable military solutions to work. South Vietnam's government was corrupt and uninspired, while most of its people just wanted to survive and live in peace. Nationalism, although distorted by communist ideology, was on the side of the North Vietnamese and Vietcong. They were fighting what they believed to be a war of independence and unification, no matter how long it took or how much they suffered.

Even so, the communist victory was a sad time for us all. In May 1975 I had the honor of giving an Armed Forces Day speech in New Orleans. Representative F. Edward Hebert of Louisiana later flattered me by calling it "one of the most profound, in-depth discussions of the situation . . . in the world today as related to our men in uniform and the ability of the military that I have ever heard." Here are some of my remarks about the impact of Vietnam.

> I believe most Americans welcome a reassessment of . . . our possible role in various parts of the world, but many frankly are tired of it all and long for the good old days when nations of the world were not so interdependent. While sympathizing with that desire . . ., I believe most of us recognize that . . . we cannot turn back the clock. . . . We may have learned several lessons from our recent experiences in Southeast Asia. There is some good in causing us to rethink our role and deciding that we do not have all the answers. However, we should not lose our convictions. We should not assume that one idea is as good as another, or that one form of government is as good as another.[15]

Turning toward Europe and the Middle East

Hand in glove with withdrawal from Southeast Asia, the Department of Defense began to reemphasize America's commitment to NATO. This was fine with me. Jim Schlesinger, being a strategic thinker, was especially worried about how to counter the growing military strength of the Warsaw Pact. Providing resources for the war in Vietnam had seriously weakened both the quantity and quality of American military forces in Europe, including those of the Air Force. Along with modernization in general, rebuilding and improving our capabilities in Europe gave the Air Force a new focus—one that served as something of a catharsis in putting the failures of Vietnam behind us. There was an obvious need to apply the lessons of the Vietnam and Middle East wars to our posture in Europe, including the importance of electronic warfare, precision-guided weapons, and survivable aircraft basing.

Despite Kissinger's policy of détente with the Soviet regime, the relentless modernization of its military capabilities helped protect us from budget cuts even more radical than those we did suffer. For example, the realities of the Warsaw Pact threat prevented Senate majority leader Mike Mansfield from passing any of his perennial amendments between 1966 and 1974 to unilaterally reduce American troop levels in Europe. Although United States Air Forces in Europe (USAFE) had declined to only 650 aircraft in 1974 and 57,000 people in 1975 (its lowest numbers during the Cold War), the command was becoming better organized and equipped—laying the foundations for a more modern and survivable NATO force after I left the Pentagon.

With Nixon's victory in the 1972 election, Henry Kissinger declared that 1973 would be the "year of Europe." Unfortunately, it quickly became the first year of Watergate. Then suddenly, in October 1973, the Middle East took center stage with the surprise attack on Israel by Egypt and Syria, known as the Yom Kippur War. The administration quickly de-emphasized NATO and our European allies, most of which refused open support to Israel. As described earlier (in the C-5 section of chap. 4), the Air Force played a key role in the course of this conflict with its long-range airlift to Israel. It also responded quickly to the administration's order to upgrade military units to Defense Condition (DefCon) level three on 25 October to counter possible Soviet intervention.[16] This action provoked some fear in Europe and suspicions at home that it was designed to draw attention away from Watergate. Despite the potential for a superpower confrontation, I did not consider the alert especially noteworthy, but the intensity of the Yom Kippur War itself was a real eye-opener for all of us in the Pentagon. It highlighted the need to counter Soviet-designed integrated air defense systems (which had taken a toll on the Israeli Air Force) and build up larger stocks of munitions and other consumables.

Although tensions with the USSR soon eased as a cease-fire took effect between Israel and its enemies, the diplomatic and economic impact of the war proved detrimental to the United States and other industrialized countries. The petroleum-rich Arab states, using their domination of the Organization of Petroleum Exporting Countries (OPEC), significantly cut back oil production. This greatly exacerbated an existing fuel shortage into a real crisis, marked by panic buying of gasoline and other disruptions to the American way of life.

For the Air Force, the most immediate and adverse impact of the OPEC action was a shortage of the jet fuel that was our lifeblood. We used about 2 percent of the nation's total petroleum consumption, and because pilots need to practice regularly to maintain their proficiency, any major cutback in flying could have serious consequences. During the worst part of the fuel crisis in late 1973 and early 1974, the Air Force reduced its fuel consumption by about one-third. We took measures ranging from shutting off hot water heaters and turning down office thermostats to retiring 400 older support aircraft (which were often flown by staff officers to maintain their flight status). We also had to cut back on flying hours for our frontline pilots. As stated at the time by Gen George Brown, "In the absence of a solution, the Air Force and the other services face a tremendous challenge to reduce fuel consumption along with the rest of American society. We must do our part with acceptable risk to combat readiness and to safety. That's a fine line to draw."[17]

Although the immediate fuel crisis gradually eased, the increased price of jet fuel added to the strains on our budget. I wrote in mid-1975, "Even though we have significantly reduced our consumption, . . . we are paying more for fuel. For example, in FY [fiscal year] '76, the Air Force plans to purchase twenty-eight percent less fuel than in FY '73, but the cost of that reduced amount will be more than $1 billion higher."[18] In my opinion, one positive result of the fuel shortage was to spur Air Force investment in flight simulators, the potential of which had previously gone unrealized. As a believer in applying new technology, I took an active interest in the simulator program as a way to save flying hours while, at the same time, improving safety and training. I remember going down to NASA's center at Langley AFB, Virginia, to check out one of their latest simulators, which was more advanced than anything being developed at our Aeronautical Systems Division at the time. This simulator could even imitate two combat aircraft in a dogfight. I went up against a real fighter pilot, who of course shot me down. Improvements in computer graphics and related technologies would steadily make simulators more effective and versatile in the future.

Less than a year after the Yom Kippur War, another crisis in the eastern Mediterranean also adversely affected the Air Force. In July 1974 the Greek-led Cypriot National Guard staged a coup and installed an anti-Turkish firebrand as the new leader of Cyprus. He promised to seek union with Greece, which was then run by a military junta. Using our NATO installation at Incirlik as a key staging base, Turkey quickly invaded northern Cyprus to protect the island's Turkish-Cypriot minority. It appeared my old fear when I worked at NATO Headquarters, of a war between Greece and Turkey, might be coming true.

Turkey, which had a strategic location bordering the Soviet Union and the Middle East, hosted a wide range of important American military and intelligence facilities during the Cold War. The majority of them belonged to the US Air Force, which operated an umbrella organization called The United States Logistics Group (TUSLOG) to support the more than 15,000 American personnel and dependents assigned to Turkey at that time. Turkey also had the third largest army in NATO, after the United States and West Germany. Rather than risk a military confrontation over Cyprus, the Greek government relied on political pressure. Despite the importance of Turkey to our national security, the US Congress defied the Ford administration and, overriding two vetoes, imposed an embargo on our military sales and assistance to Turkey that took effect in February 1975. After the House voted down a proposal to lift the embargo in July 1975, the government of Turkey retaliated by placing US forces under what it called "provisional status."

All of us in the Pentagon leadership were dismayed by Congress's actions. In his staff meeting on 11 February 1975, Jim

Schlesinger said the embargo created "a disaster area" for NATO as well as American interests in the Middle East. As regards the July vote, Gerald Ford later wrote that he "considered this the single most irresponsible, short-sighted foreign policy decision Congress has made in all the years I'd been in Washington."[19] Under provisional status, we had to shut down valuable activities for monitoring the Soviet Union, including communications, seismic, and radar intelligence facilities, as well as stop all flying operations not in direct support of NATO. Turkish officers took over control of US installations, with American flags no longer flown. CIA director Colby warned that some deficiencies from the loss of our intelligence facilities could not be corrected by other means.

Inside Turkey, American units and personnel lost many of their special privileges and had to contend with the Turkish bureaucracy, especially its infamous customs service. This caused many inefficiencies and hardships for Americans stationed there. DOD and the services withdrew personnel and dependents (to below 12,000 by the end of 1975) and did their best to work around the restrictions. We assigned Brig Gen Bill Usher, my military assistant, to the joint US military mission in Ankara partly as a signal to the Turkish General Staff and Turkish Air Force that we wanted to remain close allies despite the circumstances. I was relieved when Congress partially lifted the arms embargo a month before I left the Air Force, but it did not end the embargo until September 1978. This finally allowed a slow rebuilding of more friendly relations between the United States and Turkey, the importance of which would become apparent to even the most staunchly anti-Turkish factions in Congress after the Iranian Revolution in 1979.

I last visited Iran in October 1974. By then this large nation had become a linchpin of the Nixon doctrine to build up regional allies to counter communist expansion. The Air Force was heavily involved in equipping and training Iran's military forces with some of our latest equipment. Accompanied by Amb. Dick Helms and Maj Gen Devol "Rocky" Brett, chief of the US military advisory group in Tehran, I had a long audience with the shah of Iran. I found him very astute in his command of international affairs and knowledge of military technology. He had a vision of Iran becoming a great power under his inspired

233

leadership. Unfortunately, just five years later we would learn that he did not have as much knowledge as he needed about his own people.

The Air Force and American Society

By the late 1960s, the military's traditional culture was beginning to feel the effects of that decade's profound political and social changes. These included the growing antiwar movement, dissatisfaction with the Selective Service System that provided the military with cheap manpower, rapidly spreading use of illegal drugs among the younger generation, growing militancy among African-Americans and other minorities frustrated by discrimination, emergent demands by many women for equal opportunity, and the beginning of the gay rights movement. By the time I left the Air Force in the mid-1970s, the Department of Defense had been coping, to one degree or another, with all of these new challenges.

The Domestic Action Program

The civil rights movement and increasing concerns about poverty in America had raised my social consciousness during the 1960s, as illustrated by my work with the black community in Boston while at MITRE. So when Mel Laird created a Domestic Action Council to harmonize efforts to help needy segments of American society a month after I became undersecretary, I was happy to be named the Air Force's representative. Consistent with President Johnson's "war on poverty," Secretary McNamara had set some precedents for these efforts in 1966 when he committed DOD to help improve the status of America's underprivileged classes with such efforts as "Project 100,000." This involved lowering entrance requirements so that otherwise unqualified young men (eventually totaling 220,000) might benefit from military training and discipline. In the face of race riots and growing social unrest, DOD's programs to meet social needs expanded under Clark Clifford and continued to accelerate even with the change to a more conservative administration. Laird formalized these efforts under the name Domestic Action. This program recognized the fact that DOD, as the government's largest agency, had a wide range of resources and expertise that could greatly impact civilian communities, especially those near military installations. Laird's policy specified that such efforts were not to adversely affect the traditional roles of the services. To spread an even wider net into local communities, the program encouraged participation by Reserve, National Guard, and Civil Air Patrol units.

Roger Kelley, assistant secretary of defense for manpower and personnel from March 1969 until June 1973, was in charge of the Domestic Action Program. Roger was a true gentleman whom Mel Laird entrusted to oversee the Pentagon's people programs. Under his direction the services tried to come up with innovative ways to contribute more directly to American society, either locally or by working on a national scale with other federal agencies. Within the Air Force, I organized our own Domestic Action Policy Council in September 1969 to better convey the importance of the program to the field and act as a clearinghouse for sharing ideas and measuring progress. I believed the key to the program was the enthusiastic volunteer, so I did not favor any special funding or bureaucracy. In many cases the Domestic Action Program formalized existing efforts by Air Force bases to be good citizens of their communities.

Among DOD's goals for 1970, Mel Laird wanted every major military installation to have at least one significant cooperative project with a nearby community. Most of the initiatives were local, such as base people volunteering to help charitable and public service agencies, setting up day camps for disadvantaged children, helping counsel troubled youth, and visiting nursing homes. Officers' wives clubs were involved in many of these local projects. Units opened their recreational and athletic facilities for children from surrounding neighborhoods, and Air Force medical and dental personnel used their valuable skills to help civilians who could not afford regular health care. Judging by our early statistics, SAC's Second Air Force seemed to have an especially active program. So I invited its commander, Lt Gen David C. Jones, to come up from his headquarters at Barksdale AFB, Louisiana, to brief our policy council. When he finished, I thanked him for an excellent presentation and asked if there was anything we could do to help. I remember Dave answering, "Yes, just stay out of our way."

Projects national in scope included a Youth Opportunity Employment Program that hired students for summer jobs in cooperation with the Department of Labor and Civil Service Commission and another program for providing educational, recreational, and cultural activities for inner-city youth. In the summer of 1969, the Air Force provided summer jobs for 13,000 young people, most of them disadvantaged, and special programs for inner-city youth at 35 installations. The summer-hire program continued to grow in subsequent years. The Air Force Recruiting Service began special efforts to recruit Airmen from more than 40 areas identified with high unemployment. At the other end of the enlistment period, Project Transition helped Airmen leaving the service—especially those in military specialties without civilian counterparts—with career and education counseling, job placement, or training with private-sector employers to help these veterans become productive citizens.[20]

In May 1973 just before leaving the Pentagon, Roger Kelley said Domestic Action had done a lot of good, but it was now time to start giving it a broader approach to community relations beyond its original antipoverty mission. By mid-1974 it had been renamed the Community Services Program to better describe its purpose. (With revelations about the surveillance and disruption of various groups by the FBI and military intelligence agencies, the term *domestic action* was taking on sinister implications.) In any case I like to think that the efforts we made did some real good. Some of my colleagues warned me that field units gamed the reports to give us the data we wanted, but I was not that cynical. I believe many of the initiatives survived as elements of expanded community relations and more proactive people policies. I was therefore proud on 9 January 1973, when Mel Laird presented me the DOD Medal for Distinguished Public Service just before he left the Pentagon. In addition to mentioning the performance of my more traditional duties, such as management of space activities and procurement, Secretary Laird said, "Your work on the Domestic Action Council is not entirely understood by all members of the Department of Defense team, but I want to go out of my way to pay special tribute to you today for your interest in that program, for your leadership, for your counsel and advice in this important area."

Morale and the All-Volunteer Force

One of the most divisive aspects of the Vietnam War was the Selective Service System, which conscripted (drafted) as many young men as necessary to fill the ranks of the Army and, to a lesser extent, the Marine Corps. During his campaign for president, Richard Nixon had promised to accelerate development of an all-volunteer force, which became one of Mel Laird's top goals as secretary of defense. In March 1969 Nixon created a commission headed by former SecDef Thomas Gates to study how to achieve this goal. Its report in February 1970 recommended immediate action to raise the pay and benefits of lower ranking people and end the draft in 1971—much sooner than Nixon had anticipated. Actions by the White House and Congress later extended authority for the draft until the end of June 1973. Meanwhile, the administration replaced the often subjective decisions of local draft boards with a national lottery based on social security numbers. As the military decreased in size, this random process was used to call up fewer and fewer young men.

The Air Force's highly technical and diverse mission required somewhat better-educated and longer-serving enlisted people than the other services. Because of the Air Force's valuable technical training programs, relatively comfortable living conditions, and the fact that only a small portion of its people were exposed to combat, it had always been able to attract enough volunteers to avoid having to use the draft. Nevertheless, antimilitary sentiment appeared to be affecting the new generation of first-term Airmen as the war dragged on. For example, in a poll taken in 1965, only 13 percent of Air Force E-3s (the grade of those normally finishing their first enlistment) identified the most unfavorable aspect of an Air Force career as, "Don't like military life in general." By 1969 the proportion of E-3s naming military life in general as their main gripe had risen to 32 percent. Senior enlisted people, in contrast, continued to focus on more traditional gripes such as promotions, pay, and family separations.[21] In response to such data and other indicators, we in the Air Force leadership took seriously the need to improve and maintain morale, especially for young Airmen. Having to adapt to the all-volunteer force no doubt helped motivate us, since many young men traditionally enlisted in the Air Force mainly

237

to avoid being drafted into the Army. A survey in late 1973, for example, showed only about 42 percent of our first-term Airmen would have enlisted without the draft. There was much concern among our officers and senior NCOs as to whether the "carrot" of higher pay would be enough to replace the "stick" of the draft in prompting qualified young men to join the Air Force.

Within the Secretariat, Curtis Tarr was our point man on early issues involving the all-volunteer force. He worked closely with the deputy chief of staff for personnel, Lt Gen Robert J. Dixon, in making sure the Air Force prepared for the transition. Although we continued to meet overall manning goals while the draft was being phased out, there were at first some disturbing signs. The percentage of recruits in the top two mental categories declined from 43 percent to 36 percent and high school graduates dropped from 96 percent to 84 percent from 1970 to 1971. By 1972 more aggressive recruiting, such as guaranteeing choice of training in specific career fields, had restored the percentage of new Airmen in the top two categories back to 42.7 percent and high school graduates to almost 90 percent.[22]

In addition to cuts in the Air Force's manning (which dropped steadily from 906,000 at the end of FY 1968 to 612,000 at the end of FY 1975), better personnel management and training opportunities allowed us to keep meeting recruiting goals despite the end of the draft. For example, the Air Training Command created the Community College of the Air Force, an innovative way for Airmen to receive academic credit—including associate degrees—for their training. Better pay and benefits, a more equitable enlisted promotion system based heavily on test scores, development of an advanced personnel data system that increased chances of getting preferred assignments, and other improvements made an Air Force career seem more attractive. The reenlistment rates for enlisted people completing their first term rose from 18 percent during FY 1968 (at the peak of the draft) to 31 percent in 1974 and 40 percent in 1975. We also tried to enhance the quality of life of our young first-term Airmen in other ways. For example, we hired or contracted for civilian food service workers so that junior enlisted people need no longer perform the traditional duty of "kitchen police" (long known as KP).[23]

I worried perhaps more than necessary about the Air Force's morale in the post-Vietnam era. Unlike the Army, which went through a serious breakdown of discipline in its enlisted force and widespread disillusionment with senior leadership among its officer corps, the Air Force seemed less discouraged by Vietnam. This is not to say our officers, especially those who flew in combat, were satisfied with existing training and assignment policies.[24] On my visits to Southeast Asia, however, I always found the morale of Air Force people to be surprisingly good— better than mine would have been in their place. As a high-ranking visitor, of course, I could never be sure I was getting the unvarnished truth, but any major breakdowns in morale would surely have become known to us in the Pentagon. I also think the Air Force's generally effective performance during the Linebacker operations helped redeem its self-image.

Compared to the Army, the Air Force as a whole had suffered relatively light casualties throughout the war and so was less invested emotionally in what became a losing cause. Finally, unlike Army draftees, even the Air Force's youngest people were volunteers who kept busy supporting flying operations in peacetime as well as wartime. Their biggest gripe seemed to be a desire to fit in better with the contemporary youth culture, such as wanting to wear longer haircuts. Among career people—officers and NCOs—the main morale problems continued to involve practical issues such as reductions in force, slow promotions, pilots forced into desk jobs, moving expenses, involuntary tour extensions, inadequate housing, and lack of jobs for spouses (especially overseas). One of the more ironic situations, which I recorded in a memo on 16 September 1971, was that "the single Airmen live on base and want to be off. The married Airmen live off base and want to get on."

Despite such frustrations, Air Force people seemed fairly well satisfied with their quality of life. Because the Air Force generally trained and fought from established bases, it traditionally invested more in facilities than the Army, which deployed into the field to train and fight, or the Navy, which did so from ships. The Air Force also had what I believe to be a well-deserved reputation for taking better care of its people than the other services, and more closely matching talents to their career aspirations. I reflected some of the Air Force's philosophy when complimenting Lt Gen Bob Dixon's personnel team in remarks I made on 9 January 1973. "There are some people who will say you're wasting money educating all these people beyond where they need. . . . Well, all I can say is when we turn people back out on the civilian economy, . . . they'll do a better job for the country, and every dollar we invest is well spent."

In my opinion, one of the greatest benefits in having to adapt to the all-volunteer force was that it compelled the services to look toward the 50 percent of Americans who had previously been permitted only limited participation and second-class status in the military. I refer of course to our female population. I wholeheartedly backed Bob Seamans when, near the end of his tour as secretary in April 1973, he proclaimed a goal to triple the number of women in the Air Force by 1978. As described later, improving the status of women became one of my top priorities when I became secretary.

Dealing with Drugs

Much has been written about the post-Vietnam syndrome and its deleterious effects on the US military. Even though the Air Force did not go through what I consider a crisis, coping with the social problems spilling over from civilian society presented us with some significant concerns. One of these was the use of drugs. The 1960s saw a growing tolerance for illegal substances among American youth—the manpower pool for new military personnel. Marijuana was readily available to young servicemen assigned to Vietnam, an increasing percentage of whom ignored possible disciplinary measures by smoking it. Soon there were too many to court-martial. By the early 1970s there was also a sudden upsurge in the use of such hard drugs as opium, heroin, amphetamines, and barbiturates, especially among soldiers in Vietnam. I have to admit that initially I had blinders on when it came to recognizing the drug problem.

Under pressure from some members of Congress, who warned of a "military drug epidemic" that would release addicted veterans into society, the Department of Defense in October 1970 sanctioned new policies to deal with the problem. These included education on the dangers of drug abuse and the use of urinalyses (aptly nicknamed Project Golden Flow) to detect hard drugs and help deter their use. The OSD also encouraged the services to offer amnesty for those who voluntarily sought help and medical programs to treat and possibly rehabilitate those identified. After a visit to Southeast Asia in late February and early March 1971, I wrote to Bob Seamans, "I can't believe that if everyone knew how bad this problem was they would be sitting back and relaxing about it."

In March 1971 we in the Air Force responded to the new DOD policy with a "limited privileged communications plan" that encouraged Airmen to avoid prosecution under the *Uniform Code of Military Justice* by seeking treatment voluntarily. We followed up in June with an aggressive and systematic five-phase rehabilitation program of (1) identification, (2) detoxification, (3) psychiatric evaluation, (4) behavioral reorientation, and (5) follow-on support. Airmen judged by medical specialists to have successfully completed any of the last four phases could be returned to duty or, if not deemed amenable to treatment, be discharged from the Air Force. To conduct phases three and four, the Air Training Command established a centralized facility called the Special Treatment Center at Lackland AFB, Texas. Between 1971 and 1974, it enrolled almost 2,000 Airmen (about 64 percent of them from Pacific Air Forces) and returned more than 900 to duty. This was a much smaller number of patients than initially feared, and in early 1974 we were able to close this center. Local bases became responsible for short-term treatment of mild cases of drug abuse, with some moderate cases going to a special training group for marginal Airmen at Lowry AFB, Colorado. More serious drug cases were discharged, with responsibility for treatment shifted to the Veterans Administration.

In part, the Air Force's initial willingness to rehabilitate drug users reflected our concerns about meeting manning goals in the era of the all-volunteer force. As it became apparent that we were able to recruit and retain enough people, the Air Force became more willing to discharge Airmen with performance and discipline problems. The Air Force also began to face up to problems caused by alcohol. On a stop at Ramstein Air Base, Germany, in 1972, one of the briefers explained to me that for every Airman admitted to the hospital for drugs, about 10 more were admitted for alcohol-related problems. By 1975 the Air Force, under the leadership of General Jones, began emphasizing the identification and treatment of alcohol abuse, which was being recognized as a more widespread and pervasive social problem than drug abuse, and one that encompassed all ranks.[25]

Improving Race Relations

Another societal problem that spilled over into the armed forces was racial discrimination and unrest, to include increased militancy by many African-Americans in uniform and a backlash from some whites. I was so concerned about how to deal with race relations and better understand the underlying social environment for the Domestic Action Program that, in August 1969, I visited the White House to seek advice from Daniel Patrick Moynihan, who was then President Nixon's special assistant for urban affairs. Although the Army and Navy experienced the worst racial incidents, the Air Force was not immune. For example, 30 Airmen were arrested at Travis AFB, California, after a prolonged outbreak of violence in May 1971 during which a dormitory was set on fire.

Something had to be done. The DOD responded to growing racial tension by mandating a program of race relations awareness education and creating the Defense Race Relations Institute at Patrick AFB, Florida, to train specialists in this area. To conduct race relations training and deal with equal opportunity issues as well as such problems as drug and alcohol abuse and family violence, the Air Force established an entirely new career field and organizational element called Social Actions. We also required all sizeable installations to establish human relations councils comprised of a cross section of ranks, ages, and ethnic groups. At first some of the race relations classes tended to degenerate into "rap sessions." I attended a few early classes at the Race Relations Institute and recall one black instructor who refused to acknowledge that racism might not be exclusively a white transgression. Eventually I believe the classes became broader in scope and less divisive. One initiative I strongly supported was including equal opportunity and treatment (EOT) as a criterion in rating supervisors and managers, both military and civilian. I was happy with the emphasis that the DCS/Personnel under Bob Dixon placed on improving minority representation "to more properly reflect the population from which we draw our forces." I wrote to him in April 1972 that "this represents one more example of Air Force leadership in an area where we can be an effective agent for change."

One of the most difficult and politically sensitive issues faced by the military installations, especially in the South, was finding off-base housing for black personnel and their families. After many years of trying to tiptoe around this problem by seeking voluntary integration and limiting sanctions to individual buildings, in May 1973 DOD announced that landlords who rented segregated housing would be declared ineligible for renting to any military personnel. Although trying to avoid saying this housing would be placed "off limits" (forbidden to all military personnel), that was indeed the implication.

The Air Force led the services in racial integration during the late 1940s, but one of its most serious challenges to improving race relations in the 1970s was our small percentage of minority officers. Even by 1972, only 1.7 percent of Air Force officers were black, compared to 4.6 percent of all college-educated Americans. In that year we set a goal of 5.6 percent minority officers by 1980. By 1975 the proportion of black officers had increased to 2.5 percent. Our strategy included getting more ROTC programs at historically black colleges and offering more remedial prep school opportunities for potential Air Force Academy cadets. With Mel Laird's approval, we service secretaries also encouraged an unwritten policy to give special consideration to retaining minority officers during reductions in force.

The Air Force's enlisted ranks were more integrated than its officer corps, rising from 12.6 percent black in 1972 to 14.8 percent by the start of 1975. There were some structural problems, however, such as lower promotion rates and fewer technical job assignments for black Airmen. One symptom of apparent discrimination where we made rapid progress was punitive discharges, three-quarters of which were given to blacks in 1970. By 1972 this rate had been cut by more than half.[26] I was also happy when we named Thomas N. Barnes as chief master sergeant of the Air Force in October 1973. Chief Barnes, who served in this capacity for almost four years, was the first African-American to hold the Air Force's top enlisted position. In this role he advised George Brown, David Jones, and me on enlisted personnel policies and concerns.

With the retirement in February 1970 of Lt Gen Benjamin O. Davis, who had been commander of the famous Tuskegee Airmen of World War II, the Air Force lost one of its greatest leaders as well as a tremendous role model for African-Americans. One of his former subordinates at the end of World War II—a tall, imposing black colonel named Daniel "Chappie" James— had recently achieved some fame as commander at Wheelus Air Base in Libya. After the junta led by young Mu'ammar Gadhafi overthrew Libya's monarchy, there was an incident during which Colonel James stood at Wheelus's main gate, pistol at the ready, and defied an attempt by Gadhafi to occupy the base with Libyan troops. Mel Laird was impressed by Colonel James's performance then and in the subsequent negotiations to vacate Wheelus. He assured James's promotion to brigadier general by selecting him in 1970 as his deputy assistant secretary for public affairs, where James's outgoing personality proved to be a real asset.

Although some white officers claimed Chappie James had been promoted mainly because of his race, I thought he had certainly paid his dues with 27 years of service before becoming only the fourth African-American in the US military to achieve star rank. When OSD soon requested that James be given a second star over many more senior officers, the complaints seemed more legitimate. Mel Laird met with Bob Seamans and me to insist on the promotion. Bob pleaded that the promotion board was already too far along in its deliberations, but Mel could not be persuaded. Bob then informed General Ryan of the SecDef's orders to make it happen. So Chappie James's subsequent promotion to major general might be considered an example of affirmative action. His advance to three stars (which did not involve a promotion board) occurred at the behest of Jerry Friedheim, the assistant SecDef for public affairs, who made James his principal deputy. This happened in the spring of 1973, when Elliott Richardson was SecDef, and I was acting Air Force secretary. James then left the Pentagon to be vice-commander of the Military Airlift Command (MAC) in 1974.

When command of North American Aerospace Defense Command (NORAD) was coming open in 1975, Dave Jones told me he was thinking about recommending James for the job. I also thought this would be a good idea, but we knew there would

244

be some reservations about whether Chappie's personality and experience qualified him so soon for a four-star position. Gen P. K. Carlton, his boss at MAC, gave him a strong recommendation, and Bill Clements was also an influential supporter. Jim Schlesinger agreed that he was ready for the job and could also serve as a great example to the American public of the progress now possible for blacks in the military. Like quite a few ex-fighter pilots of that era, Chappie had a rather rakish reputation in his personal life. When I told him we were thinking of promoting him to a level of high visibility but had some concerns, he guaranteed to me that he would not do anything to embarrass the Air Force. David Jones had the honor of pinning on General James's fourth star, and I have since been proud to have played a role in appointing the first black to achieve that rank in American history. I also believe, however, that Benjamin Davis truly deserved to have been granted this distinction several years earlier. In later years I regularly used to chat with Davis at the Andrews AFB golf course. Naturally, I was delighted when, in 1998, President Clinton pinned a belated fourth star on this true American hero, and I sent a sincere letter of congratulations.[27]

Expanding Opportunities for Women

I gladly deferred to Generals Brown and Jones and their staffs on a new officer rating system and most other hot military personnel issues of that time—such as allowing navigators to command operational units—but not on the issue of women in the Air Force. Here I took the lead. The limited opportunities for females did not affect just the military. There were many women in civil service jobs, but most were found in lower graded clerical and administrative positions. With the presence of so many women in the top echelons of the government today, it's shocking to look back at Pentagon phone books and staff directories from the 1960s and 1970s and see virtually no female names except as personal secretaries to high-ranking men. Like many males in my generation, I came only gradually to realize that women deserved better career opportunities. Being increasingly exposed to the arguments of women's rights

245

advocates, to include my daughter Pam, no doubt helped raise my consciousness on this issue.

After almost becoming extinct in the early 1960s, the special personnel category designated Women in the Air Force (WAF) was growing slowly at the time Bob Seamans and I arrived on the scene. From 1965 to 1970, the size of the WAF increased by about one-third (to about 10,000), and a number of formerly all-male job specialties had opened to females. To a large extent, this reflected congressional amendments that overturned postwar legislation limiting females to only 2 percent of the military population and imposing other restrictions on their benefits, grade levels, and career opportunities. At my swearing-in ceremony as secretary on 18 July 1973, the size of the WAF stood at slightly over 16,000 (with about 3,300 more women in the Nurse Corps). This still represented only about 2.7 percent of all USAF military personnel on active duty. In my acceptance speech, I made it a point at the ceremony to emphasize my philosophy for expanding equal opportunity and our ambitious goal for recruiting more women:

> As we move into the next four years, we face a whole new set of challenges. . . . It's going to take some very imaginative solutions to recruit and retain the kind of force that we need. At the same time, we've got to provide for this force the equal opportunity that we have pledged all along and on which we have came a long way but still have much farther to go. Equal opportunity for our minorities and equal opportunity for majorities. We have more women in the country than we do men, so there's a majority group that we want to work with very closely, and we in the Air Force are going to increase the number of women [in uniform] by a factor of three in the next five years.[28]

One month before my swearing-in, DOD began a series of actions to further improve benefits for service women. These reforms were the result of a Supreme Court decision in May 1973 on a case filed by a female Air Force lieutenant who wanted to claim her civilian husband as a dependent. The most immediate step was to give military women who were married to civilians the same benefits received by servicemen married to civilian wives. Other adjustments giving females full access to other pay and benefits soon followed. Such changes obviously made military service more attractive to women. In return, female Airmen tended to have better disciplinary and attendance records than their male counterparts.

Jeanne M. Holm headed the WAF from 1965 until 1973. In July 1971 she was promoted from colonel to brigadier general. (Our only other female general officer served as chief of Air Force nurses.) In March 1973 we made Jeanne director of the Secretary of the Air Force's Personnel Council and quickly promoted her to major general. Although Jeanne disliked the need for having women recognized for "firsts" as a symptom of their restricted status, hers was the highest rank yet achieved by a female officer in any of the services. Jeanne was a strong proponent of admitting women into the Air Force Academy and opening up most military jobs to women. We couldn't accommodate all of her desires at that time, but I encouraged the Air Staff to expand female opportunities. By mid-1974, for example, 191 assignment locations worldwide were open to women, compared with just 46 six years earlier. The Air Force also assigned the first military woman to command a major field unit: in this case, our large intelligence training wing at Goodfellow AFB, Texas. Although the services suffered a large reduction in overall ROTC enrollment, we exponentially increased the number of women enrolled in ROTC programs, from only seven in 1968 to more than 1,500 in the 1972–73 school year. By the end of fiscal year 1975 (30 June), we had increased the number of women in the Air Force to 26,774 (including 1,542 officers). With the addition of another 3,439 women in medical positions, this amounted to 30,213 females in the Air Force, or more than 4.9 percent of our 612,551 people on active duty.[29] Steps were already under way to abolish the WAF as a separate category in the Air Force's organizational structure and fully integrate women into regular field units during 1976.

Two controversial and related issues posed major barriers to women seeking careers as Air Force officers. One of these, which involved all the armed forces, was allowing females to attend the service academies. The Air Force Academy at Colorado Springs is than just another engineering or liberal arts college; it provides a total immersion in military culture. The academy has a slogan, carved in stone over an entrance to the main cadet area, "Bring me men."* My uniformed colleagues had both prac-

---

*This anachronistic saying was finally removed after sexual abuse of female cadets at the Air Force Academy became a national issue in the spring of 2003.

247

tical and emotional arguments against messing with this tradition. The second issue, which was especially emotional within the Air Force, was allowing women to fly aircraft.

Both of these issues were related to the sensitive matter of sending women into combat. Most of my senior military officers opposed opening the academy to women or allowing them in the cockpit, usually citing existing restrictions against women in combat. Like other proponents of increasing opportunities for women, I did not find this argument entirely convincing. After all, about 30 percent of academy graduates were going into fields that did not involve flying, and many of those who did get their wings never flew combat aircraft. Most Air Force officers were not academy graduates, but those that were set the example. Moreover, no matter how women earned their commissions, without the chance to serve in operational jobs—most of which required rated personnel—few women would be able to rise into command positions.

The institutional opposition to granting full equal opportunities for women was illustrated in late 1973 when the Air Force joined the Army in opposing changes in the Defense Officer Personnel Management System that would require Congress to eliminate or revise existing statutes that allowed the services to treat women differently than men. In this case the Navy, reflecting the liberal social views of Elmo Zumwalt, initially favored the removal of sexual references in existing laws. For the Air Force, keeping its authority to discharge pregnant women was a major concern. A special DOD study group eventually dropped all changes in the status of women from the proposed language submitted to Congress, where some senior members were also opposed to equal treatment of women officers.

The Army and Navy, whose officers were more likely to go into combat-related assignments, seemed even more adamant against women at their academies than the Air Force. Although publicly supporting the official position, I personally was fairly comfortable about opening the Air Force Academy to women (which I believed to be inevitable) if adequate preparations were made. By serving on a committee chaired by Bill Clements that examined education at the service academies and postgraduate schools, I had become quite familiar with the issues involved. As a practical matter, however, only Congress or the courts could change the situation. The Justice Department, supported by the services, actively defended existing policy against a lawsuit filed by two females wanting to go to the Air Force and Naval academies. An appeals court gave the services a reprieve in July 1974 by denying the plaintiffs' attempts to enter the academies that year, but the issue was far from settled.

The Defense Advisory Committee on Women in the Service (DACOWITS) was one of the more influential groups lobbying to open the academies and remove other career barriers. A DOD-sponsored group, then consisting of about 40 women from various walks of life, DACOWITS met twice a year. I always found their meetings to be friendly and low-key. I especially recall addressing a DACOWITS gathering in May 1975 at Andrews AFB. Also speaking was Maj Gen Jeanne M. Holm, just before her retirement. Saying she was leaving some unfinished business behind, Jeanne made it clear that she favored admitting women to the academies, into aircraft cockpits, and aboard ships. She also said women should be allowed to serve in combat jobs that do not require brute force, and if the services didn't take the necessary actions themselves, she predicted Congress and the courts would.[30] Just a month later, one of her predictions came true when both the House and Senate voted for the service academies to open their doors to women. President Ford signed this legislation into law in October 1975. Despite some gnashing of teeth by many of my uniformed colleagues, the Air Force quickly began preparations for accommodating up to 150 women cadets in August 1976. In fact, I had made sure we already had a preliminary plan in our hip pocket for such a contingency.

A Test Case for Gays

With minorities and women making progress in the military of the 1970s, homosexuals began to demand equal rights as well. In early 1975 an Air Force NCO named Leonard Matlovich, then stationed at Langley AFB, volunteered to become a test case against the military's policy of discharging those discovered to be homosexuals. During three tours in Vietnam, he had earned various decorations, including a Bronze Star and Purple Heart. After seeking guidance from the American Civil Liberties Union, he wrote a letter admitting his sexual preferences but requesting a waiver to the rules in view of his stellar 12-year record and current job performance. Based on the advice of legal and personnel staffs, the commander at Langley decided to give him a general discharge under current regulations. A review by the Air Force's Administrative Discharge Board upheld this decision.

Meanwhile, the case received a lot of publicity, with Matlovich becoming the first avowed homosexual to be featured on the cover of *Time*. The issue provoked strong feelings within the military community. After I declined to grant a waiver to the existing regulations, he was given an honorable discharge from the Air Force on 22 October. I did not believe the American public was ready for the armed forces to begin offering complete tolerance of homosexuality in their ranks. Former sergeant Matlovich continued to press his case in court, and in 1980 a federal judge ordered the Air Force to reinstate him with back pay. Not wanting to set this kind of precedent, the Air Force negotiated a settlement with Matlovich to drop the case in exchange for a tax-free payment of $160,000.

Leonard Matlovich certainly achieved his goal of raising the issue of gays in the military. I even received a formal letter to the secretary of the Air Force from my daughter Pam, protesting our refusal to accommodate him and other gays in uniform. Yet, as Bill Clinton found out early in his presidency when he tried to honor a campaign promise to allow avowed gays into the armed services, the American military culture is not yet ready to accept openly homosexual members in uniform, nor indeed is much of American society. Whatever their personal or religious beliefs, most senior military leaders still think full integration of gays would be incompatible with the discipline and cohesion needed in the military environment. Even back in the 1970s, I had nothing against those gays in the military who kept their lifestyle private, such as is theoretically permitted by the compromise "don't ask, don't tell" policy eventually adopted by DOD in the early 1990s. But I'm afraid this is an issue that isn't going away any time soon.

My Farewell to Arms

To many Air Force officers at the time, the prospect of women pilots seemed almost as inflammatory an issue as gays in uniform. Despite the almost forgotten achievements of the Women

Airforce Service Pilots (WASP) of World War II, most male officers thought females were just not as physically, psychologically, or socially capable of flying military aircraft as men. In this regard, however, modern hydraulics had reduced the need for upper body strength, and new fly-by-wire control systems would do more so in the future. The warrior self-image of rated male officers was no doubt an underlying factor in this attitude. As with admitting women into the academies, however, the most viable argument against women in cockpits was the statutory prohibition against using them in combat.

Since the 1950s the Air Force had followed a single-track training and universal assignment policy for its pilots, meaning they all went through the same undergraduate pilot training program, after which they were ready to transition to anything from a C-9 Nightingale to an F-15 Eagle. The Navy, in contrast, used a dual-track training system in which, after basic flight training, some aviators began learning to operate jet aircraft and others trained to fly propeller aircraft or helicopters. Based on a "Z-gram" from Elmo Zumwalt, the Navy opened its propeller aircraft track to women in 1972. Then the Army too began training female pilots for support aircraft. To have the supposedly more traditionalist Navy and Army appear more progressive than the Air Force seemed embarrassing to me. In March 1975 I read a recent inspection report critical of undergraduate pilot training that convinced me the time could be ripe for changing our men-only policy. Here is what I wrote to the chief of staff, David Jones, on 2 April 1975.

> In my appearances before civilian groups, I am frequently asked why [the] USAF cannot train women pilots as the Navy is doing. I usually fall back on the statement that USAF has only one kind of pilot, and "he" must be universally assignable. It doesn't always convince. Recently, the IG reported that the concept of the universally assignable man may not be working out; and that we should take another look at dual track. I think this has merit.

My suggestion set off a new round of internal study and debate within the Air Force about its aircrew training policy. Because most of our fixed-wing aircraft were now jet-powered (the main exception being C-130s), the Air Force's two tracks would break down into fighter-type aircraft and bomber, transport, and tanker-type aircraft. Many of my blue suit colleagues argued strongly against mixing women pilots into the equation. On 15 October when President Ford asked me to leave the Air Force, consideration of two-track aircrew training was still in its early stages. Because of my imminent departure, I decided it was now time to speed things up by making a decision on the basic issue of allowing female pilots. Ironically, the November issue of *Air Force Magazine* would report that the Air Force "will stick with the longstanding ban on women becoming military flyers, even in support flying positions."[31]

On 31 October 1975, our protocol staff sent out invitations for my farewell ceremony at Andrews AFB on 7 November, with Jim Schlesinger to preside. In the interim, President Ford's announced dismissal of his secretary of defense assured the observance would attract more attention than most such affairs. Sure enough, Jim gave a hard-hitting speech on the need for a stronger defense policy that received widespread publicity. Then it was my turn. I gave a more conventional talk, reviewing my years with the Air Force. After praising our achievements in promoting equal opportunity, I announced, "We will soon open limited pilot duty to our women, who, while prohibited from participation in combat, can still serve us in transport activities and others."[32] This came as a surprise to the audience—a pleasant surprise to the recently retired Jeanne Holm, who asked others seated nearby if she had heard me correctly. It was a somewhat less pleasant surprise—indeed a shock—for many of her former colleagues on the Air Staff, which as yet had not developed a plan on how to do this. I was confident they soon would.

I had one final public forum as Air Force secretary when I held a predeparture press conference in my Pentagon office on 21 November. I gave a few more details on women pilots, explaining the Air Force would begin by training only a small number of existing female officers. As for flying combat planes, I said that would depend on future decisions, such as passage of the equal rights amendment (which seemed to be moving toward ratification at that time).

To skip forward in time: the first contingent of 10 female Air Force officers began flight screening at Hondo, Texas, in August 1976. All 10 plus one female Coast Guard officer received their wings at Williams AFB, Arizona, in September 1977. (Until moving into an apartment in 2002, I kept a picture of this pioneering group of young women on the wall of my home office.) In the months after leaving the Air Force, I received comments and letters from a number of general officers telling me what a terrible mistake I had made. In later years, some of those who had been so opposed to both female pilots and female cadets have told me that, in retrospect, I had been right. One was grateful his daughter was going to the academy and hoping to become a pilot.

Back to my final press conference: responding to a question on morale, I expressed concern that recent cuts in some benefits would be causing more manning problems if not for weaknesses in the civilian economy, but that service people tend to undervalue their remaining benefits. As for recent scandals about various DOD officials accepting favors from contractors, I responded that we had warned ranking personnel to avoid even the appearance of impropriety and now required general officers and civilian equivalents to report any stock holdings in the defense industry. When asked about not being advised on such secret operations as the bombing in Cambodia, I said I believed service secretaries should "know what the hell is going on" and had come to support a proposed bill to require the joint chiefs to keep their secretaries informed of any such activities. Even though not in the operational chain of command, I said the secretaries would at least be able to make their objections known by "kicking, screaming, or resigning."[33] In my case, however, I was leaving the Pentagon with only good feelings. I still look back at my years with the Air Force as among the most exciting and rewarding of my life.

Notes

1. "Gen Westmoreland Asks for 206,000 More US Troops," *New York Times*, 10 March 1968, 1; Lloyd C. Gardner, *Lyndon Johnson and the Wars for Vietnam* (Chicago: Ivan R. Dee, 1995), 410, 441–43, 453; and Townsend Hoopes, *The Limits of Intervention: An Inside Account of How the Johnson Policy of Escalation in Vietnam Was Reversed*, rev. ed. with epilogue (New York: D. McKay, 1973). Although Hoopes did not directly acknowledge his leak in this book, it was later revealed by Herbert Y. Schandler in *The Unmaking of a President: Lyndon Johnson and Vietnam* (Princeton, N.J.: Princeton University Press, 1977), 203.

2. For more details, see Henry Kissinger, *Ending the Vietnam War: A History of America's Involvement in and Extrication from the Vietnam War* (New York:

Simon & Schuster, 2003), 81–82; and Jeffrey P. Kimball, *Nixon's Vietnam War* (Lawrence, Kans.: University Press of Kansas, 1998), 137–39. In response to controversies over American policy in Iraq, Melvin R. Laird recently wrote an assessment of Vietnamization. See his essay, "Iraq: Learning the Lessons of Vietnam," *Foreign Affairs* 84, no. 6 (November–December 2005): 21–43.

3. For the most thorough and balanced histories of USAF activities in Vietnam during my years at the headquarters, see Wayne Thompson, *To Hanoi and Back: The U.S. Air Force and North Vietnam, 1966–1973* (Washington, D.C.: AF History and Museums Program [also Smithsonian Institution Press], 2000); and Bernard C. Nalty, *Air War over South Vietnam, 1968–1975* (Washington, D.C.: AF History and Museums Program, 2000). For a compact summary of overall Air Force participation in the war, see John Schlight, *A War Too Long: The History of the USAF in Southeast Asia, 1961–1975* (Washington, D.C.: AF History and Museums Program, 1996), reprinted as chapters 19 and 20 in Bernard C. Nalty, ed., *Winged Shield, Winged Sword: A History of the United States Air Force*, vol. 2, *1950–1997* (Washington, D.C.: AF History and Museums Program, 1997).

4. Robert C. Seamans, *Aiming at Targets* (Washington, D.C.: NASA, 1996), 161; and *Washington Post*, 26 July 1973, A1. For Kissinger's justification for the secrecy of the bombing, see Kissinger, *Ending the Vietnam War*, 65–67.

5. According to Kimball in *Nixon's Vietnam War*, total bombing tonnage increased from about 3.2 million during 1965–1968 to 4.2 million during 1969–1972 despite fewer sorties—apparently reflecting more use of B-52s.

6. John Ehrlichman, *Witness to Power: The Nixon Years* (New York: Simon & Schuster, 1982), 94–95.

7. Statement by Alexander Haig in Gerald S. Strober and Deborah H. Strober, *Nixon: An Oral History of his Presidency* (New York: HarperCollins, 1994), 189.

8. Thompson, *To Hanoi and Back*, devotes a chapter to the Lavelle affair, 199–210.

9. Senate, *Nomination of McLucas and Brown, Hearing before the Committee on Armed Services*, 93d Cong., 1st sess., 13 July 1973.

10. H. R. Haldeman, *The Haldeman Diaries: Inside the Nixon White House* (New York: Putnam's, 1994), 436–37.

11. This conclusion was later confirmed in an exhaustive account of the fighting on the ground by Dale Andradé, *America's Last Vietnam Battle: Halting Hanoi's 1972 Easter Offensive* (Lawrence: University Press of Kansas, 2001), 492, who wrote, "Clearly, the single most important factor in the North Vietnamese defeat was U.S. airpower."

12. See Thompson, *To Hanoi and Back*, 255–80, for full details on Linebacker II.

13. Tarr memo, 7 October 1969, cited by Nalty, *Air War over South Vietnam*, 177.

14. See chapter 3 for the circumstances of Dr. Seamans' statement in December 1972. Just a few weeks later, President Nixon made a secret guarantee to President Thieu promising that "we will respond with full force should the settlement be violated by North Vietnam"—a promise he was no longer

around to keep when the time came. William Bundy, *A Tangled Web: The Making of Foreign Policy in the Nixon Presidency* (New York: Hill and Wang, 1998), 362.

15. Hon. F. Edward Hebert, "Secretary McLucas Looks at the Man in Uniform Today," *Congressional Record*, 20 May 1975, E2517–18.

16. There are five defense conditions, ranging from (5) normal peacetime posture to (1) maximum force readiness, with DefCon 3 calling for increased force readiness.

17. Quoted by John L. Frisbee, "The Chief Discusses USAF's Prospects," *Air Force Magazine*, February 1974, 33.

18. John L. McLucas, "USAF's Increasing Operational Efficiency," *Air Force Magazine*, May 1975, 48–50.

19. Gerald R. Ford, *A Time to Heal* (New York: Harper & Row, 1979), 302.

20. Early publicity about the program included an article by Undersecretary McLucas, "Domestic Action—A New Challenge for the Air Force," *Air Force Magazine and Space Digest*, February 1970, 54–57. Later articles included "Air Force's Dr. John McLucas: Burgeoning Interest in Domestic Action Programs," *Government Executive*, October 1972, 36; and "Domestic Action: The Other Mission," *Air Force Times*, 18 July 1973, 12.

21. Bruce Callander, "New Poll Hits 'Military Life,'" *Air Force Times*, 17 December 1969, 13.

22. Comments by Dr. Tarr and General Dixon in John T. Correll, "The All Volunteer Force—Possible or Probable," *Air Force Magazine*, November 1971, 70–75; and Robert C. Seamans Jr., "How USAF Plans to Meet Its Personnel Needs," *Air Force Magazine*, March 1973, 45–50.

23. Data derived from "Air Force Almanac," *Air Force Magazine*, May 1976, 133.

24. For a firsthand critique of inadequate training and arbitrary assignment of aircrews in Vietnam and the reforms that followed, see C. R. Anderegg, *Sierra Hotel: Flying Air Force Fighters in the Decade after Vietnam* (Washington, D.C.: AF History and Museums Program, 2001).

25. Lawrence R. Benson documented drug problems in a monograph, *The USAF Special Treatment Center and Rehabilitation of Drug Abusers, 1971–1974* (Lackland AFB, Tex.: AF Military Training Center, 1975). For an article on problems caused by the dominant legal drug, see Ed Gates, "Turning the Spotlight on Alcohol Abuse," *Air Force Magazine*, June 1975, 77.

26. John T. Correll, "A Fair Share for USAF Minorities," *Air Force Magazine*, June 1972, 25–29; and "Black Percentage of Force Rises," *Air Force Times*, 18 October 1975, 19.

27. For more on these two pioneering black generals, see James R. McGovern, *Black Eagle: General Daniel 'Chappie' James, Jr.* (Tuscaloosa: University of Alabama Press, 1985); J. Alfred Phelps, *Chappie: The Life and Times of Daniel James Jr.* (Novato, Calif.: Presidio, 1991); and *Benjamin O. Davis, American: An Autobiography* (Washington, D.C.: Smithsonian Institution, 1991).

28. From a transcript in McLucas's Air Force papers, portions of which were also quoted by Claude Witze in "USAF's New Leaders," *Air Force Maga-*

zine, September 1973, 59. Data on WAF is from *USAF Statistical Digest, Fiscal Year 1970* (Washington, D.C.: Comptroller of the Air Force, 22 February 1971), 231, and *Fiscal Year 1973* (31 July 1974), 129.

29. *USAF Statistical Digest, Fiscal Year 1975* (Washington, D.C.: Comptroller of the Air Force, 16 April 1976), 128–31.

30. "General Holm Looks Back and Ahead," *Air Force Times*, 14 May 1975, 2. For her comprehensive study of females in uniform, see Jeanne Holm, *Women in the Military: An Unfinished Revolution*, rev. ed., (Novato, Calif.: Presidio, 1992).

31. John O. Gray, "The Bulletin Board," *Air Force Magazine*, November 1975, 81.

32. Quoted from transcript in the McLucas papers. For a summary, see "How to Liven Up the Farewell," *Air Force Times*, 26 November 1975, 12.

33. "McLucas Endorses Data Bill," *Washington Post*, 22 November 1975, A5; and "Late Thoughts by McLucas," *Air Force Times*, 10 December 1975, 9, 24.

Chapter 7

Managing Civil Aviation
and Commercial Space Programs

Fifteen October 1975: The Air Force C-140 Jetstar was descending for a refueling stop en route to California when the steward informed me I had a telephone call from Air Force One. The caller was Pres. Gerald Ford. "Can you talk?" he asked. I told him we were about to land and asked if I could please call him back from the ground. "Sure," he said. I was almost certain I knew why the president wanted to talk to me.

Heading the Federal Aviation Administration

The White House chief of staff, Donald Rumsfeld, had already informed me that the president was looking for a new chief of the Federal Aviation Administration (FAA). Ford had fired the previous administrator, Alexander Butterfield, a former Air Force colonel who had resigned his commission to run the FAA after being a member of Nixon's White House staff. It is possible that Nixon could have finished his term had it not been for the testimony Butterfield gave before Senator Samuel J. Ervin Jr.'s special Watergate Committee. In July 1973 Butterfield shocked the nation when, in response to a direct question, he revealed existence of the tape-recording system that Nixon had installed in the Oval Office.

From that day forward, Alex's effectiveness was compromised. Key people went out of their way to isolate him, including his immediate supervisor, Secretary of Transportation Claude S. Brinegar. Even so, Butterfield held on until March 1975, when President Ford asked for his resignation. Despite bad publicity over some aircraft accidents, I found his dismissal a bit perplexing. Fifteen years later my curiosity got the best of me, so I called Gerald Ford and asked exactly why he had fired Butterfield. "John," he said, "I'll tell you. I have nothing against Alex Butterfield. But it turned out that nobody would work with him. If you've got a man that people won't work with, you can't use him. I had to let him go for that reason."[1]

Crossing the River

For six months after relieving Butterfield, the White House tried hard to find a suitable replacement. The Senate was reluctant to approve another military officer, in this case former Navy astronaut Charles "Pete" Conrad. Ford couldn't get other prospective nominees to volunteer for the job, partly because he was looked upon as a lame duck. So the White House finally decided to look at appointees who had already been through the confirmation process. As I heard later, the president went down the list until he saw my name and said, "There's our man." I was called to the White House to meet with Rumsfeld, National Security Advisor Brent Scowcroft, and Special Assistant Douglas P. Bennet. They tried to interest me in volunteering for the vacancy, but I told them I would prefer to finish my tour with the Air Force rather than start over in another agency this late in the president's term in office. I remember saying as I left, "Well, I guess if the president really wants me, I'll hear from him." So that's why I got the call in my plane. President Ford was very persuasive, assuring me that I could remain at the FAA as long as I desired after the election (which he intended to win). I said, "Okay, sir, I'll seriously think about it."

After talking to Mel Laird and Secretary of Transportation William T. Coleman (who had succeeded Brinegar), I made an appointment at the White House and told the president I'd accept his offer. I did, however, mention my concern that the move across the Potomac could appear to be a demotion. "That's not really the case," he explained. "It may look that way on the surface, but the fact is, the FAA is a terribly important job. The safety of the traveling public is at stake." To help bolster my prestige, he agreed to swear me in at the Oval Office. After appearing before the Senate's Commerce Committee on 4 November, my appointment was confirmed on 13 November.[2] Soon I said good-bye at the Pentagon and went to the White House for the swearing-in ceremony, accompanied by my family and new boss, Bill Coleman (the first black Republican to serve in the Cabinet). I then moved into my new office in the southwest sector of the District of Columbia with an excellent view of the National Mall and the Smithsonian Castle.

My Roles as Administrator

Once on board, I quickly realized that the president was right; I had stepped into a very significant job. Even though the FAA (with only 58,000 employees) was a lot smaller than the Air Force, its mission was critical to the nation's economy and mobility. The FAA is entrusted with fostering air commerce, promulgating and enforcing air safety rules, certificating airmen and aircraft, establishing airways, administering grants to airports, and managing a common system of navigation and air traffic control for both civil and military aircraft.

A feeling prevailed in some quarters on Capitol Hill that the FAA's air commerce and safety missions posed a conflict of interest.[3] With all due respect, I disagreed, believing that it was very important for the agency to be both a promoter and a regulator of aviation. Otherwise, it might tend to make regulations that would accomplish relatively little to improve aviation safety and do so at an unnecessarily high cost. Dan Rather of CBS News later interviewed me in trying to make a case that my judgment on certain safety measures was compromised by my "fostering" role. When the story failed to appear, I had Joe Laitin, my public relations chief, ask when we could expect to see it. The answer was that Rather's staff had failed to find any reputable critics willing to allege that I was scheming with the airlines at the expense of safety.

Although morale at the FAA had suffered as a consequence of Butterfield's being persona non grata and the long interregnum before my appointment, a competent acting administrator, James E. Dow, had been keeping the ship on an even keel.[4] After he retired in March 1976, I appointed Jeff Cochran, the FAA's R&D chief, as my deputy and retained most of the existing staff. I especially relied on Charles E. Weithoner, the associate administrator, and Ray Belanger, director of air traffic services. I did bring in a new chief counsel, Bert Z. Goodwin, who had been my deputy general counsel at the Air Force. I also named a new assistant for public affairs, the aforementioned Joe Laitin, whom I had gotten to know when he worked for Jim Schlesinger. (After finishing his government career, Joe served as the ombudsman for the *Washington Post* from 1986 to 1988 and remained a good friend until his recent death.)

The one area of the FAA I thought seriously deficient was its procurement function. That may be why the Office of the Secretary of Transportation (OST) had set up the Transportation Systems Acquisition Review Council (TSARC), which reminded me of the DSARC at the Pentagon (see chap. 4). The TSARC could help prevent unwise or premature decisions, but it could also stifle initiative and slow down the process. One of my great frustrations was not doing more to increase the speed and efficiency of how the FAA acquired such new equipment as a much needed collision avoidance system. The FAA also needed more R&D funding. I last testified on that issue less than a month after leaving the agency, recommending to the Transportation Subcommittee of the House Appropriations Committee that it provide $25 million more in annual funding to help permit development of a more automated air traffic control system, computerized flight controls, head-up displays, and other needed equipment. I also emphasized the need to reduce the hazards of fire and toxic smoke after accidents and for research on using satellite navigation systems.[5]

Unlike the festering relationship between Brinegar and Butterfield, Bill Coleman and I enjoyed very good rapport. On the other hand, bureaucratic frictions between the OST and the FAA had built up over the years—not unlike the tensions in the Pentagon between the OSD and the services. There were still some like Gen William F. "Bozo" McKee, last head of the FAA when it was the independent Federal Aviation Agency, who believed folding it into the Department of Transportation (DOT) had been a mistake. I think the potential advantages of having all federal transportation functions under one roof outweigh the constraints this imposed on the FAA. I also maintain, however, that it's important that the FAA administrator enjoy a different status from heads of other DOT components. Although no longer serving directly as the president's advisor on aviation, he or she ought to have some kind of personal relationship with the chief executive and authority to work directly with the Departments of State, Commerce, and Defense in all areas involving aviation. I can't remember exactly how many times I saw President Ford, but there were quite a few. I felt it was not only good for my ego,

but also good for others to know that the president was interested in aviation issues and was seeking my advice on them.

Aircraft Accidents and Safety

Safety needs to be the paramount concern in building, maintaining, and operating passenger aircraft, and it is probably at
the top of every FAA administrator's agenda. The main reason
for the invention of the air traffic control system was preventing midair collisions, and I was fortunate not to have any such
accidents involving commercial air carriers during my watch.
Yet airplane collisions can also happen on the ground, as we
learned to our dismay on 27 March 1977, only four days before
I left the FAA. Two Boeing 747s collided on a runway at Tenerife in the Canary Islands, killing 583 people. This is still by far
the deadliest accident in aviation history. About the only positive thing that can be said about this avoidable collision was
that it accelerated development of a new generation of airport
surface detection equipment. Another fatal accident involving a
major air carrier occurred on 27 April 1976, when an American
Airlines Boeing 727 crashed on landing at Harry S. Truman
Airport at Saint Thomas in the Virgin Islands. The airport's
runway extended only 4,658 feet, which did not afford a very
wide margin for error, especially when combined with nearby
mountains. Thirty-seven people lost their lives. The crew had
obviously made a mistake, one that might not recur for many
years. Nevertheless, I felt Saint Thomas's runway should be
extended by 700 feet. I could not secure full funding for this
project, but Secretary Coleman did approve a matching grant
of $37 million to upgrade the airport.

Dealing with the consequences of an accident that happened
before I came to the FAA became one of my biggest headaches.
The tragic mishap occurred on 3 May 1974, when a DC-10 operated by Turkish Airlines crashed after taking off from Paris, killing all 346 people on board. A door to the baggage compartment
came unlatched, allowing air to escape from the lower deck. The
floor of the pressurized passenger compartment then collapsed,
cutting control cables leading to the tail surfaces. The manufacturer, McDonnell Douglas, was blamed for faulty design and
construction, while Turkish Airlines was blamed for not having

adequate maintenance or door latching procedures. A similar incident had occurred with an American Airlines DC-10 in 1972, but extraordinary efforts by the crew had saved the plane. The FAA took some actions at the time, but after the Paris tragedy, it was accused of not having dealt adequately with the first event and even of covering up the seriousness of the defects.

I brought myself up to speed on this issue through numerous briefings and meetings with people at McDonnell Douglas and elsewhere. The FAA implemented additional steps to improve the latching mechanism and a related monitoring device in the cockpit. We then looked at the overall airframe design to see how to strengthen the floor. Although another door failure was unlikely, there had been incidents of explosions causing a similar loss of air pressure from baggage compartments. In addition to strengthening the floor, we also required installation of an air escape mechanism with small blow-down hatches to relieve pressure differences between the upper and lower decks. Prior to taking these actions, air safety advocates accused the FAA of laxness. Now the industry accused us of overkill. Besides the cost of the modifications, adding to the weight of an airliner automatically decreases its fuel efficiency and payload, but we pressed ahead in spite of the criticism. Fortunately, the manufacturer was receiving orders for freighter versions of the DC-10 with stronger floors; so much of the design work was already being done.

Having dealt with the DC-10, we looked at other "heavies," specifically, Boeing's 747 and Lockheed's L-1011. We encountered a good deal of resistance, since these aircraft had not been involved in similar accidents. Terrorists, however, could get explosives into the cargo hold of any airliner. To avoid grounding aircraft just for this work, we allowed the modifications to be made as part of their next overhaul cycle, giving the airlines until the end of 1977 to remedy the problem. As the deadline neared, a number of airlines appealed for more time. Without my knowledge, a subordinate FAA official granted them permission to apply for a one-year extension. Word then got out that the FAA was caving in to the airlines at the cost of safety. The *New York Times* was particularly critical, with one of its top columnists accusing us of disregarding the public interest.[6] Needless to say, Secretary Coleman took note of the adverse commentary. He said something like, "John, what have you got to say about this? I realize that safety is in your area, but I don't like this bad publicity." I tried to explain that there had been an administrative glitch and that I'd straighten it out.[7] By the time the dust settled, the industry accomplished all the fixes on the DC-10 in approximately one year, on the 747 over a period of three years, and on the L-1011 somewhere in between.

Were we guilty of overkill? A few years after these upgrades, an explosion blew a large hole in the baggage area of a 747 flying over the Mediterranean. A check of the airplane after it safely landed found that there had been no distortion of the floor structures. I felt vindicated that perhaps our extra caution had been a good idea after all.

Hijackers, Bombs, and Jet Noise

As indicated by this incident, even if it were possible to prevent all accidents, passengers would not be free of peril. During the 1960s and early 1970s, terrorism usually meant hijacking. From 1973 through 1978, only one US airliner was successfully hijacked. Putting the brakes on hijackings was due in large measure to a rule promulgated by the FAA in 1972, requiring American air carriers to inspect all carry-on baggage for weapons or other dangerous objects and to scan each passenger with a magnetometer before boarding. Those and other measures helped keep metal weapons off aircraft. They did not, however, provide total security against hijacking or sabotage, nor did they protect US carriers that flew in from countries with less security.

During my second month on the job, I was rudely introduced to another face of terrorism. On 29 December 1975 a powerful bomb in a coin-operated locker at LaGuardia International Airport killed 11 people and injured more than 50. Secretary Coleman appointed me to lead a government-industry task force that developed new guidelines for safeguarding airport lockers, but checked baggage also demanded attention. Because the technology for X-ray devices and chemical "sniffers" was still immature, the Airline Transport Association (ATA) opposed 100 percent baggage screening.[8] Hence, we issued yet another rule, effective 15 April 1976, that required airlines to selectively screen checked baggage pending development of better explosive detection equipment. Keeping small explosives from luggage without slowing the airplane boarding process to a crawl

263

was a difficult challenge. This menace was later highlighted by the destruction of Pan Am Flight 103 over Lockerbie, Scotland, in 1988. But it took the disastrous events of 11 September 2001 to provide the resources needed to begin screening all baggage, using the kind of sophisticated equipment that was high on our wish list in the 1970s.

Other vulnerabilities were also hard to deal with. In September 1976 five Croatian separatists hijacked a Chicago-bound TWA flight and forced it to Paris using nothing more menacing than fake explosives. Ironically, our new procedures for screening passengers and their carry-on luggage for weapons had worked as designed, but the hijackers used several innocuous objects to construct their phony bomb. This incident posed the problem of how to spot hijackers carrying items that could not be detected by magnetometers, an issue that remained unsolved when I left office. In the hindsight afforded by the vicious use of little box cutters by the suicidal terrorists on 11 September 2001, it is apparent that our efforts to improve airline security a quarter century ago failed to anticipate such worst-case scenarios.

Although not life threatening, aircraft noise abatement was one of the more demanding problems requiring my attention. A large segment of the aviation community, however, could not acknowledge aircraft noise as a real problem. At first I believed the FAA had not been imaginative enough in pursuing this highly publicized and politicized issue. I informed our chief counsel, among others, that it was time to start looking for new solutions. The problems were real. Several months before I arrived on the scene, the US Supreme Court upheld a lower court decision compelling the city of Los Angeles to pay damages to owners of property near Los Angeles International Airport. Then, in October 1975, community activists in Boston blocked the extension of two major runways at Logan International Airport.

The first generation of turbojet-powered airliners was especially loud. Despite taking the noise problem seriously, I did not relish making operators retrofit such antiquated equipment as the 707 and DC-9 with new engines. I believed that money could be better used in purchasing new, quieter planes. Soon, however, airport operators and environmentalists put pressure to do something about older planes on President Ford, who was competing for votes with Jimmy Carter. On 21 October 1976, while campaigning in New York, he announced on the tarmac of Kennedy International Airport that all airliners not in compliance with existing noise standards for new aircraft must comply within eight years. To implement the president's promise, Secretary Coleman and I jointly issued a new noise abatement policy, effective 1 January 1977, that gave two- and three-engine jet transports six years and four-engine jets eight years to comply.

Arranging for British Airways and Air France to land their supersonic Concorde airliners in the United States handed me another political hot potato. The Concorde's thunderous sonic boom prohibited it from breaking the sound barrier over land, but there was still a lot of local resistance against subsonic flights into American airports because of the loudness of its powerful engines. We in the FAA devised a policy, announced by Secretary Coleman in February 1976, to let the Concorde operate on a temporary basis, while testing its environmental impacts on noise levels and air pollution. Concorde service at Dulles International Airport, at that time an FAA-operated facility with few nearby communities, began in May 1976. I was offered a free seat on its first flight back to Paris, but aware of how that might appear to editorial writers, I used FAA travel funds to pay for the $827 ticket. Although I considered the Concorde an impressive technological achievement, my opinion as a passenger that it was "cramped, hot, and noisy" was quoted in the press.[9] Continued opposition and legal actions by local citizens and the Port Authority of New York and New Jersey delayed Concorde passenger service at JFK until November 1977.

Negotiating with the Air Traffic Controllers

Supervising the FAA means having to deal with federal employee unions, which had been granted considerable powers during the Kennedy and Johnson administrations. By the time of my arrival, FAA employees were represented by at least nine such organizations of varying size and influence. I believed that unions should be allowed to lobby for employee benefits and participate in certain employee relations activities, but they should never go on strike against the government and the public it serves. Many FAA managers, however, did not want to deal with unions at all, an attitude that provoked unnecessary friction in our labor relations.

The most important union issue I had to deal with was reclassification of air traffic controller positions, a process that had been dragging on since 1968. On 28 July 1976 the Civil Service Commission (CSC) missed the latest deadline for completing a study on this issue. This prompted John F. Leyden, president of the Professional Air Traffic Controllers Organization (PATCO), to call for a "by the book" slowdown. The nation's airways were facing something of a crisis by Saturday, 31 July, when I began intense negotiations with Leyden. Attacking from the other side, Paul Ignatius of the ATA called me to warn that the association's members were ready to seek a court injunction against the union. Fortunately, by day's end I was able to get Leyden and his people to agree to a compromise that also satisfied the ATA. The agreement called for PATCO to end its slowdown; the CSC to complete its position classification study by the end of August; the FAA's comments on the study to reach the CSC by the end of September; and the CSC to publish the new standards as soon as possible thereafter. Upon release of the agreement, I announced that the FAA supported upgrading positions at certain facilities and that Robert Hampton, the CSC chairman, had indicated that the study should not recommend downgrades. Although hard-liners on the CSC staff who wanted a large number of positions downgraded delayed resolution of the issue until 1977, about 2,500 air traffic controllers were then upgraded out of a workforce of less than 18,000. When all was said and done, we were able to avoid a major confrontation by, in large part, earning the trust and confidence of the majority of PATCO members.

Half a decade later under more combative leadership, most PATCO members walked out on a nationwide strike. During the intervening years, my successors had used the time we bought to develop a workable strike response plan. This allowed Pres. Ronald Reagan to fire all controllers who refused to return to work, secure in the knowledge that the entire national air transportation system would not collapse.

Leaving Public Service

After Gerald Ford's loss to Jimmy Carter, Brock Adams, Carter's new secretary of transportation, kept me on board for several more months while looking for a replacement.[10] Upon leaving the FAA and, as it turned out, my career in direct government service, I sent an essay to *Aviation Week* that was printed as the op-ed piece in its edition of 25 April 1977. An editor titled it "Requiem for a Bureaucrat." (I would have preferred *technocrat.*) Toward the end of the article, I reflected on my service as a member of the Nixon and Ford administrations.

> Now those of us who came to Washington eight years ago have some new decisions to make, new lives to create, new jobs to be sought, and an evaluation to be rendered: was it worth it? Did we contribute to solving the monstrous problems of the period, or were we part of the problem? Would the country be better off if we had remained where we were? Would we be better off? It's hard to be objective. The country has lost its innocence. So have we. The country has been shaken to its foundations. But it has endured. The Bicentennial was a happy respite from the great time of troubles. Now we must go on. Many of us came to Washington with that optimism that permeates a new Administration. Many of us had been disillusioned by the promises of Camelot and the Great Society. Now we are even further disillusioned. Neither Democrats nor Republicans have had a monopoly on mistakes as we recall the Bay of Pigs, Vietnam, Cambodia, Watergate, congressional sex scandals, and all the rest.
>
> And now the new group is in town ready to clean up the mess they inherited. And those of us who came eight years ago are on the way out. We had our chance. We won some and we lost some. We have some successes for which we will get credit, some of which are called mistakes but will later be known as successes, plus some real mistakes which will never be corrected. . . . Was it worth it? I may never know, but if I hadn't tried, I'd probably regret not having done so for the rest of my life.[11]

Looking back from the perspective of another quarter of a century, I can still say that had I not accepted the challenge, I would never have known the deep satisfaction of having faithfully served my country in a leadership role. Nor would I have enjoyed that special bond of trust and friendship that exists among those Americans, both in and out of uniform, who have taken a solemn oath to preserve, protect, and defend the Constitution of the United States.

FAA Postscript: The Third Pilot Controversy

For many years thereafter, I continued to serve on various FAA panels and often went to Capitol Hill regarding the FAA's modernization requirements. In April 1982, for example, I was called upon by the congressional Office of Technology Assessment (OTA) to chair a conference that reviewed the FAA's National Airspace System Plan: a comprehensive road map for enhancing the ability of airports and the air traffic control system to meet future air transportation needs.[12] In 1985 I testified strongly in favor of divesting National and Dulles airports from federal control to a new local airport authority.[13] At almost every opportunity, I continued to advocate use of the new global positioning system as a navigation aid, most notably as a member of the FAA's R&D Committee in the early 1990s.

Probably the most significant of my undertakings on behalf of the FAA was resolving a long-standing controversy about cockpit crew complement standards.[14] Specifically, President Reagan appointed me to lead a study to determine whether long-range transport aircraft could be operated safely with no more than two pilots. Highly politicized, the issue affected the interests of aviation unions, airline operators, and aircraft manufacturers.

Since 1965 FAA workload criteria had required three pilots on most airliners with three or four engines but allowed two on smaller jet airliners, starting with the DC-9. While I was at the FAA, McDonnell Douglas was working on a stretched DC-9 known as Super 80, harbinger for a future generation of bigger and longer-range two-engine transports. Almost on my way out the door, I responded to a request by J. J. O'Donnell, president of the Airline Pilot's Association (ALPA), to establish an interagency task force to look at the relationship of pilot workload, safety, and crew size. Conducted under my successor, Langhorne Bond, this analysis found no compelling reason to revise the current standards. Almost immediately, ALPA went on the offensive, calling it "a phony in-house study." In August 1980 the FAA formally certified the DC-9-80 as safe to operate with a two-man crew. Within a few months, ALPA organized a protest in front of the White House and launched a nationwide media campaign to argue the virtues of the three-man cockpit and impugn the FAA's integrity and dedication to aviation safety. ALPA also threatened a one-day pilots' walkout in March 1981 unless the president approved a new impartial study of the crew size issue.

Reagan's new secretary of transportation, Drew Lewis, offered to appoint a three-person presidential task force if the strike was called off. O'Donnell agreed, with the proviso that

ALPA name two of the members. Its choices were Fred Drink-water, a former NASA test pilot and chief of aircraft operations at the Ames Research Center, and me, by that time president of Comsat World Systems, as task force chairman.[15] As the third member, the DOT chose Lt Gen Howard W. Leaf, inspector general of the Air Force (who supervised its safety function). Known formally as "The President's Task Force on Aircraft Crew Complement," we were assigned a staff of 45, more than half of them USAF personnel, with the rest from NASA, private industry, and academia. To help me I brought in Jack Stempler, my astute and trusted Air Force general counsel, and Robert Schwartz, a Comsat vice president.

Our charter from President Reagan was to "examine the issue of flight crew size for the 'new generation' of commercial airlines."[16] Implicit in this mission was the question of whether new airliners being introduced during the next decade—including the DC-9-80, Boeing 757 and 767, and Airbus 310—could be flown safely by a two-person flight crew. From the outset, I resolved that the investigative process would be thorough and impartial. If there is such a thing as a school for technocrats, part of my training was participating in, leading, or receiving analytical studies at all echelons of business and government. As chairman, I knew that I had to set the tone for the investigation and be visible at every stage of the process. In addition to a professional staff, we needed competent witnesses, imbued with the importance of the task, who believed that their efforts were appreciated. For the next three months, with approval of the leadership at Comsat, I made the work of the task force my top priority.

After reviewing thousands of pages of documents and listening to scores of witnesses, we published our final report on 2 July 1981. It essentially refuted all of ALPA's claims that three-member crews were inherently safer than two pilots.[17] In the interest of promoting flight safety, the report also contained some 18 other recommendations, ranging from the use of collision-avoidance systems to the reduction of cockpit distractions during the critical stages of flight.[18] Upon receipt of our report, O'Donnell was good to his word and recommended to the ALPA executive board that they accept the results. In the end, die-hards in his union got their revenge by voting O'Donnell out of office in their next general election. Not long after that, he and I had a pleasant lunch together. He confided that he had painted himself into a corner on the issue, and that I had bailed him (and the industry) out of a bad situation. Henceforth, airliner cockpit design would be driven by technical and human factors, not by archaic union agreements and specious arguments about aircraft safety. This was one of the last times I represented the United States government in an official decision-making capacity. I was pleased with the outcome and proud of the work we had done.[19]

Connecting the Global Village at Comsat Corporation

In early 1977, while waiting for Brock Adams to find a new FAA administrator, I had time to think about what to do next. Although considering a variety of offers elsewhere, both in industry and academia, in the end I decided to stay in Washington rather than start over again in another city. When the leadership at Comsat again offered me a job—this time as president of its nonregulated subsidiary, Comsat General Corporation—I was grateful for a second chance to enter the commercial space business. In addition to still being on good terms with Joe Charyk, the chairman of Comsat's board was Joseph McConnell, a fellow alumnus of Davidson College.[20]

Arthur C. Clarke, my longtime friend and soul mate, is widely credited with being first to publicly recognize the true potential of communications satellites in an article published at the end of World War II.[21] Arthur's vision of worldwide communications began to become reality just 18 years later, in 1963, with a Hughes-built satellite called Syncom II.[22] Even before this achievement, Pres. John F. Kennedy, in his first State of the Union address, excited the world when he invited all nations to join in developing satellite communications as well as weather satellites and scientific space probes.

To fulfill JFK's promise of worldwide satellite communications, his administration lobbied Congress to pass the Communications Satellite Act of 1962. This enabling legislation stated, "It is the policy of the United States to establish, in conjunction and in cooperation with other countries . . . a commercial communications satellite system, as part of an improved global communications network, which will be responsive to public communications needs of the United States and other countries, and which will contribute to world peace and understanding." Reflecting a compromise between advocates of public and private ownership, the act went on to specify that "United States participation in the global system shall be in the form of a private corporation, subject to appropriate governmental regulation."[23] The result was a regulated monopoly, the Communications Satellite Corporation, whose mission was to establish the global communications envisioned in the law and to take the lead in inviting other nations to participate. The interests of existing telecommunications companies—such as AT&T, Western Union, RCA, and General Telephone—were protected by allowing them to purchase collectively up to 50 percent of Comsat's common stock. The remaining half was sold to the public. The government (through NASA) provided launch services for Comsat's satellites and would be able to lease capacity on the satellites this new entity and its partners would develop.

Comsat was incorporated in February 1963, with Joseph V. Charyk selected as its president. Almost immediately, Comsat sent a team overseas to enlist foreign membership in a multinational consortium that would own and operate the global communications network with which Comsat was chartered to interface. Comsat's point man in this complicated international endeavor was John A. "Johnnie" Johnson, who had been chief counsel of both the USAF and NASA. After many years of negotiations and interim arrangements, the International Telecommunications Satellite Consortium (Intelsat) formally came into existence in 1973.[24]

To complement its international business, Comsat tried for many years to get permission from the Federal Communications Commission (FCC) to expand into the American market, which was already served by a relatively advanced land-based communications infrastructure. In 1974, after providing for RCA and Western Union to own and operate competing satellite systems, the FCC finally gave formal approval for Comsat to get into domestic satellite communications—if it had a separate entity conduct this business and other non-Intelsat activities. Meeting this

271

anticipated provision had already led to the establishment of Comsat General Corporation as a wholly owned subsidiary.

When I began to explore working for Comsat in early 1977, it was not entirely clear if there would still be a good place for me in the company. Joe Charyk and his team decided to restructure Comsat General's management by offering the position of chairman of its board to its current president, Johnnie Johnson. That would vacate the position of president, leaving an opening that I could fill. My first question about this offer was, "Is there enough work to justify two grown men at the top?" As it turned out, Johnnie was close to retirement and not averse to having his workload reduced. Given my long-standing interest in Comsat and my desire to stay in the local area, I agreed to take the job even before it was clear what my exact duties would be. To my relief, Johnnie and I quickly agreed on how to share responsibilities. I would focus on the internal workings of the 365-person organization, while he would concentrate on such less-stressful activities as outreach and international affairs. In my new role I directly supervised several vice presidents and some other talented people of lesser rank. Not surprisingly, I discovered that some of my new subordinates considered me an interloper, including vice presidents who had been hoping to compete for the top position someday. The only way I knew to deal with that kind of situation was to do my job the best way I knew and try to earn their respect.

One vice president who didn't seem too unhappy to see me was retired Air Force major general John L. Martin, who I already knew from my early days at the NRO. In an earlier chapter, I described how the Air Force ran into embarrassing problems developing the Navy's Fleet Satellite Communications System (FLTSATCOM, often referred to as *FleetSat*). In view of these problems, the Navy began looking for an interim solution. Two people within Comsat, both by the name of Martin, came up with an idea to help the Navy as well as the corporation. John Martin was director of Comsat General's Domestic and Aeronautical Systems Office. Edward J. Martin, a veteran of the Air Force's Cambridge Research Laboratory, was one of his top engineers. John was familiar with the kind of problems TRW was having with FleetSat, while Ed had some innovative ideas about how to provide satellite services to commercial shipping, including new satellite transmission and mobile receiver capabilities. The two Martins put their heads together and proposed a system to communicate with ships, which came to be called *Gap Filler* by the Navy, as a leased service until FleetSat was ready. Its excess capacity could also begin providing services to commercial shipping, which could be expanded when the Navy finally got FleetSat fully operational. These concepts came together to help jump-start the Marisat satellite system.

The success of the Marisat project led to the formation in 1978 of a new international satellite communications organization, the International Maritime Satellite Corporation (Inmarsat). Thanks again to the negotiating talents of Johnnie Johnson, Inmarsat became the nautical equivalent of Intelsat. I spent much of my time seeking to build a strong customer base for our new maritime communications and a related consulting service. We especially liked cruise ships, since their passengers would run up many hours of "talking time." Tankers and freighters posed more risk for less gain. It was educational for naïve folks like me, who assumed that people smart enough to operate a fleet of ships were also honest businessmen, to learn that their companies were among the most disreputable of all multinational corporations.

We also went into the component-manufacturing business in a small way by forming a subsidiary called Comsat General Telesystems to develop and manufacture signal-processing equipment and lower cost ground-station components for remote locations. Following up on ideas from my previous job, we tried to assist the FAA with an experimental satellite communications system for aircraft called Aerosat.[25] We also cooperated with the US Geological Survey and Telsat Canada to demonstrate the capabilities of satellites linked to small, unattended ground stations to collect water resource data from remote areas of North America.

Times were starting to change for the telecommunications business by the late 1970s. In part because of Comsat's enviable success, the FCC came under pressure from various private interests and sympathetic congressmen to dilute Comsat's special status. In May 1980 the FCC issued a report concluding that Comsat should carve out two separate companies, one to offer basic regulated services, and one to handle all other business operations. A few months later, as a result of these findings, I left Comsat General, which became responsible for nonregulated business, to become president of Comsat World Systems Division. My new entity offered services to companies who wanted to rent capacity on either the Intelsat or Inmarsat regulated systems. Essentially, I managed what Comsat called "jurisdictional services" as well as the Comsat Labs—activities accounting for more than half of the corporation's total earnings. The reorganization left Comsat General with nonjurisdictional satellite services, communications products, and information services. After the dust settled, World Systems Division had some 450 people and Comsat General about 100. Above both Comsat General and World Systems, the parent corporation's headquarters had about 150 people on its staff.

My transition from Comsat General to Comsat World Systems was relatively painless. I only wish I could have said the same about my domestic situation. Pat and I had been drifting ever farther apart, despite the best efforts of family, friends, and counselors to bring about some sort of reconciliation. In 1980 I moved out of our house on Lake Barcroft to an apartment in Alexandria, Virginia. In early 1981, almost 35 years to the day since our marriage, I was in court to formalize its dissolution. Several months later, in September 1981, I married Harriet Black, a charming and sophisticated woman I had first met at our church in Annandale. She has been my best friend and the pride and joy of my life ever since. With my four children and her five, we each became part of a larger extended family.

Harriet and I had a small wedding, followed by a reception with family members and close friends. Among the guests was Barbara Eisenhower, the wife of John Eisenhower (son of the former president). Harriet and Barbara had become friends during Harriet's first marriage to an Army officer. Barbara heard us describe how, on the next day, I was taking Harriet along on a "working honeymoon" to Paris, Geneva, Madrid, and London—places where I was to meet with various Intelsat executives to lobby for appointment of an American as its next director general. A few days later, at our hotel in Paris, we were surprised to get a message from Barbara. She had arranged for us to end our trip with a stay at Scotland's famous Culzean Castle, perched on a cliff overlooking the Irish Sea. In 1945 the castle's former owners had donated it to the National Trust of

Scotland with a provision that the top floor be given to General Eisenhower in appreciation of his wartime service. We gladly extended our itinerary to include Culzean and spent the final days of our European trip staying in its beautiful and historic Eisenhower suite.

As president of World Systems, I was next in line to Joe Charyk in seniority. By this time, Johnnie Johnson had retired, and John D. Harper had become the chairman of the board, replacing my friend Joe McConnell, who remained as chairman emeritus. The new president and CEO of Comsat General, Richard Bodman, was one of those strictly business types who was supposed to have a keen eye for new schemes to increase profits. Unfortunately, Comsat General lost a lot of money with what he hoped would be two new profit centers. One was a subsidiary called the Satellite Television Corporation (STC), whose satellites were designed to transmit television channels directly to home antennas. It eventually lost about $120 million. The other loser was called Satellite Business Systems (SBS), a joint venture with IBM and Aetna to provide data and voice communications for big businesses. Unfortunately, the system experienced more than usual cost growth and production slippages, making it virtually obsolete by the time it was ready to market. IBM and Aetna could easily afford their share of SBS's losses, which averaged about $25 million per month, but Comsat could not. It carried both of these losing entities on its books for years until able to sell off their assets and concentrate on core business areas.

On the regulated side of the house, I had a good run of luck. World System's operations were consistently profitable, with earnings exceeding expectations by several million dollars. World Systems was able to expand its customer base, upgrade the capabilities of existing ground stations, and build others to increase the number of available circuits. Through Comsat Labs, we maintained world leadership in R&D for commercial satellite communications, particularly in the areas of micro-electronics, microprocessor-based digital electronics, and multiple small-user communications systems. We successfully integrated Inmarsat and expanded both the types of services offered and the number of customers. We had had some failures, but overall, I felt that the division was a good operation.

I remained as president of World Systems until late November 1983, when I was elected to a new job as Comsat's executive vice president and chief strategic officer. Establishing this position was an attempt by the board to improve Comsat's ability to organize its business activities and formulate goals for the future. I had suggested to Charyk that he create this post because I felt that, despite Comsat's overall success, over time it had become less creative. The ratio of engineers to "paper pushers" kept falling, and the engineering choices that remained seemed less interesting. I thought that we had allowed the company's culture to become increasingly based on near-term financial considerations and less on innovation, which in a technological business seems like a one-way street to mediocrity. My new position also allowed more time for exploring broader professional interests and participating in community affairs.

One of my most visible community projects was raising $9 million in the Washington area and $29 million nationwide to finance rebuilding the Filene Center at the National Park Service's Wolf Trap Center for the Performing Arts in Vienna, Virginia, after it was destroyed by fire in April 1982. For the better part of two years, I wrestled with the Wolf Trap Foundation's financial, legal, and construction problems as a member of the board of directors and eventually as its chairman. The reopening of the Filene Center in 1984 gave me a great deal of satisfaction.

I held the strategic planning post at Comsat until mid-1985, when I reached the company's mandatory retirement age of 65. While in this position, I acted as Comsat's senior troubleshooter and spent a good portion of my time defending the company's interests on Capitol Hill and in the bureaucracy of the executive branch. Perhaps my crowning achievement was planning a major reorganization of the company. Its purpose was to strengthen all aspects of Comsat's business, which was facing increasing competition from other space-oriented telecommunications companies and fiber-optic cable. As stated in an earlier chapter, I have never been a believer in reorganization for its own sake. My goal in all of this was to establish a clear strategic direction and strengthen operations.[26] It seemed to work, at least for the next decade, as the company continued to grow and prosper.

When I officially retired from Comsat on the first of September 1985, I immediately hung out a shingle as an air and space consultant. After a few years of taking on various space-oriented consulting assignments, I was approached by some people at Harvard to write a book about commercial opportunities in space. I titled this book, published in 1991, *Space Commerce.* In it I explained why Comsat was such a unique and successful space venture, at least up to that time. I thought that one of the most important factors was timing. When Comsat was founded, the situation was ripe for space-based communications as soon as the technology matured enough to permit spacecraft to work as designed. There was also a pent-up demand for more reliable and affordable communications links between continents. At the same time, there was a strong desire on the part of government officials for the public to reap some tangible benefit from the large investments being made in space technology. The Communications Satellite Act, with its compromise among competing views on the government's role in this new venture, created the necessary framework for all this to work. Although Comsat's leaders fought constantly with the FCC over rates and policies, the US government stayed true to its promise not to compete, and it became a good customer for commercial satellite services.

The genius of Kennedy's concept of inviting all nations to participate, which led to Intelsat, gave Comsat's overseas partners strong incentives to support the enterprise. All shared in liberal patent and data rights policies, and some even got into the satellite-building business. All told, Comsat eventually provided technical assistance to about 50 countries in designing complete communications systems and acquiring satellites, ground stations, or both. By the time I retired, 119 countries were members of the Intelsat consortium. This level of participation spoke volumes about how Comsat met its responsibilities. For most of us, Comsat was about more than simply turning a profit for shareholders. There was a sense of adventure and wonder at the things we were able to do from orbit. We retained a keen desire to fulfill Kennedy's charge to establish a communications satellite system "which will contribute to world peace and understanding."[27]

Alas, nothing lasts forever, and today the remnants of this once proud symbol of America's technical expertise and international business acumen are being scattered to the winds. The end of Comsat came after a series of missteps and the crush of competitive pressures in the telecommunications industry that began to accelerate after the court-ordered breakup of AT&T in 1984. The final blows came when Congress amended the Communications Satellite Act to allow Lockheed Martin to complete a takeover of Comsat in 2000, even though this rapidly expanding air and space conglomerate lacked either a clear strategic plan as to where to go with this latest acquisition or a firm commitment to telecommunications.

Chairing the Arthur C. Clarke Foundation

The international aspects of Comsat's business held a special appeal to me and many of my colleagues. A number of us volunteered to support the 1983 World Communications Year (WCY), a global event sponsored by the United Nations (UN) and, in particular, the International Telecommunication Union in Geneva. The executive assistant to the director general of Intelsat, Dr. Joseph N. Pelton, was serving as the managing director of President Reagan's committee for American participation in the WCY. When Joe asked me to help set up something called the Arthur C. Clarke Foundation as part of the WCY effort, I readily agreed. Clarke was part of an informal fraternity of communications experts with extraordinary influence on how the frequency spectrum for satellite communications was allocated among various claimants. Throughout my career, I had heard many stories about him and read some of his many works.

The Arthur C. Clarke Foundation of the United States (ACCFUS) was one of Pelton's ideas to help "get things going" for the WCY. He knew that Arthur C. Clarke's famous name and personal interest would give the project instant credibility. A legend in his own time, Clarke has helped inspire two generations of space enthusiasts through his writings, including the screenplay for Stanley Kubrick's visionary 1968 motion picture, *2001: A Space Odyssey.* Joe had first met him in 1976 at his home in Sri Lanka and has been a friend and supporter ever since. My initial role was to help recruit people to serve on the foundation's board of directors. Having secured the support of the Sri Lankan Embassy, we proceeded to draft bylaws and recruited Fred Durant—a 30-year friend of Clarke who had retired as associate director of the Smithsonian Institution's National Air and Space Museum—to help us. In September 1983 we held our first board meeting at Intelsat headquarters. The founding directors included Henry Hockeimer, president of Ford Aerospace; Alfred "Bud" Wheelon, president of the Space and Telecommunications Division of Hughes Aircraft; Congressman George Brown of California, chairman of the House Science and Technology Committee; Tedson Meyers, a prominent Washington lawyer who had been Hubert Humphrey's key aide; and several other distinguished people—many of whom I had known and worked with in the past.

I first had the opportunity to meet Arthur Clarke when he came to Washington in the fall of 1983 to help Comsat celebrate various milestones, such as the twentieth anniversary of the Syncon II satellite's launch into geosynchronous orbit, and participated in various WCY events. Truly an innovative thinker, Clarke had become famous as a writer about both science facts and fiction. I used this theme when introducing him twice in the same day. In the morning, I presented him to a gathering in the Comsat building, telling them he had written 40 books. Later in the day, I introduced him again at George Washington University—this time saying, "Arthur is a very prolific writer, having completed fifty books." Suddenly I recalled the number I'd used that morning and, after correcting my mistake, admitted, "Not even someone as productive as the great Arthur C. Clarke can be that fast!" That day began a friendship that has lasted for 20 years. I found in him a kindred soul who shares my passion for using space to further the cause of world peace—to achieve, in Arthur's words "nothing less than the salvation of mankind."[28]

I have been chairman of the ACCFUS since its inception. Over the years we have used its modest budget to what I believe has been good effect. The foundation has sponsored awards and prizes, coordinated research and demonstration projects, supported travel by deserving third-world scientists to attend international conferences, supplemented salaries of some of the professors at the Clarke Centre in Sri Lanka, and sponsored the annual Arthur C. Clarke Lecture. I was honored to deliver the first of these at the University of Moratuwa, Sri Lanka, in 1987. It stressed the idea that improved space-enabled communications could be a vehicle for better relations among peoples, economic progress, and improved education, medical treatment, and health.[29] At the end of the ceremony, I received one of my most cherished and unusual mementos, the Arthur C. Clark prize for distinguished services to satellite communication. It stands about 12 inches high and takes the shape of the mysterious monolith from *2001: A Space Odyssey.*

In the summer of 1988, I was able to pay Arthur back for his hospitality when medical problems prompted him to make another visit to the United States. I met him and Hector Ikanayake, a close Sri Lankan friend, at Dulles International Airport on 13 July. Arthur was in a wheelchair, having been diagnosed two years earlier in London with amyotrophic lateral sclerosis (Lou Gehrig's disease). I drove them to Baltimore, where doctors at Johns Hopkins hospital, unaware of the earlier diagnosis, determined he was actually suffering from the recently discovered postpolio syndrome, a serious but not fatal condition. Needless to say, Arthur was relieved to learn he now had a chance to live until 2001 (and, hopefully, many years beyond).[30]

Launching the International Space University

In addition to the ACCFUS and its recent spin-off, the Clarke Institute for Telecommunications and Information (CITI),[31] perhaps my most valued collaboration with Pelton and Clarke has been our work in support of the International Space University (ISU). The ISU came to life during the flurry of activity surrounding Arthur's visit to the United States in October 1983. While in New York to give a speech to the UN General Assembly, Clarke was introduced by Pelton to three ambitious young men: Peter Diamandis from MIT; Todd Hawley from George Washington University (who was working as an intern for Pelton at Intelsat); and Robert Richards, a Canadian who was active in Students for the Exploration and Development of Space (SEDS)—another group with which Pelton was affiliated. They asked if Arthur might support their idea for an international university that would offer students from many nations an interdisciplinary curriculum focused on space science and technology. Arthur was immediately smitten by their youthful enthusiasm and the soundness of their proposal. Todd later explained that he first got the idea from seeing more than 60 different nationalities all working together at Intelsat. Transforming the ISU from this idea into reality soon began.

In addition to giving birth to the ISU, 1983 was the year that the Society of Satellite Professionals was founded to stimulate the growth of the space-based communications industry and to help satellite specialists advance their careers. Arthur agreed also to be its honorary chairman.[32] Joe Pelton, who was the founding president of this group, was thus at the center of a web connecting it, the Clarke Foundation, and the ISU. Another key figure in this organization was a mutual friend and former colleague, Burton I. Edelson.

A real space professional, Burt had become director of Comsat Labs in 1972. He worked for me there until being approached by NASA administrator James M. Beggs (one of Burt's fellow midshipmen at Annapolis in the late 1940s). With my recommendation, Beggs hired Edelson as his associate administrator for space applications, a position he held from 1982 to 1986. At NASA, Burt had some flexibility in passing out grant money, and he provided a $50,000 grant that helped get the ISU off the ground in 1984, holding its first informal session that summer at MIT. A core group that included Edelson, Clarke, Pelton, and me formed a board of trustees that spent the next 20 years garnering support.[33] At the same time, the energetic triumvirate of Diamandis, Richards, and Hawley worked out front to make their dream a reality. By 1989 the board had been successful in getting financial assistance from 70 corporations, foundations, academic institutions, and government agencies around the world.[34] Supporters of the ISU shared a conviction that it could advance science and technology while helping transcend barriers between nations. As I told one of the ISU's summer sessions:

> We have, through technology, become the masters of the world's resources. Too often, we have marshaled those resources to act out our hostilities toward each other, separating ourselves into warring tribes that don't understand just how lethal our games have become. It has been the mission of ISU to advance our understanding of each other so that wars between nations and tribes become unthinkable. Only by looking at the world as a closed system can we learn how to enjoy its

wealth without jeopardizing its ability to restore itself. Only by under-standing its physical and biological workings can we be proper guard-ians of the future.[35]

The ISU conducted its first formal summer session in 1988 at MIT with more than 100 graduate students from 21 nations. Even more attended summer sessions in Strasbourg, France, in 1989 and Toronto, Canada, in 1990.[36] In the latter year, I became chairman of the board of trustees. In that role, I an-nounced in February 1993 the choice of Strasbourg for the ISU's permanent campus.[37] In addition to summer sessions in various cities, the growing ISU began offering an intensive 11-month program leading to a master of space studies degree in 1996. The first class graduated 30 students from 14 nations. For the next five years, these classes continued to meet in tem-porary quarters while plans progressed for a new facility. In May 2002, with completion of major construction at its mag-nificent new campus in Strasbourg, the ISU has come a long way toward achieving the original vision. To date the school can boast of more than 1,800 alumni from 80 countries and a pool of 700 faculty members and lecturers. More than 400 sponsors worldwide now support the ISU. I was honored to serve as chairman and grateful to have Joe Pelton follow me in that position. ISU alumni have already become astronauts, biospherians, and top officials of air and space companies and space agencies. Perhaps one day its graduates will be among those who explore Mars or achieve other major breakthroughs in human destiny. Contributing to the success of this worthy institution has been one of my proudest achievements.

Notes

1. Although not privy to Ford's explanation, FAA historian Edmund Pres-ton wrote a good summary of Butterfield's term in *Troubled Passage: The Federal Aviation Administration during the Nixon-Ford Term, 1973–1977* (Washington, D.C.: GPO, 1987), 5–168. A reference authored by Preston, *FAA Historical Chronology: Civil Aviation and the Federal Government, 1926–1996* (Washington, D.C.: GPO, 1998), has been useful in preparing this section.

2. Senate, *Nominations, October–December: Hearings before the Committee on Commerce*, 94th Cong., 1st sess., 13 November 1975, 39–45.

3. House, Committee on Interstate and Foreign Commerce, *Air Safety: Selected Review of FAA Performance*, 93d Cong., 2d sess., 27 December 1974, 237.

4. See Preston, *Troubled Passage*, 169–200.

5. "Former FAA Head Urges More Research on Displays, Controls," *Aviation Week and Space Technology* (*AW&ST*), 25 April 1977, 100.

6. Tom Wicker, "Most Pervasive Corruption," *New York Times*, 19 October 1976, 39.

7. For a summary, see "Wide-Body Fix Extension Still under Study at FAA," *AW&ST*, 1 November 1976, 28; and "Wide-Body Modifications Pushed," *AW&ST*, 6 December 1976, 23.

8. "ATA Task Force Opposes Screening All Airline Bags," *AW&ST*, 15 March 1976, 205.

9. "Speed of Sound," *AW&ST*, 7 June 1976, 13.

10. For additional details on the FAA during McLucas's tour as administrator, see Preston, *Troubled Passage*, 201–65.

11. John L. McLucas, "Requiem for a Bureaucrat," *AW&ST*, 25 April 1977, 23.

12. Office of Technology Assessment, *Review of the FAA 1982 National Airspace System Plan* (Washington, D.C.: GPO, August 1982).

13. Senate, *Transfer of National and Dulles Airports: Hearings before the Subcommittee on Aviation, Committee on Commerce, Science, and Transportation*, 99th Cong., 1st sess., 26 June 1985, 133–40.

14. See Nick A. Komons, *The Third Man: A History of the Airline Crew Complement Controversy, 1947–1981* (Washington, D.C.: GPO, 1987), for more details.

15. O'Donnell later said of McLucas, "He was absolutely a super person to deal with" and "could pierce through all that garbage and get right to the heart of an issue very quickly." Ibid., 102–3.

16. Ronald Reagan to John McLucas, letter, 5 March 1981, Appendix A in John L. McLucas, Fred Drinkwater III, and Lt Gen Howard W. Leaf, USAF, *Report of the President's Task Force on Aircraft Crew Complement* (Washington, D.C.: GPO, 2 July 1981).

17. Ibid., 7. In fact, there was some evidence that a three-man crew might be less safe.

18. Ibid., 8–12. A few years later some third-pilot advocates complained that they would not have gone along so peacefully with the task force's conclusions had they known the FAA would be so slow and inept in its efforts to develop effective collision-avoidance devices.

19. In the summer of 2002 Dr. McLucas donated his FAA papers and records of the crew complement study to Embry-Riddle Aeronautical University.

20. Although Comsat and related companies often referred to themselves with all capital letters (e.g., "INTELSAT"), their names are not true acronyms. For editorial reasons, this chapter uses the lower case for all but the first letter.

21. Arthur C. Clarke, "Extra Terrestrial Relays," *Wireless World*, October 1945, 305–8, reprinted on the Clarke Foundation Web site, http:www.clarkeinstitute .com/vision, October 2002. Clarke later wrote a retrospective essay on this article as an introduction to Anthony M. Tedeschi, *Live via Satellite: The Story of COMSAT and the Technology That Changed World Communications* (Washington, D.C.: Acropolis Books, 1989), 9–14.

22. See McLucas, *Space Commerce*, 26–36, for a summary of early geostationary communications satellites.

23. *Communications Satellite Act of 1962*, Public Law 624, 87th Cong., 2d sess., 76 Stat. 419 (31 August 1962), sec. 102.

24. For additional details, consult Joseph N. Pelton, *Global Communications Satellite Policy, Intelsat, Politics and Functionalism* (Mt. Airy, Md.: Lomond Systems, 1974); and Joseph N. Pelton and Joel Alper, eds., *The Intelsat Global Satellite* (New York: AIAA, 1984). For a book focusing on the work of Comsat Labs through the 1980s, see Tedeschi's *Live Via Satellite*.

25. Aerosat died on the vine when Congress took away the FAA's money for the program and appropriated $1 million for a "back to the drawing boards" study.

26. *COMSAT* (Communications Satellite Corporation journal), no. 16, 1985, 2–3.

27. McLucas, *Space Commerce*, 50–53.

28. Arthur C. Clarke, "Star Peace," Address on receipt of the 10th Charles Lindbergh Award, Paris, 20 May 1987.

29. John L. McLucas, "Technology and Development," in *Future Vision: A Tribute to Arthur C. Clarke and to Space Communications in the 21st Century*, ed. Joseph N. Pelton (Boulder, Colo.: Johnson Publishing Company, 1992), 18.

30. Neil McAleer, *Arthur C. Clarke: The Authorized Biography* (Chicago: Contemporary Books, 1992), 353–57.

31. Founded in February 2000, CITI was created to facilitate "electronic networking of leading educators and researchers around the globe." See http://www.clarkeinstitute.com

32. Society of Satellite Professionals International Web site, http://www.sspi.org/html/about.html.

33. Burt Edelson passed away in January 2002, at the age of 75, before having a chance to review this section.

34. *Space Times* (Magazine of American Astronautical Society), March–April 1998, 8.

35. John McLucas, "What ISU Means to Me," in *The ISU Cosmic Pioneer* (Toronto, Canada: York University Department of Communications, 6 July 1990), 1.

36. "The International Space University," *Aerospace America*, March–April 1989, 7–11; and Past Summer Session Programs, International Space University Web site, February 2003, http://www.isunet.edu/academic_programs.htm.

37. "Toronto Loses Bid for Space University," *The Financial Post* (Toronto), 2 February 1993, 10.

Chapter 8

Promoting Space, Science, and Technology

3 December 2002:

It is with heavy heart that I inform you of the passing from our planet of Dr. John McLucas, long term Chairman of the Arthur C. Clarke Foundation, as of Sunday 1 December 2002. John McLucas was Secretary of the U.S. Air Force, FAA Administrator, Director of the National Reconnaissance Office, satellite executive, inventor, author, educator, writer, wit, father, husband and most of all a noble human being and friend. Without his support and special insight a number of key occurrences might not have happened. These include the International Space University with its $20 million campus in Strasbourg, France, the restoration of the Wolf Trap Auditorium, the Arthur C. Clarke Foundation of the US (of which he was chairman from 1983 to 2002), and the new Arthur Clarke Institute for Telecommunications and Information (CITI) that connects telecommunications research and development organizations around the world. The world suffered a great loss in the passing of John McLucas.[1]

So read a statement that Joe Pelton shared with the press and sent via e-mail, along with a message of regret from Arthur C. Clarke in Sri Lanka, to mutual friends and associates of John McLucas.[2] A large gathering of family, friends, colleagues, and government officials attended a memorial service on 19 December at the historic Old Presbyterian Meeting House in Alexandria. A few weeks later, McLucas's ashes were interred at Arlington National Cemetery with full military honors. Secretary of the Air Force James G. Roche presented Harriet McLucas with the memorial flag that had covered the casket. A flyover in "missing man" formation of sleek F-16 fighters—aircraft McLucas helped bring into existence—opened the ceremony. At a much higher altitude, some of the satellites he had once sponsored may have passed over as well.

John McLucas had devoted most of his last two decades to promoting beneficial applications for science and technology, especially in space. His knowledge, experience, and vast circle of colleagues made him a valuable contributor to various endeavors. The numerous associations and enterprises with which McLucas was affiliated reflect his penchant for networking and desire to remain active in what he called his "semiretirement." (See appendix A for a list of the more noteworthy positions he held after leaving government service.)

Defending the Free Exchange of Scientific Information

The year before McLucas retired from Comsat, he was elected president of the American Institute for Aeronautics and Astronautics (AIAA). A longtime member, he had received its highest aeronautics honor, the Reed Award, in 1982. During his term as AIAA president, from June 1984 through May 1985, McLucas did not hesitate to address controversial issues or expound his philosophy on the relationship of government with science and technology to an informed and influential audience.[3] One of his big concerns was access to space. He implored NASA to procure expendable launch vehicles and, two years before the *Challenger* disaster, pointed out "the need for a backup to the Shuttle, should it be grounded after an accident or loss."[4] Several months after this editorial, he warned the House committee overseeing NASA, "I don't know where we'd be if we lost an orbiter and didn't have a fifth one coming along." He also emphasized the need to begin planning a "next generation Shuttle," do careful advance planning on the newly sanctioned International Space Station (ISS), and expand US support of space science, including cooperation with the USSR.[5] As regards to internal AIAA affairs, McLucas backed a contentious decision to move its main office from New York to Washington, D.C.[6] He ended his presidency with a call for the government to revisit President Eisenhower's "open skies" philosophy by supporting an international Earth observation satellite and pleading to keep weapons out of space.[7] In 1987 the AIAA selected McLucas to receive its annual Goddard Award for astronautics, making him one of a select few to receive both it and the Reed Award.[8]

While AIAA president, McLucas took perhaps his boldest stand by disputing the Reagan administration's new policy, spearheaded by Assistant Secretary of Defense Richard N. Perle, to drastically curtail foreign access to American scientific research and innovations. McLucas first went on record about the potential harm of these new controls on technical data and exports as a member of the National Academies' Committee on Science, Engineering, and Public Policy (COSEPUP).[9] Two years later, his first editorial in *Aerospace America*, "Technology Transfer Scare Hurts Innovation," criticized the "paranoia" behind this policy.[10] "The U.S. cannot operate as an island," he later said. "It is absolutely necessary to engage in technology transfer . . . rather than build a wall around ourselves."[11] In March 1985 the AIAA pointed out the absurdity of not sharing research with foreigners when they were earning 50 percent of the engineering PhDs in American universities. The campaign by the AIAA and like-minded institutions had some effect. In September 1985 the White House overruled OSD by keeping the products of fundamental research unrestricted except as appropriate through standard security classification procedures.[12]

As McLucas was phasing down at the AIAA, he began seeking firsthand perspectives on technology transfer issues by visiting some of the countries most affected. In May 1985 he traveled to China with the AIAA executive director, James J. Harford, and their spouses as guests of the Chinese Society on Astronautics. Despite restrictions, they were able to gather technical and administrative information about Chinese satellite programs.[13] McLucas made a second trip to China in February 1986 under the auspices of the congressional Office of Technology Assessment (OTA) as part of a broad study of technology transfer to China. He and a China specialist, Norman Getsinger, focused on satellite communications.[14] Their study concluded "a combination of US government and private sector activities [to] establish a U.S.-China partnership in the development of space communications could provide political, economic, and strategic benefits which appear to outweigh the risks."[15]

As part of a major congressionally directed study of export controls in 1986, McLucas served on a COSEPUP panel chaired by Lew Allen Jr. McLucas again flew across the Pacific as a leader of COSEPUP's fact-finding mission to Japan, Korea, Hong Kong, Singapore, and Malaysia. A few weeks later, he joined another delegation that went to Britain, France, Belgium, Austria, West Germany, and Sweden. The panel's report, issued in January 1987, acknowledged that controls might have been too lax in the 1970s but concluded, "U.S. control policies and procedures are in danger now of overcorrecting. . . . The result is a complex and confusing control system that unnecessarily impedes U.S. high-technology exports to other countries of the Free World and directly affects relations with the CoCom allies."[16] (The Coordinating Committee for Multilateral Export Controls, or CoCom, was a means by which the industrial democracies tried to apply consistent policies on technology transfer to the Soviet Union.) Following up on the published report, McLucas represented the National Academies as an expert witness on Capitol Hill. On 11 March 1987 he recommended to a House subcommittee, "We should get rid of the U.S. unilateral approach, really work with our allies, and make the [CoCom] control system work."[17] One week later, he explained his basic philosophy to a Senate subcommittee: "Our security depends, in my view, mainly on strengthening our technology base and running faster . . . in a technological sense rather than on trying to preserve what we now have."[18]

McLucas was also cautious about using export controls as a means of economic protectionism. In November 1989, after being named chairman of NASA's Advisory Council, he denounced a proposal that NASA seek more authority for restricting the release of technical data to counter growing foreign competition in the air and space industry. "It is extremely difficult to deny unclassified R&D information to outsiders," he warned the House Committee on Science, Space, and Technology, "without interfering with its dissemination to those insiders who must have it to do their jobs." He also emphasized the special role of NASA in fostering science and technology, recalling, "President Eisenhower's desire to create an open civil space program resulted in NASA becoming one of his proudest legacies." Based on his experience, McLucas recommended five ways "of leveling the playing field on the flow of technology." These were to (1) selectively classify especially sensitive technologies, (2) negotiate technology-sharing agreements with key trading partners, (3) encourage multinational production agreements, (4) decrease the time for American companies to move from development into production, and (5) invest more for R&D in vital areas, such as space and aeronautics.[19]

Advancing International
Cooperation in Space

Besides lowering barriers to the flow of scientific information, John McLucas actively campaigned for more bilateral and
multilateral space projects, including a commercially operated
international Earth observation satellite. For him, this idea
germinated from proposals by Comsat to "privatize" Landsat
operations, a concept he advocated in repeated congressional
testimony.[20] After market studies indicated that commercial
revenue from selling its data could not pay for Landsat's costs,
McLucas and his colleagues proposed an expanded program
called Earthstar, which would also sell more financially lucrative meteorological data. Although the Department of Commerce seemed interested, Congress did not like this idea, and
Comsat withdrew from competition for the Landsat contract.[21]
The Earth Observation Satellite Company (Eosat), jointly owned
by Hughes Aircraft and General Electric, was selected to perform this mission, which McLucas later criticized as "a poor
example of an attempt to privatize a government program."[22]
Eventually, the first Bush administration had to rescue Landsat from financial meltdown with new funding arrangements.[23]
For many years, however, McLucas continued advocating an
international means of performing Earth observation. He and
some fellow activists proposed a new polar-orbiting system
that, in 1988, they began calling the Environmental Resources
Satellite (Envirosat), a multifunction system for gathering land,
ocean, and weather data.[24] He also advocated a multinational
satellite system that could detect military preparations and
hopefully prevent potential conflicts.[25]

In addition to international cooperation, McLucas endorsed
efforts to collaborate directly with the Soviets on selected space
projects, even when the Cold War intensified during the early
1980s. In the fall of 1983, while representing the AIAA at a meeting of the International Astronautical Federation (IAF) in Budapest, he met with Vladimir Kotelnikov, chairman of the USSR's
Intercosmos Council and a member of the Soviet Presidium.
One thing led to another, and in June 1984—25 years after his
previous visit to the Soviet Union—John and Harriet McLucas
along with Jim and Millie Hartford went to the USSR as guests

of the Soviet Academy of Sciences. The men visited various institutions, including the headquarters of Intersputnik (a socialist
version of Intelsat). Although noting his hosts' relatively primitive facilities and equipment, McLucas was impressed by some
of their scientific achievements, such as the Venera program in
which two satellites were radar-mapping the surface of Venus.
When Kotelnikov lamented the lack of recent contacts with NASA
(understandable after the Soviet shoot down of Korean Airlines
flight 007), McLucas confided that James Beggs had informally
encouraged him to visit "because it was difficult for Beggs himself to have such communications under the circumstances."
During a luncheon at the Bear Lake ground station that down-
linked telemetry from Venus, John followed up an eloquent toast
by its director, Alexei F. Bogmolov, by saying, "While our two
governments are not on good terms, it is very important that
scientists stay in touch, and maybe the governments will eventually improve their relations."[26]

In the spring of 1985, McLucas learned that Jim Beggs was
hesitant to celebrate the 10th anniversary of the symbolic
and historic Apollo-Soyuz docking mission in July. Believing
long-term cooperation in space should not be held hostage to
current tensions, he called a contact at the National Security
Council about the potential benefits of an Apollo-Soyuz commemoration, and the NASA administrator was soon told that
the White House thought such a remembrance would be a
wonderful idea. "As it turned out," McLucas later recalled, "it
was a great demonstration of how individuals can transcend
governments. . . . The cosmonauts made an appearance before Congress, participated in two different symposia, and went
on morning TV. I'll never forget Alexei Leonov saying on Good
Morning America [the ABC television show], when asked if he
would fly again with his American counterpart, Thomas Stafford, 'I'd go anywhere with Tom!'"[27]

With the Cold War thawing in the late 1980s, McLucas began
promoting more space ventures with the Soviet Union. In October 1986, at an IAF conference in Innsbruck, McLucas witnessed
the Soviet delegation completely upstage the Americans with new
openness about their ongoing space achievements. NASA—still in
disarray after the *Challenger* disaster—had little to say. On behalf
of the AIAA, McLucas widely distributed a rather provocative let-

ter to President Reagan lamenting that "the U.S. civil space program is no longer one in which we can all take pride" and calling for presidential leadership in reinvigorating NASA.[28] At the same time, he headed a committee of the United Nations Association of the United States on international cooperation in space that hosted a visit by Roald Sagdeev, the director of the Soviet Institute of Space Research. McLucas's group proposed various initiatives in a report entitled *The Next Giant Leap in Space: An Agenda for International Cooperation*, which also urged a new US-USSR agreement on space projects.[29]

In October 1987 McLucas returned to Moscow for the Space Future Forum, held to mark the 30th anniversary of Sputnik. While there, McLucas scored something of a coup on a visit to Soyuz Carta, an organization recently formed to market photographs taken from orbit. Using a personal check for about $900, he purchased some multispectral imagery taken over Oregon with the Soviets' KFA-1000 camera. It had a resolution of five meters—twice as good as France's *Systéme Pour l'Observation de la Terre* (SPOT). These photos, printed in the AIAA's magazine a month later, were the most detailed pictures of American territory from orbit yet made public.[30]

During the space forum, McLucas again conferred with Sagdeev on opportunities for joint space ventures. Sagdeev was a proponent of a cooperative Mars exploration program, an idea that McLucas had been promoting ever since he chaired a "Steps to Mars" conference, cosponsored by the AIAA and the Planetary Society in July 1985. McLucas believed that the USA, the USSR, and other interested nations should work together on missions to Mars, at first robotic and, if feasible, eventually manned. He thought Russian capabilities in such areas as powerful boosters and long duration space flights could well complement American expertise in systems engineering, instrumentation, and automation. After General Secretary Mikhail Gorbachev publicly endorsed this joint Mars program in May 1988, McLucas and Burt Edelson called on President Reagan to agree in principle and authorize NASA to begin planning.[31] To Sagdeev's sorrow, Russian weakness in software led to the failure of his two ambitious Phobos missions to Mars in late 1988 and early 1989.[32] In July 1989, on the 20th anniversary of *Apollo 11*'s mission to the Moon, Pres. George H. W. Bush issued a call for a return to the Moon as a stepping-stone for going to Mars, but cost estimates soon aborted this ambitious plan (which would be revived by his son in a somewhat different form 15 years later).

During the 1990s, McLucas supported those limited instances of cooperation with Russia that occurred, such as its sometimes shaky participation in the ISS, while wishing more could be done to help Russia's struggling space experts. Harriet recalls hosting various Russian scientists and engineers when they visited Washington. John continued his friendship with Roald Sagdeev, who came to America as a professor at the nearby University of Maryland in College Park. As time went on, McLucas came to realize that his earlier hopes for more extensive collaboration had been overtaken by Russia's economic and political problems and its early failure to establish a civilian space program.[33]

Overseeing QuesTech, Inc.

As John McLucas was nearing retirement from Comsat, Bruce G. Sundlun, a member of its board of directors, talked McLucas into joining him on the board of a diversified high-tech company called QuesTech, headquartered in McLean, Virginia. At the time QuesTech employed some 600 people and generated annual revenues of approximately $42 million. Its primary clients included the military services, NASA, the Department of Energy, and some commercial energy and air and space companies. What started as a part-time responsibility in 1985 became a full-time commitment in 1986. QuesTech's longtime president, CEO, and chairman, Herbert W. Klotz, perished in an auto accident on 12 November, and McLucas stepped in temporarily to fill this void until he could find a permanent replacement.[34]

When McLucas became involved with day-to-day management, he discovered QuesTech to be a troubled company, suffering from cost overruns, cash-flow problems, and a rather permissive corporate culture. Looking for a strong manager to run the company, McLucas turned to an executive retiring from Comsat named William L. Mayo. In May 1987 McLucas passed to him the job of president and CEO of QuesTech.[35] The company appeared to be on course, with net earnings growing to more than $2 million for

1987.[36] Unfortunately, Mayo's confrontational management style led to conflicts with QuesTech's four founders, and the board fired him in October 1988. In retaliation, Mayo filed a wrongful-discharge lawsuit against the company.[37] McLucas reluctantly remained as chairman while this dispute and other legal issues were pending. In May 1990 he turned over his place on the board to Sundlun, who later in the year was elected governor of Rhode Island.[38] McLucas's association with QuesTech and his struggle to restore its integrity was undoubtedly the most painful experience of his professional career.[39]

Fostering Space Commerce

John McLucas was also sorely disappointed with the actions of another former Comsat executive whose career he promoted, a man named Richard R. Colino. This affair—so contrary to McLucas's code of ethics—caused him lasting distress. Colino was the man he lobbied for as the next director general of Intelsat during his trip to Europe with Harriet in late 1981.[40] Unfortunately, Colino, who took over Intelsat in June 1983, betrayed the trust McLucas and others had placed in him. In December 1986 Intelsat's board of governors fired Colino for financial improprieties. After pleading guilty to criminal fraud and conspiracy, he was sentenced to six years in prison.[41] McLucas was among those interviewed by investigators looking into charges against Colino and his accomplices. The case expanded to include alleged kickbacks on some contracts with Hughes Aircraft and Ford Aerospace. At Hughes, former top CIA scientist Albert Wheelon (an associate of McLucas in various space endeavors) fell victim to the unfavorable publicity and prematurely resigned as its new chairman and CEO in May 1988.[42] Eleven years later, when Bud Wheelon discussed the incident with McLucas, it was obvious that John's apparently inadvertent implication of him in this affair by giving the investigators adverse information about one of Wheelon's deputies remained a painful episode in the lives of both men.

McLucas began to participate in various private space ventures after leaving Comsat. In August 1985 he was elected to the board of the Space Shuttle Corporation of America, a subsidiary of Astrotech International founded by Willard F. Rockwell Jr.

to build new space shuttles. Other members of the board included former NASA administrator James Fletcher and former Apollo astronaut Eugene Cernan. In November 1985 the new company's name was changed to General Space Corporation to reflect plans to move into other space services.[43] McLucas later described the company as "part of a conglomerate of space-related companies that Rockwell expected would soon become the space-age counterpart of General Electric, General Dynamics, and General Motors. . . . Then came the *Challenger* accident, [and] the bubble of optimism about America's fledgling space industries burst."[44] With Astrotech curtailing operations at its subsidiary, McLucas resigned from the board.

McLucas spent much of the late 1980s promoting other space enterprises. Orbital Sciences Corporation (OSC), headquartered in northern Virginia, has proven to be the most successful. In 1987 Orbital's young CEO, David Thompson, asked McLucas to join its board of directors. Since the *Challenger* accident, Thompson's company had been diversifying beyond a shuttle-launched booster called the Transfer Orbit Stage to offering complete yet affordable launch capabilities for small satellites. OSC's revised business plan was based on the concept of "Microspace," that is, taking advantage of the continual miniaturization of components to decrease the size and cost of payloads. To serve as a launch vehicle, OSC developed the truly innovative Pegasus, a winged space rocket designed for release high in the atmosphere from large aircraft. Pegasus put its first satellite into orbit from a B-52 in April 1990.[45] Orbital acquired an L-1011 transport as its own Pegasus launch platform in 1994 and adapted the Pegasus to serve as an upper stage for its ground-launched Taurus rocket.

In many ways OSC—with its reliance on private capital, its workforce of mobile young engineers, and its "dot-com"-style corporate culture—was the antithesis of the companies and government agencies McLucas had known in the past.[46] Being associated with this dynamic business seemed to stimulate him. In return the McLucas résumé helped advance the cachet of the small company with potential customers, especially at NASA, DOD, and established air and space corporations. McLucas served on Orbital's board during its expansion and diversification into a full-service space enterprise. New business areas included satellite operations, such as lightweight low-orbit data communications satellites developed by its Orbcomm division. McLucas was especially enthusiastic about the establishment in 1993 of Orbital Imaging Corporation (Orbimage), a subsidiary formed to handle remote sensing.[47] As the years went by, he was well satisfied to see Orbital compile a record of 255 successful launches in 267 space missions through 2002. Regrettably, a Taurus rocket bearing the name John McLucas on its fuselage, launched in September 2001, failed to complete its mission.[48]

During the early days of space shuttle development and for several years thereafter, one of McLucas's pet ideas was to use some of its huge external liquid fuel tanks as cost-effective space stations, each with about six times the volume of the shuttle's payload bay. To help implement this concept, McLucas in 1987 became chairman of a start-up company in Boulder, Colorado, called External Tanks Corporation. With Randolph Ware as president, the new company—founded by the University Consortium for Atmospheric Research (UCAR)—envisioned using the tanks for scientific laboratories, storage units, fuel depots, and materials or pharmaceutical processing facilities. When NASA declined to utilize a specially modified external tank with docked shuttles as the basis for a separate space station, the company proposed having one become the first component of the ISS *Freedom*—an idea McLucas tried to interest NASA in pursuing. Although External Tanks Corporation negotiated with NASA for use of 10 tanks, various issues kept these plans from becoming reality. As late as 2000, however, the US Space Foundation still considered the tanks a cheap adjunct to the ISS.[49]

A more successful opportunity for profiting from a government space program presented itself with the Navstar global positioning system, which McLucas had championed ever since its early development (see chap. 5). He became even more enthusiastic as the GPS matured. "I predict," he wrote in 1990, "that we will soon see millions of civilian users tied to these satellites, because they offer a host of opportunities that have never existed before. We have not even begun to think of all the applications for its use."[50] From 1992 to 1996, McLucas was able to advocate and evaluate new commercial applications for GPS as a technical advisor to Trimble Navigation of Sunnyvale,

California. Founded in 1978 (the year of the first GPS launch) by Charles Trimble and two fellow visionaries from Hewlett-Packard, Trimble Navigation became one of the most successful companies focusing on GPS equipment and related services.[51] Long impressed by Trimble's technological leadership, McLucas became connected with the company in his role as chairman of the board of Avion Systems of Leesburg, Virginia. Avion, which had developed a prototype collision-avoidance system for commuter airplanes, was purchased by Trimble in early 1991. To McLucas's consternation, Avion's young president later filed a lawsuit against Trimble. Fortunately, this dispute did not hurt McLucas's status as a valued technical advisor to Trimble, which continued its successful growth in satellite navigation and communications.[52]

In the public arena, McLucas argued strongly for adopting GPS for civil aviation and other nonmilitary applications, especially when he served as chairman of an FAA subcommittee on using the Global Navigation Satellite System (GNSS—the generic name for the GPS and other space-based location networks). As he told one congressional subcommittee when the FAA finally began to consider using GPS for navigation and instrument flying, "It is very gratifying to see the FAA moving into the space age."[53] McLucas also took special pride in how well GPS proved its military value during the Gulf War of 1991, even though its constellation of satellites was not yet complete.

To permit GPS to reach full potential, especially for civil aviation and other applications that required extreme precision, McLucas argued against routinely applying the "selective availability" feature of GPS that degraded its accuracy for all but specially configured military receivers. During Desert Shield in late 1990, Air Force Space Command allowed GPS satellites to begin sending unencrypted signals so that coalition soldiers could use commercial equipment, especially Trimble's small, lightweight GPS receivers (SLGRs—pronounced "sluggers" by the troops), which cost only about $3,400 each and weighed less than four pounds. Trimble's factory worked overtime to produce thousands of SLGRs for the ground forces, which were thereby able to navigate accurately across southern Iraq's almost featureless terrain during Desert Storm.[54] "One cannot help noting," McLucas wrote just after the war, "the irony of a doctrine which calls for disabling a security feature so that in wartime all friendly forces can use GPS's full capability, while as soon as peace breaks out, the security feature is enabled to prevent the enemy from benefiting from GPS's ultimate accuracy."[55] He continued to call for DOD's help in making GPS useful for civilian uses, as well for other users to share the Air Force's burden in funding the program.[56]

McLucas was always alert to possible new applications for the GPS, such as using extra stable GPS receivers to measure subtle distortions in the atmosphere and thereby collect weather-related data in a cost-effective way. The nickname of this application is GPS-MET (for meteorology).[57] The concept was championed by Randolph Ware and other associates at UCAR (where McLucas had been on the board of trustees) and its National Center for Atmospheric Research (NCAR), with additional funding from NOAA, the FAA, the National Science Foundation, and other government agencies. To McLucas's delight, Orbital Sciences Corporation was able to "piggy back" a small instrument on its MicroLab 1 satellite in 1995 to test the GPS-MET concept.[58] The results proved the concept of signal oscillation and showed considerable promise, although NASA preferred to pursue development of more specialized meteorological satellites for the immediate future.[59] Indicative of his advocacy, John complemented the "GPS-NOW" Virginia license plate he had used for several years by putting another on Harriet's car that read "GPS-MET." He continued to work behind the scenes to promote the concept. In early 2002 NASA's Jet Propulsion Laboratory, NOAA, the Air Force, and the Navy agreed to support a plan by the space agency of Taiwan (with help from OSC and UCAR) to sponsor several GPS-MET satellites in an international project known as the Constellation Observing System for Meteorology, Ionosphere, and Climate (COSMIC).[60]

In addition to McLucas's direct involvement in space entrepreneurship, he also tirelessly promoted commercial uses of space as a speaker and writer. His most tangible contribution was to author the aforementioned book *Space Commerce*, an analysis of the commercial endeavors of the recent past and the challenges he believed lay ahead. Among the space enterprises he examined were launchers, communications, remote sensing, navigation, experiments, manufacturing, and contractor services to government agencies. He concluded that for commercial space ventures to take hold, "we must be willing to engage in public/private partnership arrangements unlike anything we have known in the past."[61] Published in 1991 as part of Harvard's Frontiers of Space series, the 250-page book (with a foreword by Arthur C. Clarke) is still a valuable reference on the first quarter century of the space business. In August 2000 at McLucas's 80th birthday party, Dave Thompson of Orbital Sciences Corporation reminisced about his contributions to OSC and others in the space business.

> In the summer of 1987, John joined our board. . . . [D]uring the period that John has served on the board, we've evolved from being a small, struggling space company to now being a large, struggling space company. John's breadth and perspective, and the depth of his knowledge in the areas that matter to our business, have not only been essential to us, but they've been shared with many others in the commercial space industry. In fact, his 1991 book, entitled *Space Commerce*, really pointed the way for much of the tremendous growth that has taken place in the entire commercial space sector during the decade we have just completed. . . . John is a true pioneer in space commerce, not only as a thoughtful observer and writer about this exiting new field, but as an active practitioner himself.[62]

Mission to Planet Earth

After returning from Moscow's Space Future Forum in October 1987, John McLucas expressed his philosophy on the role that space technology could play in the future of mankind, if only it were harnessed on a truly international basis.

> Many of us recognize that the world is too small to continue playing the old games. The earth's resources are in many ways finite. The earth is a closed ecosystem. . . . We know in our hearts that we must begin to see ourselves as citizens of the Planet Earth. . . . Space surveillance, space communication, learning more about Earth's resources, and the place of the earth in the universe are areas that show promise if we join in using space for the benefit of us all. . . . Earth's resources are not ours to plunder. Our time on earth is not ours to fritter away. All are finite, but what we can do together is almost without limit. We can learn to work together, develop our talents, pool our resources, use our time to plan joint missions to Mars, organize the Mission to Planet Earth, and become the creators of the Space Age, an age of enlightenment for future generations.[63]

During the late 1980s and early 1990s, John McLucas occupied some fairly prominent pulpits from which to preach these messages, such as being chairman of the US Association for the International Space Year (ISY), chairman of the NASA Advisory Council, and once again a member of the National Academies' Committee on Science, Engineering, and Public Policy.

The International Space Year of 1992 coincided with the 500th anniversary of Columbus's first voyage to the New World. Senator Spark M. Matsunaga (D-Hawaii) introduced legislation for the project in 1985, and President Reagan sanctioned it in 1986, directing NASA to lead an interagency effort to support the ISY and asking the National Academies to mobilize the scientific community.[64] In partnership with NASA, the US Association for the ISY (US-ISY) was formed in January 1987 to publicize and coordinate participation in the effort and serve as a model for similar ISY associations abroad. Harvey Meyerson, a staff assistant to Senator Matsunaga, became president of the US-ISY Association, and John McLucas was named its chairman. Early global sponsors included the IAF and the Council of Scientific Unions' Committee on Space Research (COSPAR).

The US Association hosted the first conference on the ISY at Kona, Hawaii, in August 1987. With 150 representatives from both Pacific Rim and European nations, this meeting set in motion a broad range of multinational projects. McLucas pushed his long-standing idea for a global Landsat-type satellite, which led to a proposal to establish an international remote-sensing council.[65] August 1987 also saw publication of a major report on the future of NASA after the *Challenger* disaster submitted by Dr. Sally K. Ride, America's first female in space. With McLucas as one of its many contributors, this study proposed greater emphasis on remote sensing under an initiative dubbed "Mission to Planet Earth."[66]

McLucas followed up on the Kona meeting by hosting an international conference at the University of New Hampshire in April 1988 using "Mission to Planet Earth" as its name and theme. This was the first time leaders from so many national space agencies met to compare remote-sensing plans. When introducing NASA administrator James Fletcher as keynote speaker, McLucas called for a worldwide remote-sensing council to better harmonize national programs. To illustrate the need for such coordination, a working group compiled a list of 48 remote-sensing satellites being developed by the 17 nations represented at the conference. This number came as quite a surprise, since no one had ever consolidated such a list before. McLucas felt the prospect of so many similar satellites performing related missions proved the need for better multinational cooperation.[67]

Hurbert Curien, France's minister for research and technology (and a founder of the European Space Agency) was conference chairman. With his encouragement, the attendees agreed to form the Space Agencies Forum for the International Space Year (SAFISY) to deal with the need to coordinate space science activities.[68] They also approved several basic principles to guide the ISY as well as the need for careful design of data collection programs. As McLucas later wrote, "Space enthusiasts sometimes forget that collecting data is not the end of the exercise but only the beginning."[69] The SAFISY structure expanded as ever more nations joined in the months that followed. Addressing an audience at the National Air and Space Museum's annual Wernher von Braun memorial lecture, McLucas expressed his delight that "the conference turned out to be much more productive than we dared to hope for."[70]

McLucas next chaired a workshop of 35 space experts and advocates (ranging from author-scientist Carl Sagan to Senator Al Gore) conducted by The Planetary Society and George Washington University's Space Policy Institute in Washington, D.C., in December 1988. They made a strong pitch for Mission to Planet Earth and other international space projects to the Bush transition team. Two months later, McLucas and John Logsdon of the Space Policy Institute reiterated these recommendations to the House Committee on Science, Space, and Technology.[71] Mission to Planet Earth was the main theme of the ISY and, acquiring the acronym MTPE, became the umbrella under which the US government planned oceanographic, atmospheric, and earth sciences projects.[72]

After several SAFISY meetings in different nations and a resolution by the UN General Assembly, Pres. George H. W. Bush officially launched the ISY with a ceremony at the White House in January 1992.[73] The SAFISY held its climactic meeting in Washington on 28 and 29 August 1992, leading into the first World Space Congress, which met through 5 September. Organized in collaboration with the IAF, COSPAR, AIAA, National

Academy of Sciences, and NASA, this was the largest gathering of space experts and enthusiasts ever held, with more than 11,000 attendees from 67 nations.[74] The ISY also featured hundreds of national and private initiatives ranging from school projects to satellite launches. As summarized by Harvey Meyerson, "The International Space Year stood out from other celebratory years for the way it unfolded on several levels at once. It was part scientific research program, part international policy initiative, part global awareness project."[75] At a final SAFISY meeting, the national representatives agreed to establish a permanent space agency forum, which began convening annually in conjunction with IAF conferences.

Although most ISY activities soon faded into memory, NASA's Mission to Planet Earth program remained as a legacy. McLucas saw MTPE as a way to strengthen public appreciation of NASA's work by signaling its commitment to addressing practical "real world" concerns, such as examining the polar "holes" in the ozone layer, rates of deforestation, the El Niño cycle, the effects of greenhouse gases on climate change, signs of pollution and overpopulation, and other instances of global change.[76] In an issue featuring Mission to Planet Earth, *Aviation Week and Space Technology* reported McLucas's opinion that "White House leadership is essential" to give NASA and other participating agencies the resources needed to accomplish what "is, in my view, the most important thing we can do in space."[77]

As chairman of the NASA Advisory Council from 1989 to 1993, McLucas was in a position to help shape the agency's MTPE strategy. Early in the process, he cautioned against sinking too much money into large multisensor Earth observation satellites and recommended other less expensive means to assist in collecting data, including foreign and privately owned satellites. He also recommended looking for useful historical Earth observation data in the imagery collected by classified DOD satellite systems.[78] McLucas believed the council could provide valuable advice, especially when the administrator showed personal interest in the group and its work.[79] Besides his continued belief that NASA was relying too much on its small fleet of space shuttles, he also thought that it was neglecting some important issues involving manned space flight.[80] "In the life sciences," he complained in late 1990, "NASA still does not know . . . just

301

how people [in space] would benefit from an artificial gravity environment. . . . One wonders when NASA will decide it would be nice to know the answer."[81]

As McLucas anticipated, Congress scaled back NASA's original plans for MTPE, especially its large Earth observation satellites, during the early 1990s.[82] With the help of other nations, NASA conducted phase one of MTPE through 1997, using smaller specialized satellites and selected shuttle missions to collect vast amounts of data about various phenomena. The Goddard Space Flight Center in Maryland established the Earth Observing Data and Information System to archive and process the unprecedented volume of data being collected. Early on McLucas had envisioned some MTPE sensors being added to commercial satellites and having private companies operate some of the ground stations and analyze much of the data. NASA eventually pleased him in this regard by developing an MTPE commercial strategy in 1997.[83] In early 1998 NASA renamed the MTPE program its "Earth Science Enterprise" to better correlate with its parallel Space Science Enterprise.[84] Whatever the official name, McLucas continued until his final days to advocate learning more about our own planet by using space technology and to support plans for a more comprehensive international effort in the future.

Looking Farther into the Space Age

Although McLucas considered Mission to Planet Earth as NASA's most useful endeavor, he selectively supported other goals. Early in the planning for the ISS, he was "very happy that this space station was developed through an international approach" and, as he told the House Space Science Subcommittee, "I consider Jim Beggs' contribution to internationalizing that space station as one of his best initiatives."[85] According to one noted author, "the space station would be the subject of the longest and most bitterly fought imbroglio in the history of the US space program."[86] As NASA's design of the ISS, also known as the Space Station *Freedom* (SSF), began to mature, McLucas became more ambivalent about its scope and schedule. In February 1994 he said he favored a more cautious building block approach than currently planned.[87] McLucas made his last formal contribution to the ISS program during late 1996 and early

1997, when he served with three other experts on a special panel chaired by Beggs. Under the auspices of the Potomac Institute for Policy Studies (with a grant from NASA), they prepared a study on the commercialization of orbital space flight based on an assumption that the ISS would be operational in six years. The team saw significant opportunities for private companies to utilize the space station. The panel's findings reflected much of the philosophy about governmental encouragement of space commerce than McLucas had been preaching since the late 1970s.[88]

Although hopeful, McLucas was not naïve about the prospects for government subsidies of private space enterprises. For example, he wrote to Arthur C. Clarke in 1994, "The problem for us is to keep the proper amount of research going on so as to learn how to do various things out there without it being seen as just more pork for NASA and its rapacious contractors. I'm afraid that our whole gang of space buffs are [sic] viewed by many as a group which is very good at feeding at the government trough, but not very good at showing adequate payoff to the rest of the world."[89] Signing of the Land Remote Sensing Policy Act of 1992 made McLucas more optimistic about the future of commercial imaging satellites than when he wrote Space Commerce two years earlier (before opportunities presented by the end of the Cold War became obvious). In addition to preserving Landsat, this legislation liberalized rules for licensing private companies to develop advanced imaging technology. "I believe," McLucas predicted in 1993, "we should assume that remote sensing will become the second space success following communications."[90]

The following years saw a few risk-taking American companies launch commercial imaging satellites that began approaching the resolution of the highly classified reconnaissance systems McLucas had managed in the early 1970s. In fact, the intelligence community soon began purchasing imagery from these commercial sources for national security purposes. Following the OrbView satellites of OSC's Orbimage subsidiary (first launched in 1995), the most advanced American satellites included Space Imaging Corporation's Ikonos (1999) and DigitalGlobe's Quickbird (2001).[91] In 1996 McLucas had written of these, "We now have three companies who say they will put high resolution systems into orbit because they . . . foresee a market measured in the billions of dollars in the next five years. I think they are right."[92]

As he described earlier, McLucas devoted much time and energy during his retirement years to altruistic endeavors, most notably the International Space University and the Arthur C. Clarke organizations. In 1995 Peter H. Diamandis, one of the young cofounders of the ISU whom McLucas befriended in the early 1980s, established the X-Prize Foundation. Inspired by how cash awards encouraged air pioneers such as Charles Lindbergh to advance the aviation industry early in the twentieth century, Diamandis hoped a similar incentive might help jump-start private space travel and tourism in the next century. As gratefully acknowledged in the X-Prize's official history, "The Foundation received early seed funding from Tom Rogers and John McLucas."[93] During 1999 another band of space enthusiasts founded the Space Island Group. Its ambitious vision "is to develop a stand alone, commercial space infrastructure supporting the broadest possible range of manned business activities in Low Earth Orbit (LEO) for the 21st Century." McLucas joined Arthur Clarke as a technical advisor to the group along with a couple other old friends: Phillip Culbertson, NASA's former ISS director; and Fred Durant, former associate director of the Smithsonian Institution's Air and Space Museum.[94] If John had lived two more years, he would have been thrilled to see Burt Rutan's innovative Space Ship One fulfill Diamandis's vision by winning the $10 million Ansari X-Prize amid renewed enthusiasm for private space ventures.

McLucas would not have been as happy to witness the final fate of Comsat. By the end of 2002, Lockheed Martin was abruptly exiting the global telecommunications business, having sold what had been Comsat World Systems to Intelsat, divested Comsat International's Latin American operations to World Data Consortium, sold Comsat Mobile Communications to Norway's Telenor Satellite Services, sold part of Comsat Labs to ViaSat, and dispersed remaining remnants of the laboratory as well as Comsat General into other Lockheed Martin divisions.[95] John also worried about the privatization of Intelsat during his final two years. It was not so much the restructure and modernization of Intelsat's management and financial structure that concerned him as the fact that these "reforms" were largely ordered from the halls of Congress, often in response to lobbying by Intelsat's competitors. By late 2002 foreign investors had become primary owners of what had been another legacy of American space leadership.[96]

Despite repeated setbacks to various space programs and businesses, McLucas always tried to take a longer view. In looking ahead, he was hopeful that a new generation of practical yet idealistic scientists and technocrats, as exemplified by graduates of the International Space University, would carry the torch of progress and discovery into the future. Unfortunately, for John, his devotion to the ISU had seriously compromised his health. On 24 June 1991, he delivered the convocation at the ISU's summer session in Toulouse, France, attended by 137 students from 26 nations. He was still not feeling well after undergoing double bypass heart surgery several months earlier, but as ISU chairman, he considered it his duty to open the session. Regrettably, his condition worsened while in France. As soon as he and Harriet returned home to Alexandria, John went straight into the hospital, where doctors discovered a severe chest infection and other complications stemming from the bypass operation. He lost half his lung capacity and thereafter suffered persistent breathing and heart problems.

Final Years

Even when these and other medical complications began to slow him down, John McLucas kept up his outside interests and continued to travel overseas, now more for recreation than business. He and Harriet vacationed in such places as Baja California, Costa Rica, the Galapagos Islands, Israel, Jordan, and Canada, and he religiously attended annual AIAA and IAF-sponsored meetings in various nations. They spent his 70th birthday in Scotland visiting the ancestral homeland of the McLucas clan on the Isle of Mull. He celebrated his 75th birthday with a two-week trip to Scotland and Ireland. By the late 1990s, John's breathing became increasingly difficult and his mobility more restricted, but he sought to remain active in various business and professional activities, especially in the Washington area. As late as 1998, for example, he continued to serve on an advisory panel to the Senate Select Committee on Intelligence. In the aftermath of the hijackings on 11 September 2001, McLucas and Burt Edelson, with some help from Ken Alnwick, submitted a proposal to FAA administrator Jane Garvey on using advanced telemetry techniques developed for the space program to monitor commercial aircraft in flight and report anomalies, whether mechanical or man-made.

One of the highlights of John McLucas's last few years was an 80th birthday party, organized by family and friends at the Belle Haven Country Club in Alexandria. More than 120 people attended. His children and some of the more notable guests told stories about his past. In tribute to McLucas's pride in his Scottish heritage, a rather loud bagpipe concert preceded the after-dinner remarks. James Schlesinger then drew much laughter from the audience when opening his tribute by advising John that "the definition of a gentleman is somebody who knows how to play the bagpipes—but doesn't."[97]

Like many Americans, in his later years McLucas began to delve more deeply into his family's genealogy and to renew contacts with distant relatives and childhood friends. After thinking about it for years, he finally started work on this autobiography. He began to record memories and interview former associates about past events and current issues. He also became more introspective about his personal life. Among the autobiographical essays John left behind was an imaginary letter written in 1996 to the father he never knew. He wished his father could have been part of his life, but he expressed gratitude for the other family members who raised him. Explaining that he was now "thinking more about what we leave behind us when we die," he ended this essay as follows. "I know that I've had lots of good breaks in life and should be thankful to all those who helped me along the way. I'm sorry for my mistakes, but all in all, I think I haven't done too badly, and I hope you would approve of the record I've left."

Ten months after the late John McLucas was honored for his national service by the formal burial ceremony at Arlington National Cemetery, a more humble observance took place in South Carolina. On 9 November 2003, Harriet McLucas and John's four children joined his sister Jean and several others for a short memorial service at the little Reedy Creek Presbyterian Church, where young John had spent many a Sunday morning while growing up on Harold Cousar's nearby farm. They then gathered with other relatives and family friends outside the town of Clio at the old McLucas cemetery, which donations from John had helped refurbish. After tributes to a local boy who made good, a portion of his ashes that had not been interred at Arlington were buried near the 80-year-old grave of his father. The mortal remains of John Luther McLucas now rested in both the company of fellow veterans and that of his ancestors.

Notes

1. Joseph Pelton to Ken Alnwick et al., e-mail, subj.: Arthur C. Clarke's Message, 3 December 2002. After working at Comsat and Intelsat from 1969 to 1989, Dr. Pelton became a professor at George Washington University's Institute for Applied Space Research.

2. For a detailed obituary, see Bart Barnes, "John McLucas Dies: Oversaw Air Force, FAA," *Washington Post*, 5 December 2002, B6. Because John passed away before we could start writing about his activities after retiring from Comsat, this chapter has been prepared purely as a biography using available sources.

3. John L. McLucas, "Goals and Objectives," *AIAA Bulletin*, June 1984, B46; and "Aerospace America and Controversial Subjects," *AIAA Bulletin*, October 1984, B32.

4. John L. McLucas, "At the Least, Some ELVs," *Aerospace America*, October 1984, 4.

5. House, *1986 NASA Authorization*, vol. 2, *Hearings before the Subcommittee on Space Science and Applications, Committee on Science, Space, and Technology*, 99th Cong., 1st sess., February–March 1985, 1031–53.

6. Nelson Friedman, "New Headquarters Approved," *AIAA Bulletin*, January 1986, B6.

7. John L. McLucas, "Whither Landsat?" *Aerospace America*, January 1985, 6; "Open Skies a Fresh Challenge," *Aerospace America*, April 1985, 6; and "Space: Sanctuary or Menace," *Aerospace America*, May 1985, 6.

8. AIAA Major Awards, http://www.aiaa.org/about/index (accessed February 2003).

9. House, *Impact of National Security Considerations on Science and Technology: Hearing before the Subcommittees on Science, Research, and Technology and on Investigations and Oversight, Committee on Science, Space, and Technology*, 97th Cong., 2d sess., 29 March 1982, 111–58; and Panel on Scientific Communication and National Policy, Committee on Science, Engineering, and Public Policy, *Scientific Communications and National Security* (Washington, D.C.: National Academy Press, 1982), also known as "the Corson Report."

10. John L. McLucas, "Technology Transfer Scare Hurts Innovation," *Aerospace America*, June 1984, 6.

11. Ellen Marzullo, "McLucas Hosts Section Luncheon," *AIAA Bulletin,* February 1985, B7.

12. Eric J. Lerner, "DoD Information Curbs Spread Fear and Confusion," *Aerospace America,* March 1985, 76–80; and National Security Decision Directive 189, "National Policy on the Transfer of Scientific, Technical, and Engineering Information," 21 September 1985, http://www.fas.org/irp/offdocs/nsdd/nsdd-189.htm (accessed March 2003).

13. John McLucas and James Harford, "Visit to Satellite and Launch Vehicle and Supporting Technology Facilities in Beijing, Xian, and Shanghai, China, May 5–19, 1985," report (New York: AIAA, 21 June 1985).

14. John McLucas, "Summary of Trip to China in Connection with OTA Study of Technology Transfer to China, 15–25 February 1986" (draft).

15. Case study, John McLucas and Norman Getsinger, "Satellite Telecommunications Technology Transfer to China" (Alexandria, Va.: China Business Development Group, July 1986), 4.

16. Panel on the Impact of National Security Controls on International Technology Transfer, Committee on Science, Engineering, and Public Policy, *Balancing the National Interest: U.S. National Security Export Controls and Global Economic Competition* (Washington, D.C.: National Academy Press, 1987), also known as "the Allen Report," quotation from page 2.

17. House, *Omnibus Trade and Competitiveness Act of 1988,* vol. 2, *Hearing before the Subcommittee on International Economic Policy and Trade, Committee on Foreign Affairs,* 100th Cong., 1st sess., 11 March 1987, 73–74. (For his written statement, see 76–82.)

18. Senate, *Export Controls: Hearing before the Subcommittee on International Finance and Monetary Policy, Committee on Banking, Housing, and Urban Affairs,* 100th Cong., 1st sess., 17 March 1987, 48.

19. Transcript, "Testimony of John L. McLucas before the House Committee on Science, Space, and Technology re NASA's Request for Changes in Certain Laws Dealing with the Release of Technical Data," 12 November 1989.

20. Senate, *Earth Resources and Environmental Information System Act of 1977: Hearings before the Subcommittee on Science, Technology, and Space, Committee on Commerce, Science, and Transportation,* 95th Cong., 1st sess., 13 June 1977, 202–55; Senate, *U.S. Civilian Space Policy: Hearings before the Subcommittee on Science, Technology, and Space, Committee on Commerce, Science, and Transportation,* 96th Cong., 1st sess., 1 February 1979, 188–220; Senate, *Operational Remote Sensing Legislation, Part 1: Hearings before the Subcommittee on Science, Technology, and Space, Committee on Commerce, Science, and Transportation,* 11 April 1979, 179–214; House, *Earth Resources Information System: Hearings before the Subcommittee on Space Science and Applications, Committee on Science, Space, and Technology,* 95th Cong., 1st sess., 22 June 1977, 156–93; and House, *Earth Resources Data and Information Service: Hearings before the Subcommittee on Space Science and Applications, Committee on Science, Space, and Technology,* 96th Cong., 1st sess., 2 May 1979, 2–42, 216–19.

21. "Comsat Withdraws from Sparx Venture," *AW&ST,* 5 March 1984, 16.

22. John L. McLucas, chairman, NASA Advisory Council, *Mission to Planet Earth: The Wernher von Braun Memorial Lecture for 1989* (Pasadena, Calif.: Planetary Society, 1989), 14.

23. The *Land Remote Sensing Policy Act of 1992* established an interagency office to assure continued operations, but funding and management problems persisted. See "Landsat Program Chronology," http://geo.arc.nasa.gov/sge/landsat/html (accessed February 2003).

24. John L. McLucas, "Create a Global Landsat," *Aerospace America*, July 1987, 5; J. L. McLucas and P. M. Maughan, "The Case for Envirosat: A Space Remote Sensing Initiative in Keeping with the International Space Year," *Space Policy*, August 1988, 230–39; and McLucas, *Space Commerce*, 112–30.

25. For a comprehensive study of these proposals, see Walter Dorn, *Peacekeeping Satellites: The Case for International Surveillance and Verification*, Royal Military College of Canada, http://www.rmc.ca/academic/gradrech/dorn19-4_e.html (accessed April 2003), originally published in *Peace Research Review* 10, nos. 5 and 6 (1987).

26. Report, "Visit to the Soviet Union as Guests of the Soviet Academy of Sciences," John L. McLucas, president, and James J. Harford, executive director, AIAA, 15 July 1984, 21, 40; James Harford, "Soviets Renew Calls for Space Cooperation," *Aerospace America*, August 1984, 22–26; John L. McLucas, "Could Space Thaw the Freeze?" *AIAA Bulletin*, January 1985, B32; and Harriet McLucas, interview by Lawrence Benson, 10 March 2003.

27. John McLucas to Hugh Wachter, letter, in *Washington Technology*, subject: "Op Ed Piece on Soviet Offers of Space Hardware and Services," 6 September 1991.

28. John L. McLucas, past president, AIAA, to the president (with copies to the vice president and 22 others), letter, 16 October 1986.

29. United Nations Association of the United States of America (UNA-USA), *The Next Giant Leap in Space: An Agenda for International Cooperation*, Final Report of the 1986 UNA-USA Multilateral Project, October 1986.

30. "Soviets Refine Space Cooperation Overtures," *Aerospace America*, November 1987, 6–9.

31. John L. McLucas and Burton I. Edelson, "Let's Go to Mars Together," *Issues in Science and Technology*, Fall 1988, 52–55; and Edelson and McLucas, "US and Soviet Planetary Exploration," *Space Policy*, November 1988, 337–49.

32. William Burrows, *This New Ocean: The Story of the First Space Age* (New York: Random House, 1998), 573–76.

33. John L. McLucas, personal essay on space cooperation, ca. 1996, 7; and Harriet McLucas, interview.

34. M. B. Regan, "McLucas Is Appointed Temporary President and CEO of Questech, Inc.," *Washington Post*, 24 November 1986, F8.

35. M. B. Regan, "Retired Comsat General Chief Is New CEO at Questech Inc.," *Washington Post*, 15 June 1987, F10.

36. QuestTech, Inc., *Annual Reports, 1984–1987* (McLean, Virginia: 1985–1988).

37. The *Washington Post*'s financial section covered this affair in the following articles: Elizabeth Tucker, "Ousted CEO Mayo Ran Afoul of Firm's Found-

ers," 9 November 1988, B1; Sandra Sugawara, "Some Steamy Allegations by Fired QuesTech President: Suit Alleges Sexual Misconduct by Directors," 31 January 1989, C1; Sandra Sugawara and Malcolm Gladwell, "QuesTech Tries to Pick Up the Pieces: Lawsuit by Ex-President Is Latest in Series of Difficulties," 6 February 1989, F1; and Sandra Sugawara, "Former President's Lawsuit against QuesTech Dismissed: Court Said to Have No Jurisdiction in Dispute," 26 April 1989, F3.

38. "Roundup," *Washington Post*, 23 May 1990, C2; and Biography of Bruce G. Sundlun, National Association of Governors, http://www.nag.org/governors (accessed February 2003).

39. QuesTech was eventually acquired by CACI International for $42 million. See CACI news release, "CACI Completes Acquisition of QuesTech, Inc.," 16 November 1998, http://www.caci.com/about/news/11_16_98NR.html (accessed March 2003).

40. "New Chairman for Intelsat," *Financial Times*, 22 August 1983, 12. Colino, a former FCC lawyer, had retired as a vice president at Comsat in 1979.

41. Elizabeth Tucker, "Officials Fired amid U.S. Investigation of Finances; Criminal Probe Set as Allegations Mount," *Washington Post*, 5 December 1986, A1; "Former CEO of Intelsat Pleads Guilty; Colino, Associates Charged in $4.8 Million Fraud," *Washington Post*, 15 July 1987, D1; Lee Hockstader, "Ex-Official of Intelsat Sentenced to Prison," *Washington Post*, 6 December 1988, B3; and "Digest" (Finance Section), *Washington Post*, 16 April 1992, B11.

42. "U.S. Subpoenas Hughes and Ford in Improper Payments Probe," *AW&ST*, 16 May 1988, 20; and Ronald Grover et al., "Is a Big Scandal Brewing at Hughes Aircraft," *Business Week*, 23 May 1988, 58.

43. Astrotech news release, *PR Newswire*, 30 August 1985; "Astrotech Reorganizes Space-Related Business," *AW&ST*, 25 November 1985, 18; and "Astrotech Curtails Space Unit Operations," *AW&ST*, 3 November 1986, 44.

44. McLucas, *Space Commerce*, 2.

45. A story on page 1 of the *Wall Street Journal* predicting failure of Pegasus forced OSC to withdraw its initial public stock offering, but a few weeks later the *Journal* buried news of the successful launch on page 18, an affair described sardonically in McLucas, *Space Commerce*, 107.

46. For an irreverent account of OSC's struggle for success in the 1990s, focusing on its Orbcomm division, see Gary Dorsey, *Silicon Sky: How One Small Start Went over the Top to Beat the Big Boys into Satellite Heaven* (Reading, Mass.: Perseus Books, 1999).

47. "Orbital Sciences," *Aerospace America*, October 1992, 8–10; Orbital Sciences Corp., *1993 Annual Report* (Dulles, Virginia, 1994); and "About Orbital," http://www.orbital.com/ (accessed February 2003).

48. Barron Beneski, OSC vice president for corporate communications, to Lawrence R. Benson, e-mail, subject: "Re John McLucas," 24 February 2003.

49. Randolph Ware and Phil Culbertson, "STS-Lab: A Low Cost Shuttle-Derived Space Station," 1 February 1992, http://www.space-frontier.org/projects/external tanks.../html (accessed March 2003). Culbertson had been general manager of NASA's shuttle program. John McLucas to Dr. Joseph Shea, NASA assistant deputy administrator for space station analysis, letter, 23 March 1993; and

"Shuttle Continues to Waste External Tanks," *Space Daily*, 8 September 2000, http://www.spacedaily. com/news/shuttle-00y.html (accessed March 2003).

50. McLucas, *Space Commerce*, 149.

51. Web site, Trimble Navigation Limited, www.trimble.com, March 2003.

52. "Avion Systems, Inc." *AW&ST*, 20 June 1988, 117; "Trimble Navigation Reports Record First Quarter Results," *PR Newswire*, 24 April 1991; Philip J. Klass, "Bendix, BF Goodrich, Trimble Vie for TCAS-1 Business," *AW&ST*, 11 January 1993, 45–47; "Heller Navigates Clear Path in Licensing Feud: Avion Systems v. Trimble Navigation," *Recorder* (American Lawyer Media), 24 August 1994, 2; and "Trimble Settles Avion Dispute," *PR Newswire*, 3 October 1994.

53. House, *Fiscal Year 1993 FAA R,E,&D Authorization: Hearings before the Subcommittee on Technology and Competitiveness, Committee on Science, Space, and Technology*, 100th Cong., 2d sess., 10 March 1992, 77; and John McLucas, "Report of the GNSS Technology Subcommittee" (final draft), 24 January 1994.

54. David Spires, *Beyond Horizons: A Half Century of Air Force Space Leadership*, rev. ed. (Maxwell AFB, Ala.: Air University Press, 1998), 248–50, 256–57.

55. Essay by John McLucas, "New Space Order," 4 March 1991.

56. "DoD Support of WDGPS 'Critical' to Civil Use, McLucas Says," *AW&ST*, 7 February 1994, 204.

57. For scientific background, see Thomas P. Yunck, Chao-Han Liu, and Randolph Ware, "A History of GPS Sounding," *COSMIC*, January 2000, in Genesis OnLine Repository, http:genesis.jpl.nasa.gov/html/publications (accessed April 2003).

58. Microsat's main cargo was a NASA-funded instrument package to measure lightning.

59. John L. McLucas, essay on space applications, ca. 1996, 16: GPS/MET Home Page, September 1998, http://www.cosmic.ucar.edu/gpsmet/ (accessed April 2003).

60. David Thompson, interviewed by Lawrence Benson, 15 April 2003.

61. McLucas, *Space Commerce*, 110, 213.

62. Transcription of remarks by David Thompson at John McLucas's 80th birthday party, Belle Haven Country Club, Alexandria, Virginia, 26 August 2000.

63. Dr. John L. McLucas, "Perspective on Sputnik," *Washington Technology*, 9 January 1988, 4.

64. Ronald Reagan, "Message to the Congress Transmitting a Report on Establishing an International Space Year in 1992," 15 May 1986, www.reagan.utexas.edu/resource/speeches/1986/51586.htm (accessed March 2003).

65. US-ISY Association, *ISY News* 1, no. 1 (Fall 1987).

66. *NASA Leadership and America's Future in Space: A Report to the Administrator by Dr. Sally K. Ride*, August 1987, http://history.nasa.gov/riderrep.htm (accessed March 2003).

67. As an example of improving American-Russian relations, Susan Eisenhower and Roald Sagdeev married after attending the conference. Harriet McLucas was a friend of Susan's parents while John McLucas had developed a close professional relationship with Sagdeev.

68. John L. McLucas, "ENVIROSAT: An ISY Initiative, a Global Change Imperative," manuscript, 12 September 1988.

69. McLucas, *Space Commerce*, 134.

70. McLucas, "Mission to Planet Earth," 26 January 1989 (Planetary Society Paper no. 2; condensed in the American Astronautical Society's *Space Times*, March–April 1989, 3–6).

71. House, *Review of Major Space Programs and Recommendations Made to Bush Transition Team: Hearings before Subcommittee on Space Science and Applications, Committee on Science, Space, and Technology*, 101st Cong., 1st sess., 9 February 1989, 107–29.

72. NASA/XID, "Participation in the International Space Year by the National Aeronautics and Space Administration," unpublished paper, April 1988.

73. George H. W. Bush, "Message to the Congress Submitting a Report on the International Space Year," 31 January 1991, http://www.bushlibrary .tamu.papers/1991/91012104.html (accessed March 2003); UN General Assembly Resolution 46/45, "International Cooperation in the Peaceful Uses of Outer Space," 9 December 1991, http://www.un.org/documents/ga/res/46/ a46r045.htm (accessed March 2003); and NASA, News Release 92-12, "President Bush Launches International Space Year," 24 January 1992, http:// spacelink.nasa.gov/Previous.News.Releases/92.01.14/htm (accessed March 2003).

74. The next World Space Congress was held 10 years later in Houston.

75. Harvey Meyerson and Danelle K. Simonelli, *Launchpad for the 21st Century: Yearbook of the International Space Year* (San Diego: Univelt, Inc. for the American Astronautical Society, 1995), ix. This book summarizes hundreds of projects conducted during the ISY. SAFISY's files were later retired to the National Archives and Records Administration.

76. John L. McLucas, "A Plea for Global Cooperation: Remote Sensing for International Security," *Via Satellite*, January 1990, 20–23.

77. Craig Covault, "Major Space Effort Mobilized to Blunt Environmental Threat," *AW&ST*, 13 March 1989, 37.

78. "Statement by John L. McLucas, chairman, NASA Advisory Council, to the Augustine Committee, 20 November 1990," 4–5. Norm Augustine led this in-depth evaluation of NASA, documented in his *Report of the Advisory Committee on the Future of the U.S. Space Program* (Washington, D.C.: GPO, 1990).

79. John McLucas to Dan Goldin [NASA administrator] via Sylvia Fries, fax, "Utility of NAC, NASA Advisory Council, to the Administrator," 19 March 1992.

80. Diane L. Nitschle, ed., *The Sixth National Space Symposium* (Colorado Springs: Space Foundation, 1990), 34–35.

81. Statement to the Augustine Committee (note 78 above), 2. In other sources, McLucas suggested a revolving external tank as a good way to test artificial gravity in orbit.

82. Garrett Culhane, "Mission to Planet Earth," *Wired Magazine*, December 1993, http://www.wired.com/archive/1.06/mission.earth_pr.html (accessed March 2003).

83. Goddard Space Flight Center, News Release, "Major Review of Mission to Planet Earth Endorses Flexible Approach to Future Satellites, Steers Data

System Development," 21 August 1997, http://www.gsfc.nasa.gov/news-releases/1997/97-180.htm (accessed April 2003); and RAND Corp., "Workshop on Commercial MTPE," September 1997, http://www.rand.org/publications/db247/pdf (accessed April 2003).

84. NASA News Release 98-12, "Mission to Planet Earth Enterprise Name Changed to Earth Science," 21 January 1998, http://geo.arc.nasa.gov/sge/landsat/mtpe.html (accessed April 2003). NASA later consolidated its earth and space science enterprises into a new science directorate. NASA News Release 04-205, "Administrator Unveils Next Steps of NASA Transformation," 24 June 2004, http://www.nasa.gov/home/nqnews/2004/jun/1-1Q-04205_transformation.html.

85. House, *Review of Major Space Program Studies*, 107.

86. Burrows, *This New Ocean*, 591.

87. Brenda Forman, "Quo Vadimus?: An Unscientific Survey," *AIAA San Francisco Section Newsletter*, February 1994, http://www/aiaa-sf.org/files/forman/94-02.txt (accessed March 2003).

88. *The International Space Station Commercialization Study* (Arlington, Va.: Potomac Institute for Policy Studies, March 1997).

89. John McLucas to Arthur C. Clarke, fax cover sheet, 15 August 1994.

90. John McLucas to Randolph Ware, letter, subject: "Op Ed Material Related to Remote Sensing Systems," 1 December 1993, with attached paper, "Is There a New Commercial Opportunity in Space? Converting Spy Satellite Technology to Civil Use," 11.

91. Office of Technology Assessment, *The Future of Remote Sensing from Space: Civilian Satellite Systems and Applications* (Washington, D.C.: GPO, July 1993); John C. Baker, "Commercial Observation Satellites: A Catalyst for Global Transparency," *Imaging Notes*, July–August 2001 and "Commercial Satellite Imagery: Progress and Promise," *Imaging Notes*, March–April 2002, http://www.imagingnotes.com; and J. D. Wilson, "The Satellite Imaging Race Intensifies," *GeoWorld*, January 2002, http://www.geoplace.com/gw (accessed May 2003).

92. John McLucas, essay on space applications, 1996, 14. On 25 April 2003, Pres. George W. Bush signed an even more liberal space imaging policy that gave priority to commercial companies in meeting US government remote sensing requirements: Robert Wall and James R. Asker, "Unrolling the Welcome Mat," *AW&ST*, 19 May 2003, 35–36. In September 2003 NIMA approved a contract to fund a commercial satellite with better than .5 meter resolution: Robert Wall, "DigitalGlobe Wins: NIMA Seeks Funding to Sustain Space Imaging as a Second Provider," *AW&ST*, 6 October 2003, 50–51.

93. History of the X-Prize Foundation, http://www.xprize.org/press/history.html, June 2004. Thomas F. Rogers was president of the Space Transportation Association, a major advocate for commercial participation in orbital missions. McLucas had worked with him at the DDR&E in the 1960s and on various space-related issues since the late 1970s.

94. Space Island Group Web site, http://www.spaceislandgroup.com (accessed January 2003).

95. Tom Butash, "Communications" column, *Aerospace America*, December 2002, 46; ViaSat History, http://www.viasat.com/about/history .htm (accessed March 2003); Telenor Chronology, http://www.telenor.com/ about/history/chronology/ (accessed March 2003); Edward J. Martin, executive secretary of Comsat Legacy Project and former Comsat VP of engineering and operations, interviewed by Lawrence Benson, 12 March 2003; and Joseph N. Pelton to Lawrence R. Benson, e-mail, subject: "Comments on Manuscript," 7 March 2003.

96. Pelton e-mail cited above. Not until August 2004 did the now privately owned Intelsat Ltd. attempt to start selling itself to outside investors as had been required by law: Frank Morring Jr., "Intelsat Sale," *AW&ST*, 23/30 August 2004, 39. Zeus Holdings Ltd. later completed purchase of Intelsat for $5 billion, *AW&ST*, 7 February 2005, 17.

97. Transcript of remarks by Dr. James R. Schlesinger at John McLucas's 80th birthday party, 26 August 2000.

Appendix A

Selected Organizational Affiliations

Professional Associations

American Institute of Aeronautics and Astronautics (AIAA): president, 1984–85; associate fellow, later fellow, 1971–90; honorary fellow, 1991–2002

Armed Forces Communications and Electronics Association: chairman, 1979–81

Committee on Science, Engineering, and Public Policy: member, 1981–84, 1990–93 (a combined unit of the National Academy of Engineering [NAE], National Academy of Sciences [NAS], and Institute of Medicine)

Institute of Electrical and Electronic Engineers (IEEE): fellow, 1962–2002

National Academy of Engineering (NAE): member of the council, 1987–93[1]

Advisory Groups

Air Force Space Division Advisory Group: chairman, 1979–82

Air Force Studies Board: member, 1990–92

Boeing Technical Advisory Committee: member, 1985–ca.95

FAA Research, Engineering, and Development Advisory Committee: member, 1992–96

NASA Advisory Council: chairman, 1989–93

Space Vest: member, Board of Advisors, ca.1992–2000[2]

Trimble Navigation, Ltd.: member, Advisory Committee, 1992–96

Commercial Boards of Directors

Avion Systems, Inc.: chairman, ca. 1988–92

Dulles Access Rapid Transit (DART): director, ca.1985–ca.95[3]

External Tanks Corporation: chairman, 1988–ca.92

C-COR Electronics: director, 1979–92; director emeritus, 1993–2002

General Space Corporation, director, 1985–86

Orbital Sciences Corporation: director, 1987–2000; director emeritus, 2000–2002

QuesTech, Inc.: director, 1985; chairman 1986–90

Educational and Cultural Institutions

Arthur C. Clarke Foundation of the United States: chairman, Board of Directors, 1983–2002

International Space University: chairman, Board of Trustees, 1990–93; member, Board of Advisors, 1994–2002

University Corporation for Atmospheric Research (UCAR): member, Board of Trustees, 1987–93[4]

US Space Foundation: member, Board of Trustees, ca.1980–2000

US Association for the International Space Year: chairman, 1987–92

Wolf Trap Foundation for the Performing Arts: director, 1982–85; chairman, 1985–88

Notes

1. McLucas was also a fellow of the American Association for the Advancement of Science and a member of the International Academy of Astronautics, American Physical Society, and Operations Research Society.

2. Space Vest is a venture capital firm of limited partners, investing in space-related start-up companies, founded in 1991, with headquarters in Reston, Virginia.

3. DART is a private company that proposed a light rail connection to Dulles International Airport.

4. University Corporation for Atmospheric Research is a consortium of universities and affiliated organizations, founded in 1959 under the primary sponsorship of the National Science Foundation to operate the National Center for Atmospheric Research in Boulder, Colorado, and conduct related environmental and education programs.

Letter on Declassifying the Existence of the National Reconnaissance Program

John L. McLucas
1213 Villamay Boulevard
Alexandria, VA 22307

October 31, 1991

The Honorable Richard Cheney
Secretary of Defense
The Pentagon
Washington, D.C. 20331

Dear Mr. Secretary:

The last time you and I had a substantive discussion was in 1975 when President Ford met with you and me to discuss my move to the FAA from the Air Force. The subject I want to raise now relates to my involvement in managing the military space program going back to those Air Force days. Because of my connection with the program and subsequent service on various pro bono advisory groups, I have kept up with the progress of system evolution and the giant steps the Services have made in using space for military purposes.

Recently, as a member of the National Academy of Sciences management structure, I have been asked to advise the Academy in its work with PBS on a series of documentary videos called SPACE AGE. There are 8 planned videos covering various aspects of space services—mostly civil but to include military also. Subjects such as communications, surveillance, navigation, [and] remote sensing are all planned for a certain amount of coverage. Our people are already meeting with military officials to insure that the coverage on communications, etc. is accurately presented to the public.

There is one sensitive area which we hope to cover in a general way which falls under the heading of remote sensing—specifically overhead photography. We have no intention to get into detail which of course would involve classified information. Although the total time in the series devoted to military activity is quite limited, we think it is essential to cover remote sensing because it is one of the most significant parts of the military program. So we are searching for an appropriate way to cover overhead photography.

When we began this exercise some months ago, the way I proposed to cover it was to have two former secretaries of defense discuss very generally how they had used such material when they were in office. They agreed to appear in the videos for a few minutes each to state that they found the material especially valuable for certain specific events, for example to help discriminate a new Soviet ICBM from an old one when we were trying to negotiate limits of one kind or another. These two men found this approach acceptable and asked for a DIA briefing on what they proposed to cover more as a refresher than anything else because the events took place a long time ago and their memories might be faulty. Of course in the process of the refresher, they would find out if the Pentagon had any problem with them doing this particular coverage of the overhead program.

I have spoken with Denis Clift, Chief of Staff of DIA, about this. He has told me that he can easily brief Messrs. Laird and Brown on the two or three incidents they would propose to cover but he would need to be instructed to do so by higher authority. I was not surprised at his answer. At the moment, as he pointed out, the approved language for use in public discussions of satellite photography is extremely limited.

The reason for this letter to you is to ask that this situation be reviewed and if possible changed to allow discussion of what is clearly already in the public record. Specifically, I propose that the programs be treated like other secret programs. In such cases, we maintain security but the obvious aspects of the program where the public already knows a great deal are acknowledged openly. For years we have followed a policy of "no comment" and for many years—perhaps decades—the policy made sense.

In today's world, I believe the policy is counterproductive. It preserves the fiction that if we don't acknowledge it, we can keep people from talking and writing about it. We long since lost control of the public discussion. Witness the following facts: 1) the trade press announces every launch and usually identifies the payload; 2) several widely circulated books by respected authors have been published and, while not totally accurate, give a correct overall picture; 3) the Federation of American Scientists prides itself on its ability to brief the press on the latest status of the program; 4) several spies have been prosecuted for selling instruction books to the Soviets; who knows how many have not been found out?; 5) a man named Morison, a relative of the famous naval historian Samuel Eliot Morison, gave to *Janes*, the publisher, a photograph of a Soviet aircraft carrier which clearly shows what the quality of our photography is; 6) Bob Woodward's recent book on the Gulf War refers to overhead photography as though it were routine and discusses its use in analyzing the situation in the area; 7) many officials now involved with the program consider the policy an anachronism which should be changed; 8) half a dozen former high officials with experience in the program with whom I have recently discussed the situation have said they think a more open policy is in order. Many former officials believe they have no obligation to avoid discussing the program openly since the existence and success of the program long since became public knowledge.

As a person who worked in managing the creation and analysis of this material during the years 1966-75 and who has served as an unpaid consultant for all the years since then, I am quite familiar with all the reasons why the resulting products have been tightly held insofar as official release is concerned. Our close-hold policy has a long history, dating as it does to the days of Gary Powers. As we all know, President Eisenhower proposed the Open Skies Policy, based on using existing aircraft, in 1955; I was in the audience when he referred to that policy in a speech at Penn State shortly after the event. Mr. Krushchev vehemently rejected the idea so we built the U-2 and did the job unilaterally. In May 1960, Gary Powers was shot down and after a brief period of dissembling by administration officials who claimed it was a NASA mission gone astray, President Eisenhower announced what had happened and promised never to overfly Soviet territory again. Of course he was assisted in making that statement by knowing that we had developed a way to get coverage by satellite; three months later, we began the successful recovery of satellite photography. Seventeen years later, we began operating electronic systems and will soon celebrate 15 years of such operations.

One of the most compelling reasons why we have kept these programs under close control for so many years is that the systems themselves have represented the most advanced technology we were capable of building at the time; naturally, we did not want to give away any of that advantage to our enemies. Over the years, the Soviets and more recently the Chinese and Europeans have developed systems of their own, but in all cases we have continued to benefit by having more sophisticated and capable systems. Most people including myself believe that there is no reason to change our policy on keeping the specific technology under wraps.

A second strong reason to keep the program under close control was that the Soviets were extremely upset that we were getting coverage and they could do nothing to prevent our doing so. We agreed that we would keep such activity out of the press to avoid confronting them with public announcement or even comment on the subject. As they developed the ability to operate similar satellites, their complaints about our coverage eased up. But we felt it to our advantage to continue to hold the "fact of" satellite coverage on a close-hold basis. Beginning about 20 years ago, we considered abandoning the policy several times but reasons were always found as to why it was easier to say nothing than to decide how much to say. The controversial nature of discussing such coverage has continued. Some presidents have chosen to say nothing about it, but both President Johnson and President Carter decided to make statements about the "fact of" our satellite capability—President Johnson to make the point that without it, we would "harbor fears we didn't need to harbor" about Soviet capabilities and President Carter to assure the public that we knew what was going on inside the Soviet Union. But whether a given president decides to make a statement about it, the rest of us have been constrained by the rules which say we don't discuss it.

In spite of the fact that we have always had what appeared to be good reasons to maintain the close-hold policy, I think the attempts we have made to deny even the existence of such activity have been counterproductive for many years. We get credit for either looking stupid about how widely known and discussed is the existence of such activity or we get credit for not caring about public opinion, our relations with the press, or even our relations with Congress. There are many people on the Hill who deeply resent our unwillingness to talk to them about what actually goes on, how much money goes to carry out such programs, etc. In fact, I have been told by several key people that the tendency which seemed to peak in former recent administrations to paint more and more programs black did a lot to sour our Congressional friends on their former support of the defense budget.

An interesting anomaly exists in that while we have continued to stonewall on any public discussion of our programs, the Soviets have actually begun offering for sale the results of their early military overhead photography. In fact, I was a member of the U.S. delegation which attended the Space Future Forum in Moscow in October 1987—the 30th anniversary of the first Sputnik—when the Soviets made their first such sale. Our group paid by personal check. I brought the photo home, and our AIAA magazine scooped the American press by publishing the first such photo to appear in the U.S. Its quality exceeded Landsat quality by a factor of 6, not surprising knowing the origin of their coverage. China has also gone commercial and is offering round-trip microgravity rides to space in what were originally return capsules for satellite photography.

The world has changed dramatically in the past few years. The Cold War has been transformed in ways we couldn't have imagined just a few years ago. That does not mean we should assume everything is sweetness and light from now on but it does mean we ought to review which policies still make sense versus which have been overtaken by events. One of the most important reasons why we avoided public discussion of our program was to avoid confrontation with the Soviets. While there are certainly international ramifications to a change in policy, I do not believe they justify maintaining the policy. In addition to what we do, there are the Chinese; also the Canadians soon will be flying Radarsat. The French system SPOT is known to be the prototype for a higher quality military system, which will be in operation shortly. Its products will be shared by the French and a number of their friends in Europe.

But in addition to these reasons, I believe we should change our policy on discussion of satellite photography for philosophical reasons. When we pretend that things are different than they are, we kid ourselves and our friends, and we even begin to believe our own stories about a make-believe world that we assume we can control. And as said earlier, we risk alienation of our friends on the Hill whom we ask to appropriate money for programs that don't exist or on which we refuse to brief them. It is hard to measure such loss of faith in dollar terms, but it is known to cause erosion of support, as I have been told recently by people closely involved in seeking such support. As usual, we need all the help we can get—especially on programs which some people may think are no longer as high priority as they once were.

There is one final reason why I hope we can find a more open policy with respect to satellite photography. This reason has nothing to do with whether the government should change its policy but it does have to do with finding sponsors for PBS programs. No corporation with defense contracts is going to sponsor a program if there is any question that government is not happy about that program being put on the air. As long as the suspicion exists that this PBS series is somehow tainted by security violations or even skating on thin ice in that regard, we will not be able to find sponsors. While not a factor in your decision, it does affect our ability to get public support for the nation's space program. We expect our series on PBS to be very positive in that regard. As we all know, there has been an erosion of support for space, partly because of overall budget problems but also because of NASA's problems with Shuttle, etc. Successes in either the civil program or the military program can be helpful in countering the image of expensive and unproductive systems.

For roughly 20 years we have found ways to discuss "national technical means" for verification of arms control agreements. I hope that you can take action soon to review the present policy on preventing discussion of overhead photography—not to open the floodgates to reveal the true secrets of the program but to permit discussions which are normal to other classified programs.

Sincerely,

/s/ John L. McLucas

cc: General Brent Scowcroft
DCI

Abbreviations

ACCFUS	Arthur C. Clarke Foundation of the United States
AC&W	air control and warning
AEW	airborne early warning
AFB	Air Force base
AFOTEC	Air Force Operational Test & Evaluation Center
AF/RD	deputy chief of staff for research and development (Air Force)
AFSATCOM	Air Force Satellite Communications System
AFSC	Air Force Systems Command
AGARD	Advisory Group for Aeronautical Research and Development (NATO)
AGM	air-to-ground missile
AIAA	American Institute of Aeronautics and Astronautics
AIM	air intercept missile
ALPA	Air Line Pilot's Association
AMSA	Advanced Manned Strategic Aircraft (evolved into B-1 bomber)
AMST	Advanced Medium Range Short Takeoff and Landing (STOL) Transport
ARPA	Advanced Research Projects Agency (DOD)
ASD	assistant secretary of defense
ATA	Air Transport Association
ATIC	Air Technical Intelligence Center (USAF)
AWACS	Airborne Warning and Control System (E-3 Sentry)
AW&ST	*Aviation Week and Space Technology*
BDA	bomb damage assessment
C^2	command and control
C^3I	command, control, communications, and intelligence
CAS	close air support
CCD	charge-coupled device
CDTC	Combat Development and Test Center (DOD)
CECO	Community Engineering Corporation (later C-COR)
CEO	chief executive officer

CIA	Central Intelligence Agency
CINCPAC	commander in chief Pacific Command
CITI	Clarke Institute for Telecommunications and Information
CNO	chief of naval operations
CoCom	Coordinating Committee for Multilateral Export Controls
COIN	counterinsurgency
COMINT	communications intelligence
COMIREX	Committee on Imagery Requirements and Exploitation
COSEPUP	Committee on Science, Engineering, and Public Policy
COSMIC	Constellation System for Meteorology, Ionosphere, and Climate
COSPAR	Committee on Space Research
CSAF	chief of staff, Air Force
CSC	Civil Service Commission
CX	heavy cargo aircraft concept that led to C-5
DAB	Defense Acquisition Board (replaced DSARC)
DACOWITS	Defense Advisory Committee on Women in the Service
DACT	dissimilar air combat tactics
DARPA	Defense Advanced Research Projects Agency
DART	Dulles Area Rapid Transit
DCA	Defense Communications Agency
DCI	director of central intelligence
DCL	Digital Computer Laboratory (MIT)
DCPG	Defense Communications Planning Group
DCS	deputy chief of staff
DDR&E	director of defense research and engineering (OSD)
DEW	Distant Early Warning (Line)
DIA	Defense Intelligence Agency
DMSP	Defense Meteorological Satellite Program
DMZ	demilitarized zone
DOD	Department of Defense
DOT	Department of Transportation
DRDC	Defense Research Directors' Committee (NATO)
DSARC	Defense System Acquisition Review Council
DSB	Defense Science Board

DSCS	Defense Satellite Communications System
DSP	Defense Support Program
DT&E	developmental test and evaluation
ECM	electronic countermeasures
ELINT	electronic intelligence
ELV	expendable launch vehicle
Envirosat	Environmental Resources Satellite
EO	electro-optical
EoSat	Earth Observation Satellite Company
EOT	equal opportunity and treatment
ESD	Electronics Systems Division (USAF)
ESL	Electronic Systems Laboratory (company)
EUCOM	European Command (short for USEUCOM)
ExCom	Executive Committee (National Reconnaissance Program)
FAA	Federal Aviation Administration (originally Agency)
FCC	Federal Communications Commission
FCRC	federal contract research center
FFRDC	federally funded research and development center
FLTSATCOM	Fleet Satellite Communications System
FY	fiscal year
FYDP	five-year defense plan (later future years defense plan)
GCA	ground controlled approach
GD	General Dynamics
GNSS	Global Navigation Satellite System
GPO	Government Printing Office
GPS	global positioning system
H-P	Hewlett-Packard
HRB	Haller, Raymond, and Brown
IAF	International Astronautical Federation
IDA	Institute for Defense Analyses
IEEE	Institute of Electrical and Electronics Engineers
ILS	instrument landing system
Inmarsat	International Maritime Satellite Consortium
Intelsat	International Telecommunications Satellite Consortium
IONDS	Integrated Operational Nuclear Detonation Detection System

IOT&E	initial operational test and evaluation
IR	infrared
IRBM	intermediate-range ballistic missile
ISS	International Space Station
ISU	International Space University
ISY	International Space Year (1992)
ITU	International Telecommunications Union
JCS	Joint Chiefs of Staff
KGB	Committee on State Security (translated)
LARA	Light Armed Reconnaissance Aircraft
LAV	Light Attack Aircraft (Navy)
LGB	laser-guided bomb
LORAN	Long-Range Navigation (System)
LTV	Ling Temco Vought
LWF	Lightweight Fighter (YF-16 and YF-17)
MACV	Military Assistance Command Vietnam
MEW	microwave early warning (radar)
MHz	megahertz
MIDAS	Missile Defense Alarm System
MIRV	multiple independently targeted reentry vehicle
MIT	Massachusetts Institute of Technology
MITRE	MIT Research and Engineering
MMRBM	Mobile Medium-Range Ballistic Missile
MOL	Manned Orbiting Laboratory
MTPE	Mission to Planet Earth
MX	Missile Experimental (later Peacekeeper)
NADGE	NATO Air Defense Ground Environment
NAE	National Academy of Engineering
NAS	National Academy of Sciences
NASA	National Aeronautics and Space Administration
NATO	North Atlantic Treaty Organization
NDRC	National Defense Research Council
NGA	National Geospatial-Intelligence Agency (formerly NIMA)
NIMA	National Imagery and Mapping Agency
NKP	Nakhon Phanom (Royal Thai AFB)
NMCC	National Military Command Center
NOAA	National Oceanic and Atmospheric Administration
NORAD	North American Air (later Aerospace) Defense Command

NPIC	National Photographic Interpretation Center (CIA)
NRO	National Reconnaissance Office
NRP	National Reconnaissance Program
NSA	National Security Agency
OD&E	Office of Development and Engineering (CIA)
OMB	Office of Management and Budget
ONR	Office of Naval Research
OPEC	Organization of Petroleum Exporting Countries
ORT	Overland Radar Technology
OSAF	Office of the Secretary of the Air Force
OSC	Orbital Sciences Corporation
OSD	Office of the Secretary of Defense
OSP	Office of Special Projects (CIA)
OST	Office of the Secretary of Transportation
OTA	Office of Technology Assessment (US Congress)
OT&E	operational test and evaluation
OTH-B	over-the-horizon backscatter (radar)
PACAF	Pacific Air Forces
PACOM	Pacific Command
PAL	permissive action link
PATCO	Professional Air Traffic Controllers Organization
PAWS	phased array warning system
PFIAB	President's Foreign Intelligence Advisory Board
PGM	precision-guided munitions
POM	program objective memorandum
PPBS	Planning, Programming, and Budgeting System
PR	public relations
PSAC	President's Scientific Advisory Committee
RAAF	Royal Australian Air Force
RAF	Royal Air Force
Rafax	Radar Facsimile (trademarked HRB image processing system)
RAND	Corporate name (initially Research and Development)
R&D	research and development
RDT&E	research, development, test, and evaluation
Reconofax	trademarked HRB reconnaissance video scanning and transmission system
RFP	request for proposal
ROTC	Reserve Officer Training Corps

RPV	remotely piloted vehicle
SAC	Strategic Air Command (USAF)
SAF/AQ	assistant secretary of the Air Force for acquisition
SAFISY	Space Agencies Forum for the International Space Year
SAF/RD	assistant secretary of the Air Force for research and development
SAFSP	Secretary of the Air Force Office of Special Projects (NRO Program A)
SAGE	Semi-Automatic Ground Environment (System)
SALT	Strategic Arms Limitation Talks/Treaty
SAMSO	Space and Missile Systems Organization (Air Force)
SBS	Satellite Business Systems
SCAD	Subsonic Cruise Armed Decoy
SCTV	State College Television
SDS	Satellite Data System
SEA	Southeast Asia
SecAF	secretary of the Air Force
SecDef	secretary of Defense
SEDS	Students for the Exploration and Development of Space
SIGINT	signals intelligence
SIOP	Single Integrated Operational Plan
SLBM	submarine-launched ballistic missile
SPADATS	Space Detection and Tracking System
SPO	system program office
SPOT	Systéme Pour l'Observation de la Terre
SRAM	Short Range Attack Missile
S&T	science and technology
STC	Satellite Television Corporation
STS	Space Transportation System (shuttle)
TACS	Tactical Air Control System (407-L)
TDY	temporary duty
T&E	test and evaluation
TFX	Tactical Fighter Experimental (later F-111)
TISEO	Target Identification System Electro Optical
TOA	total obligating authority
TRW	Thompson-Ramo-Wooldridge (corporate name)

TSARC	Transportation Systems Acquisition Review Council (DOT)
TWP	Tactical Warfare Programs office (OSD)
UAV	unmanned (later uninhabited) aerial vehicle
UCAR	University Corporation for Atmospheric Research
UHF	ultrahigh frequency
USAF	United States Air Force
USAFE	United States Air Forces in Europe
USEUCOM	United States European Command
USS	United States ship
VHF	very high frequency
V/STOL	vertical and/or short take off and landing
WAF	Women in the Air Force
WCY	World Communications Year (1983)
WWMCCS	World-Wide Military Command and Control System
YPO	Young Presidents' Organization

THIS PAGE INTENTIONALLY LEFT BLANK

Selected Bibliography

Books, Monographs, and Published Reports

Anderegg, C. R. *Sierra Hotel: Flying Air Force Fighters in the Decade after Vietnam.* Washington, D.C.: Air Force History and Museums Program, 2001.

Andradé, Dale. *America's Last Vietnam Battle: Halting Hanoi's 1972 Easter Offensive.* Lawrence: University Press of Kansas, 2001.

Aronstein, David C., and Albert C. Piccrillo, *The Lightweight Fighter Program: A Successful Approach to Fighter Technology Transition.* Reston, Virginia: American Institute of Aeronautics and Astronautics (AIAA), 1996.

Art, Robert J. *The TFX Decision: McNamara and the Military.* Boston: Little, Brown, & Co., 1968.

Augustine, Norman (chairman). *Report of the Advisory Committee on the Future of the U.S. Space Program.* Washington, D.C.: US Government Printing Office (GPO), 1990.

Bamford, James. *Body of Secrets: Anatomy of the Ultra-Secret National Security Agency.* New York: Doubleday, 2001.

Beggs, James (chairman). *The International Space Station Commercialization Study.* Arlington, Virginia: Potomac Institute for Policy Studies, March 1997.

Benson, Lawrence R. *Acquisition Management in the United States Air Force and Its Predecessors.* Washington, D.C.: Air Force History and Museums Program, 1996.

———. *History of Air Force Operational Test and Evaluation.* Kirtland AFB, New Mexico: Air Force Operational Test & Evaluation Center, 1992.

———. *The USAF Special Treatment Center and Rehabilitation of Drug Abusers, 1971–1974.* Lackland AFB, Texas: AF Military Training Center, 1975.

Berger, Carl, ed. *The United States Air Force in Southeast Asia, 1961–1973: An Illustrated Account.* Rev. ed. Washington, D.C.: Office of Air Force History, 1984.

Bilstein, Roger E. *Orders of Magnitude: A History of the NACA and NASA, 1915–1990.* Washington, D.C.: NASA, 1989.

331

Bundy, William. *A Tangled Web: The Making of Foreign Policy in the Nixon Presidency.* New York: Hill and Wang, 1998.

Burrows, William E. *Deep Black: Space Espionage and National Security.* New York: Random House, 1986.

———. *This New Ocean: The Story of the First Space Age.* New York: Random House, 1998.

Cole, Ronald H. et al. *The History of the Unified Command Plan, 1946-1993.* Washington, D.C.: Joint Chiefs of Staff, Joint History Office, 1995.

Columbia Accident Investigation Board. *Report, Vol I.* Washington, D.C.: GPO, August 2003.

Commission to Assess United States National Security Space Management and Organization. *Report Pursuant to Public Law 106–65*, 11 January 2001.

Comptroller of the Air Force. *USAF Statistical Digests, Fiscal Years 1968–1976.* Washington, D.C.: Comptroller of the Air Force, 1969–1977.

Crickmore, Alison J., and Paul F. Crickmore. *F-117 Nighthawk.* Osceola, Wisconsin: MBI Publishing Co., 1999.

Davis, Benjamin O. *Benjamin O. Davis, Jr., American: An Autobiography.* Washington, D.C.: Smithsonian, 1991.

Day, Dwayne A. *Lightning Rod: A History of the Air Force Chief Scientist's Office.* Washington, D.C.: Headquarters, USAF, 2000.

Day, Dwayne A., John Logsdon, and Brian Latell, eds. *Eye in the Sky: The Story of the Corona Spy Satellites.* Washington, D.C.: Smithsonian, 1998.

Dorsey, Gary. *Silicon Sky: How One Small Startup Went over the Top to Beat the Big Boys into Satellite Heaven.* Reading, Massachusetts: Perseus Books, 1999.

Dyer, Davis, and Michael Aaron Dennis. *Architects of Information Advantage: The MITRE Corporation since 1958.* Montgomery, Alabama: Community Communications, 1998.

Ehrlichman, John. *Witness to Power: The Nixon Years.* New York: Simon and Schuster, 1982.

Fitzgerald, A. Ernest. *The High Priests of Waste.* New York: Norton, 1972.

———. *The Pentagonists: An Insider's View of Waste, Mismanagement, and Fraud in Defense Spending.* Boston: Houghton Mifflin, 1989.

Ford, Gerald R. *A Time to Heal.* New York: Harper & Row, 1979.

Francillon, René J. *Vietnam: The War in the Air.* New York: Crown Publishers, 1987.

Gaffney, Timothy R. *Secret Spy Satellites: America's Eyes in Space.* Berkley Heights, New Jersey: Enslow Publishers, 2000.

Gardner, Lloyd C. *Lyndon Johnson and the Wars for Vietnam.* Chicago: Ivan R. Dee, 1995.

Geiger, Clarence J. *History of the F-16: Prototype to Air Combat Fighter, 1971–1975.* Wright-Patterson AFB, Ohio: Aeronautical Systems Division, n.d.

Getting, Ivan A. *All in a Lifetime: Science in the Defense of Democracy.* New York: Vantage Press, 1989.

Gorn, Michael H. *The Universal Man: Theodore von Kármán's Life in Aeronautics.* Washington, D.C.: Smithsonian, 1992.

Haines, Gerald K. *The National Reconnaissance Office: Its Origins, Creation, and Early Years.* Chantilly, Virginia: NRO, ca. 1997.

Halberstam, David. *The Best and the Brightest.* New York: Random House, 1972.

Haldeman, H. R. *The Haldeman Diaries: Inside the Nixon White House.* New York: Putnam's, 1994.

Hall, R. Cargill. *A History of the Military Polar Orbiting Meteorological Satellite Program.* Chantilly, Virginia: NRO, September 2001.

———. *Missile Defense Alarm: The Genesis of Space-Based Infrared Early Warning.* July 1988. Reprint, Chantilly, Virginia: NRO CD-ROM, 2001.

Hall, R. Cargill, and Jacob Neufeld, eds. *The U.S. Air Force in Space: 1945 to the 21st Century.* Proceedings of Air Force Historical Foundation Symposium, Andrews AFB, Maryland, 21–22 September 1995. Washington, D.C.: Air Force History and Museums Program, 1998.

Hoffman, Richard A. *The Fighting Flying Boat: A History of the Martin PBM Mariner.* Annapolis: Naval Institute Press, 2004.

Holm, Jeanne. *Women in the Military: An Unfinished Revolution.* Rev. ed. Novato, California: Presidio, 1992.

Hoopes, Townsend. *The Limits of Intervention: An Inside Account of How the Johnson Policy of Escalation in Vietnam Was Reversed.* Rev. ed. with epilogue. New York: D. McKay, 1973.

Jacobs, John F. *The Sage Air Defense System: A Personal History.* Bedford, Massachusetts: The MITRE Corporation, 1986.

Johnson, Stephen B. *The United States Air Force and the Culture of Innovation.* Washington, D.C.: Air Force History and Museums Program, 2002.

Keller, Edward R. *The History of HRB: 50 Years of Excellence.* State College, Pennsylvania: HRB-Singer, 1997.

Killian, James R. *The Education of a College President: A Memoir.* Cambridge, Massachusetts: Massachusetts Institute of Technology, 1985.

Kimball, Jeffey P. *Nixon's Vietnam War.* Lawrence: University Press of Kansas, 1998.

Kinnard, Douglas. *The Secretary of Defense.* Lexington: University of Kentucky, 1981.

Kissinger, Henry. *Ending the Vietnam War: A History of America's Involvement in and Extrication from the Vietnam War.* New York: Simon & Schuster, 2003.

Knaack, Marcelle Size. *Post World War II Fighters, 1945–1973.* Washington, D.C.: Office of Air Force History, 1975.

Komons, Nick A. *The Third Man: A History of the Airline Crew Complement Controversy, 1947–1981.* Washington, D.C.: GPO, 1987.

Laurie, Clayton D. *Congress and the National Reconnaissance Office.* Chantilly, Virginia: NRO History Office, 2001.

Lewis, Jonathan A. *Spy Capitalism: ITEK and the CIA.* New Haven: Yale, 2002.

Lindsey, Robert. *The Falcon and the Snowman: A True Story of Friendship and Espionage.* Rev. ed. New York: Simon & Schuster, 1980.

Lonquest, John C., and David F. Winkler. *To Defend and Deter: The Legacy of the United States Cold War Missile Program.* DOD Legacy Project. Rock Island, Illinois: Defense Publishing Service, 1996.

McAleer, Neil. *Arthur C. Clarke: The Authorized Biography.* Chicago: Contemporary Books, 1992.

McGovern, James R. *Black Eagle: General Daniel 'Chappie' James, Jr.* Tuscaloosa: University of Alabama Press, 1985.

McLucas, John L. *Space Commerce.* Cambridge, Mass.: Harvard, 1991.

McLucas, John L., Fred Drinkwater III, and Lt Gen Howard W. Leaf, USAF. *Report of the President's Task Force on Aircraft Crew Complement.* Washington, D.C.: GPO, 2 July 1981.

McLucas, John L., and James Harford. "Visit to Satellite and Launch Vehicle and Supporting Technology Facilities in Beijing, Xian, and Shanghai, China, 5–19 May 1985." AIAA Report, New York: AIAA, 21 June 1985.

McLucas, John L., and Norman Getsinger. Case Study, *Satellite Telecommunications Technology Transfer to China.* Alexandria, Virginia: China Business Development Group, July 1986.

———. "Visit to the Soviet Union as Guests of the Soviet Academy of Sciences." New York: AIAA Report, 15 July 1984.

McMaster, H. R. *Dereliction of Duty: Lyndon Johnson, Robert McNamra, the Joint Chiefs of Staff, and the Lies That Led to Vietnam.* New York: HarperCollins, 1997.

McNamara, Robert S., with Brian Van DeMark. *In Retrospect: The Tragedy and Lessons of Vietnam.* New York: Random House, 1995.

Mets, David R. *NATO: Alliance for Peace.* New York: J. Messner, 1981.

Meyerson, Harvey, and Danelle K. Simonelli. *Launchpad for the 21st Century: Yearbook of the International Space Year.* San Diego: Univelt Inc. (for the American Astronautical Society), 1995.

MITRE Corp. *MITRE: The First Twenty Years: A History of the MITRE Corporation (1958–1978).* Bedford, Massachusetts: The MITRE Corporation, 1979.

Moore, Lt Gen Harold, and Joseph L. Galloway. *We Were Soldiers Once . . . and Young.* New York: Random House, 1992.

Morison, Samuel Eliot. *History of United States Naval Operations in World War II.* Boston: Little, Brown, 1962.

Nalty, Bernard C. *Air War over South Vietnam, 1968–1975.* Washington, D.C.: AF History and Museums Program, 2000.

———, ed. *Winged Shield, Winged Sword: A History of the United States Air Force.* Vol. 2, *1950–1997.* Washington, D.C.: AF History and Museums Program, 1997.

National Commission for the Review of the National Reconnaissance Office. *The NRO at the Crossroads.* Washington, D.C.: 15 November 2000.

Naval Historical Center. *Dictionary of American Naval Fighting Ships.* Vol. 6. Washington, D.C.: GPO, 1976.

Neufeld, Jacob. *Ballistic Missiles in the United States Air Force, 1945–1960.* Washington, D.C.: AF Office of History, 1989.

Neufeld, Jacob, George M. Watson, and David Chenoweth, eds. *Technology and the Air Force: A Retrospective Assessment.* Washington, D.C.: Air Force History and Museums Program, 1997.

Nitschle, Diane L., ed. *The Sixth National Space Symposium.* Colorado Springs: Space Foundation, 1990.

Nixon, Richard. *RN: The Memoirs of Richard Nixon.* New York: Grosset and Dunlap, 1978.

North Atlantic Treaty Organization. *NATO and Science: Facts about the Activities of the Science Committee of the North Atlantic Treaty Organization, 1959–1966.* Paris: NATO Scientific Affairs Committee, 1967.

———. *The NATO Handbook.* 16th ed. Paris: NATO Information Office, 1963.

Oder, Frederic C. E., James Fitzpatrick, and Paul Worthman. *The Corona Story.* 1987. Reprint, Chantilly, Virginia: NRO, CD-ROM, 2001.

Office of the Secretary of Defense, Historical Office. *Department of Defense Key Officials, 1947–2000.* Washington, D.C.: OSD Historical Office, 2000.

Panel on the Impact of National Security Controls on International Technology Transfer, Committee on Science, Engineering, and Public Policy. *Balancing the National Interest: U.S. National Security Export Controls and Global Economic Competition* ("Allen Report"). Washington, D.C.: National Academy Press, 1987.

Panel on Scientific Communication and National Policy, Committee on Science, Engineering, and Public Policy. *Scientific Communications and National Security* ("Corson Report"). Washington, D.C.: National Academy Press, 1982.

Parkinson, Bradford, and James Spilker Jr., eds. *Global Positioning System: Theory and Applications.* Vol. 1. Washington, D.C.: AIAA, 1996.

Pedlow, Gregory W., and Donald E. Welzenbach. *The CIA and the U-2 Program.* Langley, Virginia: CIA Center for the Study of Intelligence, 1998.

Peebles, Curtis. *The Corona Project: America's First Spy Satellites.* Annapolis: US Naval Institute, 1997.

———. *Dark Eagles: A History of Top Secret Aircraft.* Novato, California: Presidio, 1995.

———. *High Frontier: The United States Air Force and the Military Space Program.* Washington, D.C.: Air Force History and Museums Program, 1997.

Pelton, Joseph N. *Global Communications Satellite Policy, Intelsat, Politics and Functionalism.* Mt. Airy, Maryland: Lomond Systems, 1974.

Pelton, Joseph N. and Joel Alper, eds. *The Intelsat Global Satellite.* New York: AIAA, 1984.

Perkins, Courtland D. *Recollections.* Vol 11. Privately published, n.d.

Perry, Robert L. *Management of the National Reconnaissance Program, 1960–1965.* 1969. Rev. ed. Edited by R. Cargill Hall. Chantilly, Virginia: NRO History Office, 2001.

Phelps, J. Alfred. *Chappie: The Life and Times of Daniel James Jr.* Novato, California: Presidio, 1991.

Preston, Edmund. *FAA Historical Chronology: Civil Aviation and the Federal Government, 1926–1996.* Washington, D.C.: GPO, 1998.

———. *Troubled Passage: The Federal Aviation Administration during the Nixon-Ford Term, 1973–1977.* Washington, D.C.: GPO, 1987.

Puryear, Edgar F. *George S. Brown, General, U.S. Air Force: Destined for Stars.* Novato, California: Presidio, 1983.

Reed, Thomas C. *At the Abyss: An Insider's History of the Cold War.* New York: Ballantine Books, 2004.

Rich, Ben R., and Leo Janos. *Skunk Works: A Personal Memoir of My Years at Lockheed.* Boston: Little, Brown, 1994.

Richelson, Jeffrey T. *America's Secret Eyes in Space: The U.S. Keyhole Spy Satellite Program.* New York: Harper Business, 1990.

———. *America's Space Sentinels: DSP Satellites and National Security.* Lawrence: University Press of Kansas, 1999.

———. *The US Intelligence Community.* Boulder, Colorado: Westview, 1999.

———. *The Wizards of Langley: Inside the CIA's Directorate of Science and Technology.* Boulder, Colorado: Westview, 2001.

Schandler, Herbert Y. *The Unmaking of a President: Lyndon Johnson and Vietnam.* Princeton University Press, New Jersey: Princeton University, 1977.

Schlight, John. *A War Too Long: The History of the USAF in Southeast Asia, 1961–1975.* Washington, D.C.: Air Force History and Museums Program, 1996.

Schneider, James G. *The Navy V-12 Program: Leadership for a Lifetime.* Boston: Houghton Mifflin, 1987.

Seamans, Robert C., Jr. *Aiming at Targets.* Washington, D.C.: NASA, 1996.

Space Awareness Initiative. *It Came from Space: A Down-to-Earth Inventory of Dividends Produced by America's Space Endeavors.* Washington, D.C.: Space Foundation, n.d.

Spires, David N. *Beyond Horizons: A Half Century of Air Force Space Leadership.* Rev. ed. Maxwell AFB, Alabama: Air University Press, 1998.

———. *Orbital Futures: Selected Documents in Air Force Space History.* 2 vols. Peterson AFB, Colo.: Air Force Space Command, 2004.

Stever, Guy. *In War and Peace: My Life in Science and Technology.* Washington, D.C.: John Henry Press, 2002.

Strober, Gerald S., and Deborah H. Strober. *Nixon: An Oral History of his Presidency.* New York: HarperCollins, 1994.

Sweetman, Bill, et al. *The Great Book of Modern Warplanes.* New York: Portland House (for Salamander Books Ltd.), 1987.

Taubman, Philip. *Secret Empire: Eisenhower, the CIA, and the Hidden Story of America's Space Espionage.* New York: Simon & Schuster, 2003.

Tedeshi, Anthony M. *Live via Satellite: The Story of COMSAT and the Technology That Changed World Communications.* Washington, D.C.: Acropolis Books, 1989.

Thompson, Wayne. *To Hanoi and Back: The U.S. Air Force and North Vietnam, 1966–1973.* Washington, D.C.: Air Force History and Museums Program, 2000.

Trask, Roger R. *The Secretaries of Defense: A Brief History, 1947–1985.* Washington, D.C.: Office of the Secretary of Defense Historical Office, 1985.

Trask, Roger R., and Alfred Goldberg. *The Department of Defense, 1947–1997: Organization and Leaders.* Washington, D.C.: OSD Historical Office, 1997.

United Nations Association of the United States of America. *The Next Giant Leap in Space: An Agenda for International Cooperation.* Final Report of the 1986 UNA-USA Multilateral Project. New York, October 1986.

Van der Bliek, Jan, ed. *AGARD: The History, 1952–1997.* Paris: NATO Research and Technology Office, 1999.

Wagner, Ray. *American Combat Planes.* 3d enl. ed. Garden City, New York: Doubleday, 1982.

Wagner, William. *Lightning Bugs and other Reconnaissance Drones.* Fallbrook, California: Aero Publishers, 1982.

Watson, George. *The Office of the Secretary of the Air Force, 1947–1965.* Washington, D.C.: Center for Air Force History, 1993.

———. *Secretaries and Chiefs of Staff of the United States Air Force.* Washington, D.C.: Air Force History and Museums Program, 2001.

Weinert, Richard P., Jr. *A History of Army Aviation – 1950–1962.* Historical Monograph Series. Hampton, Virginia: Training and Doctrine Command [US Army], 1971.

Zumwalt, Adm Elmo. *On Watch.* New York: Quadrangle, 1976.

Articles and Essays

"Air Force Declassifies Weather Satellite Data for NOAA." *Aerospace Daily,* 7 March 1973.

Asker, James. "U.S. Declassifies More Cold War Recce Satellite Imagery." *Aviation Week and Space Technology,* 4 November 2002, 68–70.

"Astrotech Curtails Space Unit Operations." *Aviation Week and Space Technology,* 3 November 1986, 44.

"Astrotech Reorganizes Space-Related Business." *Aviation Week and Space Technology,* 25 November 1985, 18.

"ATA Task Force Opposes Screening All Airline Bags." *Aviation Week and Space Technology,* 15 March 1976, 205.

"Avion Systems, Inc." *Aviation Week and Space Technology,* 20 June 1988, 117.

Barnes, Bart. "John McLucas Dies; Oversaw Air Force, FAA." *Washington Post,* 5 December 2002, B6.

Butash, Tom. "Communications." *Aerospace America,* December 2002, 46.

Callander, Bruce. "New Poll Hits 'Military Life.'" *Air Force Times*, 17 December 1969, 13.

Clark, Evert. "Satellite Spying Cited by Johnson." *New York Times*, 17 March 1967, 13.

"Comsat Withdraws from Sparx Venture." *Aviation Week and Space Technology*, 5 March 1984, 16.

Correll, John T. "The All Volunteer Force – Possible or Probable." *Air Force Magazine*, November 1971, 70–75.

———. "A Fair Share for USAF Minorities." *Air Force Magazine*, June 1972, 25–29.

Covault, Craig. "Major Space Effort Mobilized to Blunt Environmental Threat." *Aviation Week and Space Technology*, 13 March 1989, 34–44.

"DoD Support of WDGPS 'Critical' to Civil Use, McLucas Says." *Aviation Week and Space Technology*, 7 February 1994, 204.

Dornheim, Michael A. "Affordable Spaceship: Burt Rutan's Quest for Space." *Aviation Week and Space Technology*, 21 April 2003, 64–73.

———. "SpaceShipOne Flies." *Aviation Week and Space Technology*, 26 May 2003, 74.

———. "SpaceShipOne Solos." *Aviation Week and Space Technology*, 18 August 2003, 32–33.

Edelson, Burton I., and John L. McLucas. "US and Soviet Planetary Exploration." *Space Policy*, November 1988, 337–49.

"Former CIA Officer Arrested in Secret Satellite Manual Sale: The Case of William P. Kampiles." *Aviation Week and Space Technology*, 28 August 1978, 22–23.

"Former FAA Head Urges More Research on Displays, Controls." *Aviation Week and Space Technology*, 25 April 1977, 100.

Friedman, Nelson. "New Headquarters Approved: NOFORN Policy Revised . . . ." *AIAA Bulletin*, January 1986, B6.

Frisbee, John L. "The Chief Discusses USAF's Prospects." *Air Force Magazine*, February 1974, 32–35.

Fulgrum, David. "Targets Become UAVs." *Aviation Week and Space Technology*, 28 July 2003, 54–55.

Gates, Ed. "Turning the Spotlight on Alcohol Abuse." *Air Force Magazine*, June 1975, 77.

Geddes, William G. "The 'Live via Satellite' Era." In *Intelsat Memoirs*. Washington, D.C.: International Telecommunications Satellite Organization, 1979.

Harford, James. "Soviets Renew Calls for Space Cooperation." *Aerospace America*, August 1984, 22–26.

Hebert, F. Edward. "Secretary McLucas Looks at the Man in Uniform Today." *Congressional Record*, 20 May 1975, E2517–18.

"Heller Navigates Clear Path in Licensing Feud: Avion Systems v. Trimble Navigation." *The Recorder* (American Lawyer Media), 24 August 1994, 2.

Hessman, James, et al. "Federal Contract Research Centers: DoD's Cerebral Reserve." *Armed Forces Journal*, 28 September 1968.

"How to Liven Up the Farewell." *Air Force Times*, 26 November 1975, 12.

Hsichun, Mike Hua. "The Black Cat Squadron." *Air Power History*, Spring 2002, 4–19.

"The International Space University." *Aerospace America*, March–April 1989, 7–11.

Klass, Philip J. "Bendix, BF Goodrich, Trimble Vie for TCAS-1 Business." *Aviation Week and Space Technology*, 11 January 1993, 45–47.

Laird, Melvin R. "Iraq: Learning the Lessons of Vietnam." *Foreign Affairs*, 84, no. 6 (November–December 2005): 21–43.

"Late Thoughts by McLucas." *Air Force Times*, 10 December 1975, 9, 24.

Lerner, Eric J. "DoD Information Curbs Spread Fear and Confusion." *Aerospace America*, 76–80.

Martin, Brig Gen John L., Jr. "Specialized Incentive Structure for Satellite Project Contracts." *Aerospace Management*, General Electric Company, Winter 1967, 47–53.

Marzullo, Ellen. "McLucas Hosts Section Luncheon." *AIAA Bulletin*, February 1985, B7.

"McLucas Endorses Data Bill." *Washington Post*, 22 November 1975, A5.

"McLucas Firmly against Force Cuts." *Air Force Times*, 12 March 1975, 5.

McLucas, John L. (chronological listing)

———. "My Two Years in the Pentagon." *Short Circuit* (HRB-Singer), July 1964, 7–10.

———. "The New Look in R&D Management." Part 1. *Armed Forces Management*, December 1969, Part 2, January 1970, 12–15.

———. "Domestic Action—A New Challenge for the Air Force." *Air Force Magazine and Space Digest*, February 1970, 54–57.

———. "An Outline of Air Force Views on RPV Potentials." *Teledyne Ryan Aeronautical Reporter*, Summer 1972, 9–13.

———. "A New Look from USAF's Weather Satellites." *Air Force Magazine*, June 1973, 64–67.

———. "Achieving an Effective Air Force in Peacetime." *Air Force Policy Letter for Commanders*, December 1973, 10.

———. "USAF's Role in Space." *Aviation Week and Space Technology*, 18 February 1974, 20–22.

———. "The Role of RPVs in the Air Force." *Commanders Digest* (DOD), 16 January 1975, 12–16.

———. "USAF's Increasing Operational Efficiency." *Air Force Magazine*, May 1975, 48–50.

———. "Requiem for a Bureaucrat." *Aviation Week and Space Technology*, 25 April 1977, 23.

———. "Technology Transfer Scare Hurts Innovation. *Aerospace America*, June 1984, 6.

———. "Goals and Objectives." *AIAA Bulletin*, June 1984, B46.

———. "At the least, some ELVs." *Aerospace America*, October 1984, 4.

———. "Aerospace America and Controversial Subjects." *AIAA Bulletin*, October 1984, B32.

———. "Whither Landsat?" *Aerospace America*, January 1985, 6.

———. "Could Space Thaw the Freeze?" *AIAA Bulletin*, January 1985, B32.

———. "Open Skies a Fresh Challenge." *Aerospace America*, April 1985, 6.

———. "Space: Sanctuary or Menace." *Aerospace America*, May 1985, 6.

———. "Create a Global Landsat." *Aerospace America*, July 1987, 5.

———. "Perspective on Sputnik." *Washington Technology*, 9 January 1988, 4.

———. "Mission to Planet Earth: The Wernher von Braun Memorial Lecture for 1989." Pasadena, California: Planetary Society, 1989.

———. "A Plea for Global Cooperation: Remote Sensing for International Security." *Via Satellite*, January 1990, 20–23.

————. "The Opportunity in Soviet Space." *Washington Technology,* ca. 15 September 1991, 15.

————. "Technology and Development." In *Future Vision: A Tribute to Arthur C. Clarke and to Space Communications in the 21st Century.* Edited by Joseph N. Pelton. Boulder, Colorado: Johnson Publishing Company, 1992.

McLucas, John L. and Burton I. Edelson. "Let's Go to Mars Together," *Issues in Science and Technology,* Fall 1988, 52–55.

McLucas, John L. and P. M. Maughan. "The Case for Envirosat: A Space Remote Sensing Initiative in Keeping with the International Space Year." *Space Policy,* August 1988, 230–39.

Mintz, John, and Don Phillips. "Military Ordered to Update Air Safety Equipment on All Passenger Planes." *Washington Post,* 17 April 1996, A7.

Mooring, Frank, Jr. "Instelsat Sale." *Aviation Week and Space Technology,* 23/30 August 2004, 39.

Noguchi, Yuki. "Intelsat Plans to Hold Long-Delayed Stock Sale before July." *Washington Post,* 4 February 2004.

————. "Intelsat Renegotiates Terms of Its $5 Billion Sale to Zeus." *Washington Post,* 19 January 2005, E5.

"Orbital Sciences." *Aerospace America,* October 1992, 8–10.

Regan, M. B. "McLucas Is Appointed Temporary President and CEO of Questech, Inc." *Washington Post,* 24 November 1986, F8.

————. "Retired Comsat General Chief Is New CEO at Questech Inc." *Washington Post,* 15 June 1987, F10.

"Research Council Appoints Space Station Review Panel." *Aviation Week and Space Technology,* 8 June 1987, 22.

Richelson, Jeffrey. "Out of the Black: The Disclosure and Declassification of the National Reconnaissance Office." *International Journal of Intelligence and CounterIntelligence,* Spring 1998, 1–25.

Seamans, Robert C., Jr. "How USAF Plans to Meet Its Personnel Needs." *Air Force Magazine,* March 1973, 45–50.

"Soviets Refine Space Cooperation Overtures." *Aerospace America,* November 1987, 6–9.

"Speed of Sound." *Aviation Week and Space Technology,* 7 June 1976, 13.

Stern, Laurence. "A $1.5 Billion Secret in the Sky." *Washington Post,* 9 December 1973, A1, 9.

343

Sugawara, Sandra. "Some Steamy Allegations by Fired QuesTech President: Suit Alleges Sexual Misconduct by Directors." *Washington Post*, 31 January 1989, C1.

Sugawara, Sandra, and Malcolm Gladwell. "QuesTech Tries to Pick Up the Pieces: Lawsuit by Ex-President Is Latest in Series of Difficulties." *Washington Post*, 6 February 1989, F1.

———. "Former President's Lawsuit against QuesTech Dismissed: Court Said to Have No Jurisdiction in Dispute." *Washington Post*, 26 April 1989, F3.

Thimmesch, Nick. "Schlesinger: The Early Years." *Air Force Times*, 5 March 1975, 12–14.

Toma, Sue. "Pat McLucas: An Interview." *Air Force Times*, 13 March 1974, 33, 35.

"Toronto Loses Bid for Space University." *Financial Post* (Toronto), 2 February 1993, 10.

"Trimble Navigation Reports Record First Quarter Results." *PR Newswire*, 24 April 1991.

"Trimble Settles Avion Dispute." *PR Newswire*, 3 October 1994.

Tucker, Elizabeth. "Ousted CEO Mayo Ran Afoul of Firm's Founders." *Washington Post*, 9 November 1988, B1.

Ulsamer, Edgar. "A Hard Look at the US Technological Posture." *Air Force/Space Digest*, May 1969, 72–76.

———. "Secretary McLucas Looks at Pressing Air Force Needs." *Air Force Magazine*, January 1974, 38–44.

Wall, Robert. "DigitalGlobe Wins: NIMA Seeks Funding to Sustain Space Imaging as a Second Provider." *Aviation Week and Space Technology*, 6 October 2003, 50–51.

Welles, Benjamin. "Foreign Policy: Disquiet over Intelligence SetUp." *New York Times*, 22 January 1971, 1, 8.

———. "H-L-S of the C.I.A." *New York Times Magazine*, 18 April 1971, 34–54.

Wicker, Tom. "Most Pervasive Corruption." *New York Times*, 19 October 1976, 39.

"Wide-Body Fix Extension Still under Study at FAA." *Aviation Week and Space Technology*, 1 November 1976, 28.

"Wide-Body Modifications Pushed." *Aviation Week and Space Technology*, 6 December 1976, 23.

Wilson, George C. "Soviets Learned of Spy Satellite from U.S. Manual." *Washington Post*, 23 November 1978, 1, 12.

Witze, Claude. "USAF's New Leaders." *Air Force Magazine*, September 1973, 58–59.

Internet Documents and Web Sites

Air Force Biographies. AFLink Web site, January 2004. http://www.af.mil/lib/bio/.

Air Force Space and Missile Pioneer Awards. AFSPC Web site, November 2003. http://www.spacecom.af.mil/hqafspc/history/pioneers.htm.

Baker, John C. "Commercial Observation Satellites: A Catalyst for Global Transparency." *Imaging Notes*, July–August 2001. http://www.imagingnotes.com.

———. "Commercial Satellite Imagery: Progress and Promise." *Imaging Notes*, March–April 2002. http://www.imagingnotes.com.

Baugher, Joe. "General Dynamics F-111 History," August 1999. http//www.f-111.net/JoeBaugher.html.

Bush, George [H. W.]. "Message to the Congress Submitting a Report on the International Space Year," 31 January 1991. http://www.bushlibrary.tamu.papers/1991/91012104.html.

Bush, Peter. "The Story behind the McNamara Line." www.chss.montclair.edu/english/furr/pbmcnamara.html (reprinted from *Vietnam* magazine, February 1996).

C-COR.net Inc. Web site, May 2003. http://www.c-cor.net.

Clarke, Arthur C. "Extra Terrestrial Relays." *Wireless World*, October 1945, 305–8, reprinted at http:www.ClarkInstitute.com/vision.

Culhane, Garrett. "Mission to Planet Earth." *Wired Magazine*, December 1993. http:www.wired.com/archive/1.06/mission.earth_pr.html.

Day, Dwayne A. "NIMA and NASA Agree to Image Shuttles Henceforth While in Orbit," 1 April 2003. http://www.friends-partners.org/fpspace/2003-April/008337.html.

Defense Industries-Army, February 2001. http://www.army-technology.com.

Department of Defense news releases. http://www.defenselink.mil/news.

Dorn, Walter. *Peacekeeping Satellites: The Case for International Surveillance and Verification.* Royal Military College of Canada.

http://ww.rmc.ca/academic/gradrech/dorn19-4_e.html. (originally published in *Peace Research Review*, 10, nos. 5 and 6 [1987].

Forman, Brenda. "Quo Vadimus?: An Unscientific Survey." *AIAA San Francisco Section Newsletter*, February 1994. http://www/aiaa-sf.org/files/forman/94-02.txt.

Goddard Space Flight Center. News Release. "Major Review of Mission to Planet Earth Endorses Flexible Approach to Future Satellites, Steers Data System Development," 21 August 1997. http:www.gsfc.nasa.gov/news-releases/1997/97-180.htm.

GPS/MET home page, September 1998. http://www.cosmic.ucar.edu/gpsmet.

Hanley, Timothy C., and Harry N. Waldron. *Historical Overview: Space and Missile Systems Center 1954–1995*, November 1997. http://www.losangeles.af.mil/SMC/HO/Smchov.htm.

Hirshberg, Michael J. "V/STOL: The First Half Century," December 2001. http://www.aiaa-ncs.org/vstol/VSTOL.html.

International Space University Web site, February 2003. http://www.isunet.edu.

Landsat Program Chronology. Landsat Web site. http://geo.arc.nasa.gov/sge/landsat/html.

Molczan, Ted. "Could Columbia Have Been Imaged by a Keyhole?" Section 6. "Skylab Encounters with KH 8-38," 3 April 2003. http://satobs.org/columbia/KeyHolesattosat.html.

NASA Leadership and America's Future in Space: A Report to the Administrator by Dr. Sally K. Ride, August 1987. http://history.nasa.gov/riderrep.htm.

NASA News Release 92-12. "President Bush Launches International Space Year," 24 January 1992. http://spacelink.nasa.gov/Previous.News.Releases/92.01.14/htm.

NASA News Release 98-12. "Mission to Planet Earth Enterprise Name Changed to Earth Science," 21 January 1998. http://geo.arc.nasa.gov/sge/landsat/mtpe.html.

NASA News Release 04-205. "Administrator Unveils Next Steps of NASA Transformation," 24 June 2004. http://www.nasa.gov/home/hqnews/2004/jun/HQ_0425_transformation.html.

National Association of Governors Web site, February 2003. http://www.nag.org/governors.

National Imagery and Mapping Agency (NIMA) News Release PA 03-01. "NIMA Sponsors Historical Imagery Declassification Conference," 3 October 2002. http://www.nima.mil/general/2002/presrel/3oct02.html.

National Reconnaissance Office Web site, November 2003. http://www.nro.gov.

National Security Archives. "The NRO Declassified," July 2002. www.gwu.edu/nsarch1/NSAEBB35.

National Security Decision Directive 189. "National Policy on the Transfer of Scientific, Technical, and Engineering Information," 21 September 1985. http://www.fas.org/irp/offdocs/nsdd/nsdd-189.htm.

Nau, Evan D. "The Bumblebee Project," 1998. http://www.peronal.umich.edu/~buzznau/bmblbee.html.

NIMA. "Historical Imagery Declassification – America's Eyes: What We Were Seeing, Historical Background," 25 September 2002. http://www.nima.mil/pa/newsroom/history/history/html.

North Atlantic Treaty Organization. http://www.nato.int/.

Orbital Sciences Corporation. "About Orbital," February 2003. http://www.orbital.com/.

"Orbital's Launch of Taurus Rocket Unsuccessful." 21 September 2001. http://www.orbital.com/...Press Releases.

Plumlee, Thurman Ray. "History of USS Norton Sound, AV-11/AVM-1," 2 June 2003. http://rayplumlee.com/ships/nortonsoundAVM1.phtml.

RAND Corporation. "Workshop on Commercial MTPE," September 1997. http:www.rand.org/publications/db247/pdf.

Reagan, Ronald. "Message to the Congress Transmitting a Report on Establishing an International Space Year in 1992," 15 May 1986. http://www.reagan.utexas.edu/resource/speeches/1986/51586.htm.

Richelson, Jeffrey T., ed. "The U-2, OXCART, and the SR-71: U.S. Aerial Espionage in the Cold War and Beyond." National Security Archive Electronic Briefing Book No. 74, 16 October 2002. http://www.gwu.edu/-nsarchiv/NSAEBB74.

"Shuttle Continues to Waste External Tanks." *Space Daily*, 8 September 2000. http://www.spacedaily.com/news/shuttle-00y.html.

Society of Satellite Professionals International Web site, July 2002. http://www.sspi.org/html/about.html.

Space Island Group Web site, January 2003. http://www.space islandgroup.com.

TCI Corporation Web site, May 2003. http://www.tcibr.com.

"Telenor Chronology," March 2003. http://www.telenor.com/about/history/chronology/.

Trimble Navigation Limited Web site, March 2003. www.trimble.com.

UN General Assembly Resolution 46/45. "International Cooperation in the Peaceful Uses of Outer Space," 9 December 1991. http://www.un.org/documents/ga/res/46/a46r045.htm.

ViaSat. "History," March 2003. http://www.viasat.com/about/history.htm.

Ware, Randolph, and Phil Culbertson. "STS-Lab: A Low Cost Shuttle-Derived Space Station." 1 February 1992. http:www.space-frontier.org/projects/external tanks.../html.

Whalen, David J. "Communications Satellites: Making the Global Village Possible," November 2002. http://www.hq.nasa.gov/office/pao/History/satcomhistory.html.

Wilson, J. D. "The Satellite Imaging Race Intensifies." *GeoWorld*, January 2002. http://www.geoplace.com/gw.

X-Prize Foundation. "History of the X Prize Foundation," April 2003. http://www.xprize.org/history.html.

Yunck, Thomas P., Chao-Han Liu, and Randolph Ware. "A History of GPS Sounding," *COSMIC*, January 2000. Genesis OnLine Repository, http://genesis.jpl.nasa.gov/html/publications.

US Congressional Documents (in chronological order)

Senate. *Nominations: Beal, . . . McLucas, . . . and Clarke: Hearing before the Committee on Armed Services.* 91st Cong., 1st sess., 26 February 1969.

Senate. *Nomination of McLucas and Brown: Hearing before the Committee on Armed Services.* 93d Cong., 1st sess., 13 July 1973.

Senate. *Questions Related to Secret and Confidential Documents.* 93d Cong., 1st sess., S.Rept. 93-456, 12 October 1973.

House. Committee on Interstate and Foreign Commerce. *Air Safety: Selected Review of FAA Performance.* 93d Cong., 2d sess., 27 December 1974.

Congressional Record. Rep. F. Edward Hebert. "Secretary McLucas Looks at the Man in Uniform Today." 94th Cong., 1st sess., 20 May 1975, E2517–18.

Senate. *Nominations, October–December: Hearings before the Committee on Commerce.* 94th Cong., 1st sess., 7 October; 4, 5, 13 November; 3–5 December, 1975.

Senate. *Earth Resources and Environmental Information System Act of 1977: Hearings before the Subcommittee on Science, Technology, and Space, Committee on Commerce, Science, and Transportation.* 95th Cong., 1st sess., 24–25 May, 13–14 June, 1977.

House. *Earth Resources Information System: Hearings before the Subcommittee on Space Science and Applications, Committee on Science, Space, and Technology.* 95th Cong., 1st sess., 21–23 June 1977.

Senate. *U.S. Civilian Space Policy: Hearings before the Subcommittee on Science, Technology, and Space, Committee on Commerce, Science, and Transportation.* 96th Cong., 1st sess., 25, 31 January, 1 February 1979.

Senate. *Operational Remote Sensing Legislation, Part 1: Hearings before the Subcommittee on Science, Technology, and Space, Committee on Commerce, Science, and Transportation.* 96th Cong., 1st sess., 9, 11 April 1979.

House. *Earth Resources Data and Information Service: Hearings before the Subcommittee on Space Science and Applications, Committee on Science, Space, and Technology.* 96th Cong., 1st sess., 2, 3 May 1979.

House. *Impact of National Security Considerations on Science and Technology: Hearing before the Subcommittees on Science, Research, and Technology and Subcommittee on Investigations and Oversight, Committee on Science, Space, and Technology.* 97th Cong., 2d sess., 29 March 1982.

Office of Technology Assessment. *Review of the FAA 1982 National Airspace System Plan.* Washington, D.C.: GPO, August 1982.

House. *1986 NASA Authorization.* vol. 2, *Hearings before the Subcommittee on Space Science and Applications, Committee on*

Science, Space, and Technology. 99th Cong., 1st sess., 6 February–26 March 1985.

Senate. *Transfer of National and Dulles Airports: Hearings before the Subcommittee on Aviation, Committee on Commerce, Science, and Transportation.* 99th Cong., 1st sess., 26 June, 11 July 1985.

House. *Omnibus Trade and Competitiveness Act of 1988* (vol. 2), *Hearing before the Subcommittee on International Economic Policy and Trade, Committee on Foreign Affairs.* 100th Cong., 1st sess., 11–12 March 1987.

Senate. *Export Controls: Hearing before the Subcommittee on International Finance and Monetary Policy, Committee on Banking, Housing, and Urban Affairs.* 100th Cong., 1st sess., 12, 17 March 1987.

House. *Review of Major Space Programs and Recommendations Made to Bush Transition Team: Hearings before the Subcommittee on Space Science and Applications, Committee on Science, Space, and Technology.* 101st Cong., 1st sess., 8–9 February 1989.

House. *Fiscal Year 1993 FAA R,E,&D Authorization: Hearings before the Subcommittee on Technology and Competitiveness, Committee on Science, Space, and Technology.* 100th Cong., 2d sess., 10–11 March 1992.

Office of Technology Assessment. *The Future of Remote Sensing from Space: Civilian Satellite Systems and Applications.* Washington, D.C.: GPO, July 1993.

Oral Histories of John L. McLucas

Ahman, Hugh N., and Scott S. Thompson. Air Force Historical Research Center. 13–14 September 1978.

Garber, Steve. NASA History Office, 9 January 2001.

Goldberg, Alfred, and Roger Trask. OSD Historical Office, 10 November 1998.

Haines, Gerald. NRO Historian, 8 May 1997. Redacted transcript.

Hall, R. Cargill. NRO Historian, 24 August 1999. Redacted transcript.

Moody, Walton, and Walter Poole. DOD Acquisition History Project, 5 June 2001.

Puryear, Edgar R., 21 February 1980. (Research for biography of Gen George Brown.)

Watson, George W. Air Force History Support Office, 10 April and 7 May 1996.

Interviews by John L. Mclucas

Transcripts of John McLucas's interviews and recorded conversations with the following former colleagues and government officials provided background information and insights.

Dr. Lew Allen Jr. (Gen, USAF, retired)
Norm Augustine
Langhorne Bond*
Jeff Cochran*
Roger G. DeKok (Lt Gen, USAF, retired)
Russell E. Dougherty (Gen, USAF, retired)
Robert R. Everett*
Ronald R. Fogleman (Gen, USAF, retired)
Jimmie D. Hill
Glenn A. Kent (Lt Gen, USAF, retired)
Joseph Laitin*
Lester L. Lyles (Gen, USAF)
Robert T. Marsh (Gen, USAF, retired)
Gregory S. Martin (Gen, USAF)
Keith D. McCartney (Maj Gen, USAF, retired)
George Monroe (Col, USAF, retired)*
George K. Muellner (Lt Gen, USAF, retired)
F. Whitten Peters
Edwin "Spec" Powell (Brig Gen, USA, retired)*
Joseph W. Ralston (Gen, USAF, retired)
Dr. Richard Raymond
William B. Robinson
Thomas F. Rogers
Michael E. Ryan (Gen, USAF, retired)
Dr. James R. Schlesinger
Bernard A. Schriever (Gen, USAF, retired)
Frank A. Shrontz
William Y. Smith (Gen, USAF, retired)
Jack L. Stempler
William R. Usher (Maj Gen, USAF, retired)

BIBLIOGRAPHY

Gene Weithoner*
Larry D. Welch (Gen, USAF, retired)
Dr. Albert "Bud" Wheelon
William W. Woodruff

———

   *Conducted with Ken Alnwick

Index

40 Committee, 172, 181–82, 209

2001: A Space Odyssey, 278, 280

Abrams, Creighton, 214
accidents (aircraft), 257, 261
acquisition process, 35, 101, 104, 138, 148, 177
Adams, Brock, 266, 270
Advanced Manned Strategic Aircraft (AMSA), 110, 130–31. *See also* aircraft: B-1
Advanced Research Project Agency (ARPA), 40. *See also* Defense Advanced Research Project Agency
Advisory Group for Aeronautical Research and Development (AGARD), 47, 49, 61
Aegis, 45
Aeronautical Systems Division, 121, 140, 232
Aerosat, 273, 284
Aerospace America, 212, 284, 287, 307–10, 314
Aerospace Corp., 211
Aetna, 275
Afghanistan, 112, 122
Air Combat Fighter (ACF), 125. *See also* aircraft: YF-16
Air Force. *See* US Air Force
Air Force Academy. *See* US Air Force Academy
Air Force Historical Foundation, 207
Air Force Magazine, 95, 148–49, 202, 208, 211, 252, 255–56
Air Force Recruiting Service, 236
Air Force Satellite Communications System (AFSATCOM), 200
Air Force Scientific Advisory Board, 53, 149, 205, 211, 215
Air Force Security Service, 23
Air Force Space Command, 206, 208, 296
Air Force Studies and Analysis Office, 118

Air Force Systems Command (AFSC), 41, 51, 62, 98, 101, 103–4, 116, 121, 148, 177–78, 192–93, 204
Air Force (Operational) Test and Evaluation Center (AFOTEC), 104
Air Force units (numbered)
1st Tactical Fighter Wing, 120
Second Air Force, 235
Seventh Air Force, 57, 70, 88–89, 142–43, 217, 219, 221
8th Tactical Fighter Wing, 142
Thirteenth Air Force, 217
1130th Aerospace Technical Development and Training Group, 209
Air Materiel Command, 17
Air National Guard (ANG), 18–19
Air Staff, 44, 67, 69, 71–72, 83, 88, 101, 120–22, 128, 132, 136, 138, 148, 178, 193, 202, 225–26, 233, 247, 252
Air Technical Intelligence Center (ATIC), 22
air traffic control, 259–61, 268
Air Training Command, 128, 238, 241
Airborne Warning and Control System (AWACS), 57–58, 110–111, 120, 134–36. *See also* E-3
Aircraft by designation
147. *See* Ryan 147 Firebee
707, 264
727, 261
747, 44, 116, 145, 262–63
757, 269
767, 269
A-1 Skyraider, 97, 227
A-4 Skyhawk, 43
A-7 Corsair II, 44, 97
A-10 Thunderbolt II, 111, 121–23
A-12 Oxcart, 42, 180
A-37 Dragonfly, 227
A-310 Airbus, 269
AC-119 Shadow, 227
AC-130 Specter, 121–22
AV-8 Harrier, 39
A-X, 99, 110, 121–23. *See also* YA-9/10

B-1 Lancer, 84, 111, 130–33, 144–46, 225
B-2 Spirit, 100, 133
B-24 Liberator, 10, 88
B-52 Stratofortress, 130, 145, 215–16, 218, 294
BQM-34, 149
C-5 Galaxy, 98, 102, 110, 115–19, 130, 145, 148–49, 230
C-7 Caribou, 227
C-9 Nightingale, 251
C-17 Globemaster III, 100, 130
C-123 Provider, 227
C-130 Hercules, 114, 129, 137
C-133 Cargomaster, 44
C-140 Jetstar, 257
C-141 Starlifter, 44, 110–11, 115, 118
DC-9, 104, 264, 268
DC-10, 261–63
DC-130, 138
E-3 AWACS, 111, 136–37
E-4 National Airborne Ops Center, 116
EC-121 Warning Star, 56, 138
EC-137D, 135
EF-111A, 113
F/A-18 Hornet, 127
F-4 Phantom II, 97, 118, 128, 227
F-5 Freedom Fighter, 127–28, 227
F-5E/F Tiger II, 127–28
F-8 Crusader, 44
F-14 Tomcat, 43, 127
F-15 Eagle, 110–11, 118–21, 123–24, 126, 130–31, 225, 251
F-16 Fighting Falcon, 126–27, 136, 285
F-104 Starfighter, 126
F-111, 43–44, 60, 98, 110, 112–15, 119, 198. See also TFX
F-117A Nighthawk, 133
F-X, 118–20
F-XX, 123–24
L-1011 Tristar, 116, 262–63, 294
MiG-21 Fishbed, 128
MiG-23 Flogger, 43
MiG-25 Foxbat, 121
Mirage F1, 126, 310
Nimrod, 136
O-1 Birddog, 227
O-2 Skymaster, 227
OV-10 Bronco, 38, 97

P.1127 Kestrel, 39
PBM-3 Mariner, 10
QU-22, 56
RQ-1A Predator, 141
RQ-4A Global Hawk, 141
Ryan 147, 137, 139
SR-71 Blackbird, 42, 119, 133, 138, 181–82, 209
T-28 Trojan, 38
T-37 Dragonfly, 227
TFX, 33, 40–43, 60, 98. See also F-111
U-2, 22, 176, 180–82, 188, 209
V-22 Osprey, 39
Viggen 37E, 126
X-20. See Dyna-Soar
X-32. See Joint Strike Fighter
X-35. See Joint Strike Fighter
XC-142, 39
YA-9, 99, 122
YA-10, 99, 122. See also A-10
YC-14, 99, 129
YC-15, 99, 129
YF-16, 99, 123, 125–29. See also lightweight fighter
YF-17, 99, 123, 125, 127. See also lightweight fighter
YF-22, 99
YF-23, 99
Airline Pilot's Association (ALPA), 268–69
Airline Transportation Association (ATA), 263, 266, 283
Aldridge, Edward, 188, 211
Alexandria, Va., 308
All-Volunteer Force, 75, 84, 237–38, 240–41
Alnwick, Kenneth J., 29
Alvarez, Luis, 12, 37, 54
American Airlines, 261–62
American Civil Liberties Union, 249
American Institute of Aeronautics and Astronautics (AIAA), 60, 211–12, 284, 286–87, 289–91, 300, 305, 307–9, 313
Andrews AFB, 42, 88, 93, 120, 207, 245, 249, 252
antisatellite system (ASAT), 203, 207
Apollo 11, 169, 291
Apollo–Soyuz mission, 290
Arlington National Cemetery, 61, 285, 306

Arlington, Va., 313
Arms Control and Disarmament Agency, 52
Armstrong, Robert, 10
Arthur C. Clarke Foundation, 278, 285
Astrotech International, 293
AT&T, 21, 40, 106, 133, 271, 278
atomic bomb, 11, 147. *See also* nuclear weapons
Augustine, Norm, 87, 94, 312
Australia, 42, 114–15, 198
Aviation Week and Space Technology, 134, 140, 149, 283, 301
Avion Systems, 296, 311

Bailey, Bernard, 181–82
Bangkok, 56
Barksdale AFB, 235
Barnes, Thomas N., 243
Bedford, Mass., 61
Beech Aircraft, 56
Beggs, James M., 106, 281, 302
Beijing, 308
Belanger, Ray, 259
Belenko, Viktor, 121
Belgium, 49, 126–27, 287
Belle Haven Country Club, 210, 306, 311
Bellis, Benjamin, 119–20
Bennet, Douglas P., 258
Bennett, Donald, 74
Benson, Lawrence R., 147–48, 255, 310, 314
Berlin Airlift, 118
Beverly, William, 37
Big Safari, 138
Bisplinghoff, Raymond, 131, 211
Bissell, Richard, 171
Black, Harriet, 274, 285, 289, 292–93, 297, 305–6, 309, 311. *See also* McLucas, Harriet
Blackburn, Al, 37, 42
Blue Ribbon Commission on Defense Management, 105
Bodman, Richard, 275
Boeing, 41–42, 44, 87, 106, 116, 129, 134–35, 140–41, 149, 176, 261–62, 269
Bogmolov, Alexei F., 290
bomb damage assessment (BDA), 218
Bond, Langhorne, 268
Boone, Nathan, 2

Booneford, N.C., 2
Borda, Richard J., 67
Bork, Robert, 95
Boston, 9, 16, 29, 48–50, 53, 60, 65, 148, 208, 234, 264
Boswell, Marion L., 87
Boulder, Colo., 207, 284
Boyce, Christopher, 189
Boyd, John, 118, 123
Bradburn, David, 178, 208
Brett, Devol, 233
Brinegar, Claude S., 257
Britain. *See* United Kingdom
Brosio, Manlio, 48
Brown, George S., 83, 89, 95, 121
Brown, Harold, 31–32, 34, 37, 41, 44–47, 49, 55, 61, 64, 66, 68, 115, 117, 132–33, 144, 194
Brown, Ron, 203
Brown, Walter, 18
Brownman, Harold L., 179
Buckley Air National Guard Base, 199
Budapest, 289
Bunker, Ellsworth, 218
Bush, George H. W., 95, 291, 300, 312
Bush, George W., 95, 291, 300, 312–13
Butterfield, Alexander, 257
Byers, Jeffrey, 90

Callaway, Howard, 74
Cambodia, 55, 215, 217, 219, 224, 253, 267
Cambridge Field Station, 16, 50
Cambridge, Mass., 148, 207
Canada, 273, 282, 284, 305, 309
Canary Islands, 261
Cape Canaveral, 23, 191, 197
Carlton, P. K., 245
Carter administration, 130, 134
Carter, Jimmy, 89, 186, 264, 266
Castro, Fidel, 181
CBS News, 259
C–COR Corporation, 28–29
Central Intelligence Agency (CIA), 64
Cernan, Eugene, 294
Chafee, John, 74, 130
Challenger, 188, 195, 206, 286, 290, 294, 299
Chandra X–Ray Telescope, 187
Charles, Robert H., 116
Charleston, S.C., 9, 16
Charyk, Joseph V., 82–83, 171, 271–72

Cheney, Richard, 190
chief of naval operations (CNO), 89, 180
chief of staff (Air Force), 71, 120–21,
    132, 136
China, 38, 87, 100, 137–38, 140, 182,
    184, 209, 287, 308
China Lake, Calif., 87, 100
Chirac, Jacques, 127
Citadel, The, 4
Civil Air Patrol, 235
Civil Service Commission, 117, 236, 266
Clarke Institute for Telecommunications
    and Information (CITI), 280, 285
Clarke, Arthur C., 207, 270, 278–79,
    283–85, 298, 303–4, 307, 313
Clements, William, 81, 85, 95, 125,
    127–28, 173, 245, 248
Clemson University, 4
Clifford, Clark, 52, 208, 234, 280
Clinton, William J., 134
Clio, S.C., 2, 307
close air support (CAS), 39, 41, 121
Cloudcroft, N.Mex., 22, 24
Cochran, Jeff, 259
Colby, William E., 95, 174
Coleman, William T., 258
Colino, Richard R., 293
Columbia, 32, 90, 259
Combat Dawn, 138
Combat Development and Test Center
    (CDTC), 40
Combat Lightning, 54, 57
Commando Hunt, 56
Commerce. See Department of
Commission on Government Procure-
    ment, 103
Committee on Imagery Requirements
    and Exploitation (COMIREX), 172
Committee on Science Engineering and
    Public Policy (COSEPUP), 287
communications intelligence (COMINT),
    23
Communications Satellite Act, 270,
    277–78, 284
Communications Satellite Corp. (Com-
    sat), 82–83, 170, 178, 201–2, 211,
    269–79, 281, 283–84, 286, 289, 292–
    93, 304, 307–10, 314
Community College of the Air Force, 238
Community Engineering Corporation
    (CECO), 24–25, 28
Compass Arrow, 138–39, 149

Compass Dwell, 139
comptroller (DOD), 33
Comsat General Corporation, 270, 272
Concord, Mass., 50, 58, 65–66
Concorde, 265
concurrency, 35, 98–99, 103
Conley Ridge, N.C., 2, 4
Conley, Viola (mother), 2
Connolly, John, 116
Conrad, Charles "Pete," 258
Convair, 123, 191
Cook, Charles W., 177
Coordinating Committee for Multilateral
    Export Controls (CoCom), 288
Cornell University, 38
Corning Glass Company, 19
Corona (conference), 205
Corona (satellite), 209
Costa Rica, 305
counterinsurgency, 37–38, 40, 97, 224–25
Cousar, Charlton, Jr., 4
Cousar, Donella, 3–5, 16
Cousar, Harold, 1, 3, 307
Cox, Archibald, 95
Crowley, John J., 179
Culbertson, Phillip, 304
Culloden, Battle of, 2
Culzean Castle, 274
Curien, Hubert, 63, 279
Currie, Malcolm R., 86, 133, 202
Cyprus, 48, 232

DaNang, 54
Dassault, 127
David, Edward E., 173
Davidson College, 6, 270
Davis, Benjamin O., 244, 255
Davis, Harry, 68, 77, 112
Dayton, Ohio, 17, 149, 188
Defense Acquisition Board (DAB), 99
Defense Advanced Research Project
    Agency (DARPA), 133. See also Ad-
    vanced Research Project Agency
Defense Advisory Committee on Women
    in the Service (DACOWITS), 249
Defense Communications Agency (DCA),
    55
Defense Communications Planning
    Group (DCPG), 55–56, 220
Defense Intelligence Agency (DIA), 53,
    74, 184

Defense Meteorological Satellite Program (DMSP), 196–97, 211
Defense Race Relations Institute, 242
Defense Reorganization Act of 1986, "Goldwater-Nichols," 94, 148
Defense Research Directors' Committee, 46
Defense Satellite Communications System (DSCS), 200
Defense Science Board, 53
Defense Support Program (DSP), 197–99, 211
Defense Systems Acquisition Review Council (DSARC), 99, 104–5, 123, 132, 136, 202, 260
De Gaulle, Charles, 48
DeMarines, Victor, 56
Denmark, 126
Department of Commerce, 289
Department of Defense (DOD), 33, 35, 51–52, 55, 57–58, 61, 70, 72–73, 75–77, 80, 90–92, 101–9, 113, 123, 148–49, 170–74, 186–87, 191, 193, 195–96, 200, 202–4, 206, 210, 212, 214, 218, 224, 233–36, 241–43, 246, 248, 250, 253, 294, 297, 301, 308, 311. *See also* secretary of defense (SecDef), Office of the Secretary of Defense (OSD)
Department of Energy, 95, 292
Department of Health, Education, and Welfare, 52
Department of Labor, 236
Department of Transportation (DOT), 260, 269
depression (1930s), 3, 6, 198
Desert Shield, 220, 296
Desert Storm, 296. *See also* Gulf War of 1991
Detzer, Gus, 24, 27
Diamandis, Peter, 280, 304
Digital Computer Laboratory, 50
Dillon, S.C., 5
Director of Central Intelligence (DCI), 173–74, 176, 182, 184, 190, 206, 210
Director of Defense Research and Engineering (DDR&E), 31, 33–34, 36, 38–40, 45, 47–48, 60, 86–87, 94, 106, 118, 121, 133, 135, 173–74, 176–77, 187, 194, 202, 220, 313
Dirks, Leslie C., 179

dissimilar air combat tactics (DACT), 128
Dixon, Robert J., 238, 240–42
Domestic Action Program, 234–35, 242
Dominitz, Jack, 55–56
Doolin, Dennis J., 87
Dougherty, Russell E., 45
Douglas, Donald, Jr., 49
Dr. Strangelove, 146
Draper, Charles Stark, 108
Drinkwater, Fred, 269, 283
drones, 137, 139, 149, 222. *See also* remotely piloted vehicles
drug abuse, 240–42
DuBridge, Lee A., 173
Duckett, Carl E., 179
Duke Law School, 79
Dulles International Airport, 265, 280
Dumont Television Corporation, 20
Durant, Fred, 279, 304
Dyna-Soar (X-20), 194

Earth Observation Satellite Company (Eosat), 289
Earth Resources Satellite, 184, 209
Earthstar, 289
East Germany. *See* Germany
Eastman Kodak, 176
Edelson, Burton I., 281, 291, 304, 306, 309
Edwards AFB, 119, 126, 129, 141, 182
Eggleton, Reginald, 19
Eglin AFB, 122
Egypt, 117, 181, 230
Eisenhower, Barbara, 274
Eisenhower, Dwight D., 25
Eisenhower, John, 274
Eisenhower, Susan, 311
El Segundo, Calif., 177–78, 180
electronic countermeasures (ECM), 18
Electronic Industries Association, 139
electronic intelligence (ELINT), 18
Electronic Systems Laboratory (ESL), 20, 133
Electronics Systems Division (ESD), 48, 51, 61, 135–36
Ellender, Allen, 186
Elliott, Daniel, 8
Ellsberg, Daniel, 109
England. *See* United Kingdom
Enhance Plus, 227
Eniwetok, 10, 13

Enthoven, Alain, 34, 41
Environmental Protection Agency, 52
Environmental Resources Satellite (Envirosat), 289, 309, 312
equal opportunity, 84, 234, 242, 246, 252
Ervin, Samuel J., Jr., 257
European Space Agency, 300
Evans, William J., 101
Everett, Robert R., 50
Executive Committee (ExCom), 171–74, 184, 186, 206
External Tanks Corporation, 295

Faga, Martin, 189–90, 210
Falls Church, Va., 66
Farnborough Air Show, 182
Federal Aviation Administration (FAA), 52, 58, 68, 93, 96, 108, 203, 257–68, 270, 273, 282–85, 296–97, 306–7, 311
Federal Communications Commission (FCC), 20, 271, 273, 277, 310
federal contract research center (FCRC), 49, 51, 59, 108–9, 148
Ferguson, James, 51, 62, 116
Ferret, 18
Filene Center, 276
Firebee (Ryan 147), 137, 139
Fitzgerald, A. Ernest, 117, 148
Fitzhugh, Gilbert W., 77, 103
Five-Year Defense Plan (FYDP), 34, 60
Flax, Alexander H., 170–71, 174
Fleet Satellite Communications System (FLTSATCOM), 200–1, 272
Fletcher, James, 294, 299
flexible response, 31, 48
flight simulators, 231
Ford Aerospace, 279, 293
Ford, Gerald, 90, 93, 95, 233, 255, 257, 266
Fort Belvoir, 101
Fort Benning, 40
Fort McArthur, 193
Fort Riley, 123
Fort Schuyler, 9
Fort Worth, 43, 125–26
Foster, John S., 55, 64
France, 3, 7, 45–46, 48, 126–27, 265, 282, 285, 287, 291, 300, 305
Fraser, Malcolm, 114
Freedom. See International Space Station

Friedheim, Jerry, 197, 244
Frosch, Robert, 74
Fubini, Eugene, 31

Gabriel, Charles, 220
Gadhafi, Mu'ammar, 244
Galapagos, 305
Garvey, Jane, 306
Gates, Robert, 190
Gates, Thomas S., 88
gays. See homosexuals
Gemini, 194
General Accounting Office (GAO), 51, 206
General Dynamics (GD), 41, 43, 125, 127
General Electric (GE), 129
General Space Corporation, 294
George Washington University, 279–80, 300, 307
Germany, 7, 48, 57, 79, 232, 241, 287
Getsinger, Norman, 287, 308
Getting, Ivan A., 192–93, 201, 211, 253
Gilpatric, Roswell L., 34
Gingrich, Newt, 92
Giraudo, John C., 67
Glasser, Otto J., 101, 109
Global Navigation Satellite System (GNSS), 296, 311
global positioning system (GPS), 60, 202–3, 295–97, 311
Goddard Space Flight Center, 302, 312
Goddard, George, 21
Goldwater-Nichols. See Defense Reorganization Act of 1986
Goodfellow AFB, 247
Goodwin, Bert Z., 259
Goodyear Aircraft, 19
Gorbachev, Mikhail, 291
Gore, Al, 300
Gorton, John, 198
Great Depression. See depression
Greece, 232
Griffith, Andy, 3
Grumman, 41, 43
Guam, 14, 215
Gulf War of 1991, 112, 199, 206, 296
guided missiles. See missiles

Haig, Alexander, 95, 205, 218, 223, 254
Haiphong, 140, 221–23
Hairston, Guy E., 87

Haldeman, H. R., 81, 95, 114, 254
Hall, Albert C., 173, 196–97
Haller, George, 17–18, 24
Halligan, Clair W., 50
Hampton, Robert, 266
Hanoi, 140, 223–24, 254
Hanscom AFB, 48, 51, 64
Hansen, Grant, 67, 100, 103, 123, 128, 193, 198
Harford, James J., 287, 309
Harlfinger, Frederick J., II, 180
Harper, John D., 275
Hartley, Frank W., 181
Harvard University, 18
Have Blue, 133
Hawker P. 1127 Kestrel, 39
Hawley, Todd, 280–81
Hebert, F. Edward, 92, 229, 255
Heilmeier, George, 133
Helms, Richard, 64, 172–73, 177, 181, 184, 186, 188, 233
Hewlett, William, 78
Hewlett-Packard, 78, 296
Hickam AFB, 220
Higdon, Robert, 19, 27
high-low mix, 124–25
Hill, Jimmie D., 179–80, 208
Hiroshima, 14
Hitch, Charles J., 33
Ho Chi Minh Trail, 55–56, 122, 142, 216
Hogan, Henry L., 67
Holloman AFB, 23
Holloway, Bruce, 70
Holm, Jeanne M., 247, 249, 252, 256
homosexuals, 249
Hondo, Tex., 252
Hong Kong, 287
Hoopes, Townsend, 66, 213, 253
Howze, Hamilton H., 39
HRB-Singer, 28–29, 32, 60, 107
Hue, 40
Hughes Aircraft, 49, 279, 289, 293, 310
Hughes, Harold, 84, 219
Humphrey, Hubert H., 63

IBM, 56, 275
Igloo White, 54, 56
Ignatius, Paul, 266
Ikanayake, Hector, 280
Ikonos, 303
Infiltration Surveillance Center, 56
infrared (IR), 21

Initial Defense Communications Satellite, 191
Institute of Defense Analyses (IDA), 55
Institute of Electrical and Electronics Engineers (IEEE), 52, 62
intercontinental ballistic missile (ICBM), 23, 144–45
Internal Revenue Service, 52
International Astronautical Federation (IAF), 289–90, 299–301
International Fighter Aircraft (IFA), 128
International Maritime Satellite Corporation (Inmarsat), 273–75
International Space Station (ISS), 286, 292, 295, 302–4
International Space University (ISU), 280–82, 284, 304–5
International Space Year (ISY), 299–301, 311–12
International Telecommunications Satellite Consortium (Intelsat), 271, 273–74, 277–81, 283–84, 290, 293, 304–5, 307, 310, 314
Intersputnik, 290
Iran, 111, 119–20, 173, 233
Ireland, 305
Israel, 53, 55, 111, 117–18, 181, 197, 230–31, 305
Israel, David, 53
Itek, 176, 209

Jacobs, John, 53, 61
James, Daniel "Chappie," 244, 255
Jason Committee, 55
Jet Propulsion Laboratory, 178, 297
Johnson, John A., 271
Johnson, Lyndon B., 59
Johnston Island, 203
Johnstown, Pa., 20
Joint Chiefs of Staff (JCS), 69, 73, 88–89, 94, 122, 132, 182, 194–95, 213, 217–18, 224
Joint Defence Space Communications Station. See Nurrungar
Joint Strike Fighter (JSF), 43
Jones, David C., 89, 235
Jones, Joseph, 109
Jones, Thomas, 106
Jordan, 305

kamikaze, 14
Kampiles, William, 189, 210

Keller, Edward, 29
Kelley, Roger, 235–36
Kennedy administration, 31, 43
Kennedy International Airport, 265
Kennedy, John F., 31, 270
Kennedy, Robert, 63
Kenneth Whiting, USS, 10
Kent, Glenn, 109
Kerama Retto, 13
KGB, 189
Khe Sanh, 56
Killian, James R., 48, 64, 148
Killpack, Larry, 142
Kimmel, Chester, 13
King, Martin Luther, Jr., 63
King, William G., 178
Kissinger, Henry, 76, 90–91, 172, 181, 197, 218, 221, 223, 230, 253
Kleindienst, Richard, 82
Klotz, Herbert W., 292
Knapp, Patricia, 16–17, 27, 32, 49, 58, 65–66, 77, 89–90, 95, 274. *See also* McLucas, Pat
Komons, Nick, 283
Kona, Hawaii, 299
Korea, 71, 128, 138, 287
Korean Airlines, 290
Korean War, 19, 39, 118
Kosovo, 123, 143
Kotelnikov, Vladimir, 289
Kubrick, Stanley, 278
Kulpa, John E., 178

LaBerge, Walter B., 87, 100
Lackland AFB, 241, 255
LaGuardia International Airport, 263
Laird, Melvin R., 63–64, 70, 72–82, 84–85, 98, 103–5, 109–11, 113, 116, 144, 172–73, 181, 185, 195, 214, 217–19, 222–23, 225, 234–37, 243–44, 254, 258
Laitin, Joseph, 85
Lake Barcroft, 66, 77, 274
Land, Edwin, 106, 186
Landsat, 209, 289, 303, 307, 309, 313
Laos, 122, 215, 217, 224
Latta, S.C., 5–6
Laughlin AFB, 181
launchers (space). *See* missiles and launchers
LAV (light attack aircraft), 43–44. *See also* A-7

Lavelle, John D., 219
Lawrence Livermore National Laboratory, 64
Leaf, Howard, 269, 283
Lear Siegler, 140
Leesburg, Va., 296
Legget, Robert F., 77
Leonov, Alexei, 290
Lewis, Drew, 268
Lexington, Mass., 60, 107
Leyden, John F., 266
Libya, 244
Light Armed Reconnaissance Aircraft (LARA), 38. *See also* OV-10 Bronco
lightweight fighter (LWF), 123–26. *See also* YF-16/17
Lincoln Laboratory, 50, 107
Linebacker, 139, 221–23, 239, 254
Linebacker II, 223, 254
Ling Temco Vought (LTV), 43
Lockerbie, Scotland, 264
Lockheed, 42, 44, 87–88, 98, 114–16, 128–29, 133, 176, 183, 191, 204, 262, 278, 304
Lockheed Martin, 278, 304
Lockheed Skunk Works, 42, 204
Logan International Airport, 264
Logsdon, John, 207, 300
London, 182, 274, 280
Loomis, Henry, 54
Los Angeles, 25, 49, 177, 182, 192–93, 211, 264
Los Angeles Air Force Station, 177, 192
Los Angeles International Airport, 264
Love, John, 81
Lowry AFB, 199, 241
LTV-Hiller-Ryan, 39
Luke AFB, 119

MacArthur, Douglas, 13
Maglione, Ralph J., 87
Mahon, George, 91
Manned Orbiting Laboratory (MOL), 194–95
Mansfield, Mike, 108, 230
Marianas, 11–12
Marisat, 201, 273
Martin, Edward J., 272, 314
Martin, John L., Jr., 177
Martin-Marietta, 94, 176
Massachusetts Institute of Technology (MIT), 9, 16, 48, 50, 53, 64, 107–8, 148–49, 280–82

Matlovich, Leonard, 249–50
Matsunaga, Spark M., 299
Mayo, William L., 292
McCartney, Keith, 86
McClellan, John, 148
McColl, S.C., 2–3, 5
McConnell, John P., 70
McDonnell, Sanford, 106
McGovern, George, 77, 79
McKee, William F., 260
McLaurin, Donnie, 3
McLean, Va., 57, 292, 309
McLucas, Archibald, 2
McLucas, Effie, 3
McLucas, Harriet (née Black), 274, 285,
    289, 292–93, 297, 305–6, 309, 311
McLucas, Hugh, 2–3, 309
McLucas, Jean, 2–3, 306
McLucas, Luther, 1–2, 63, 307
McLucas, May, 286, 291–92, 313
McLucas, Pamela, 16, 65
McLucas, Pat (née Knapp), 16–17, 27,
    32, 49, 58, 65–66, 77, 89–90, 95,
    274
McLucas, Rod (son), 27, 65–66, 90, 208,
    211
McLucas, Susan, 26, 65, 90, 311
McLucas, Willie, 3
McNamara, Robert S., 36, 60, 216–18
McNamara Line, 55, 62
McNaughton, John T., 34
McNickle, Marvin L., 101
Mettler, Rube, 106
Meyer, John C., 71
Meyers, Tedson, 279
Meyerson, Harvey, 299, 301, 312
MicroLab, 297
Microwave Early Warning (MEW), 12
Middendorf, William, 74
Middle East War (1973). See Yom Kip-
    pur War
Miles, Raymond, 19
Military Airlift Command (MAC), 196,
    244–45
Military Assistance Command Vietnam
    (MACV), 216–17, 221
Missile Defense Alarm System (MIDAS),
    191, 197
missiles and launchers
    Agena, 183, 191
    Atlas, 60, 191
    Condor, 44
    Hellfire, 141
    Jupiter, 60
    Lance, 44
    Maverick (AGM-65), 143
    Minuteman, 58, 144–45
    Missile Experimental (MX), 145
    Peacekeeper, 145
    Phoenix, 120
    Roland, 47
    Short Range Attack Missile (SRAM),
        144–45
    Sidewinder (AIM-9), 97
    Subsonic Cruise Armed Decoy
        (SCAD, AGM-86), 146
    Thor, 60, 191, 203
    Titan, 60, 144, 191, 197
    TOW, 44
    Typhon, 45
    Walleye, 44, 142
Mission to Planet Earth (MTPE), 300–2,
    313
Mitchell, John, 82
MITRE Corporation, 49, 52, 59, 61–62,
    107, 192
Moorer, Thomas, 74
Morison, Samuel Elliot, 14, 29
Morris, Joseph, 8
Morton, Rogers, 95
Mount Nittany, Pa., 16, 20, 26
Moynihan, Daniel P., 242
Muellner, George K., 120
Mull, Scotland, 2, 305
multiple independently targetable reen-
    try vehicle (MIRV), 144–45
Murphy, John R., 67
mutual assured destruction (MAD), 147,
    217

Nagasaki, 14–15, 147
Naka, F. Robert, 53, 175–77, 179–80,
    184
Nakhon Phanom, 56, 221
National Academies, 287–88, 299
National Aeronautics and Space Admin-
    istration (NASA), 52–53, 58, 64, 87,
    94, 108, 125, 149, 170, 178, 184,
    187–88, 192, 194–95, 209, 232, 254,
    269, 271, 281, 286, 288, 290–92,
    294–95, 297, 299, 301–4, 307–13
National Air and Space Museum, 185,
    279, 300

National Archives and Records Administration, 185, 312
National Center for Atmospheric Research (NCAR), 297
National Imagery and Mapping Agency (NIMA), 210, 313
National Oceanic and Atmospheric Administration (NOAA), 196–97, 211, 297
National Photographic Interpretation Center (NPIC), 174, 185
National Reconnaissance Office (NRO), 65–66, 68, 74, 78–79, 82–83, 87, 94, 101, 106, 112, 116, 137–38, 170–91, 193, 195–96, 204–12, 217, 222, 272
National Reconnaissance Program (NRP), 171, 174, 190, 206
National Science Foundation, 108, 131, 208, 297
National Security Agency (NSA), 21, 74, 174, 178, 192, 204, 208
National Security Council (NSC), 172–74, 190, 209
National Weather Service, 52, 196
Naval Indoctrination Center, 9
Naval Observatory, 56
Naval Research Laboratory, 201
Naval Weapons Center, 87, 100. See also China Lake
Navstar, 202, 295. See also global positioning system
Navy. See US Navy
Netherlands, 57, 126
Neumann, Gerhard, 106
New Orleans, 8, 229
New York City, 31
New York Times, 109, 208–9, 213, 253, 262, 283
Ngo Dinh Diem, 40
Nguyen Van Thieu, 223
Nickel Grass, 118
Nitze, Paul H., 34
Nixon, Richard M., 59
North American Air/Aerospace Defense Command (NORAD), 244
North American/Rockwell, 130, 202
North Atlantic Treaty Organization (NATO), 33, 37, 45–49, 61, 75, 89, 93, 125–27, 136, 147, 200, 229–30, 232–33
North Korea, 138
North Vietnam. See Vietnam

Northrop, 106, 122–24, 127–28, 133
Norway, 126, 304
nuclear weapons, 64, 144–47. See also atomic bomb
Nurrungar, 198–99

O'Donnell, J. J., 268
O'Malley, Jerome, 220
O'Neill, John, 37, 192
Office of Management and Budget (OMB), 83
Office of Technology Assessment (OTA), 268, 287, 308
Office of the Secretary of Defense (OSD), 32–36, 39, 51, 53, 60, 64, 67, 73, 75–77, 86, 88–89, 96, 102, 104–5, 109, 119, 125, 130, 135–36, 171, 173–74, 195, 199, 202–3, 241, 244, 260, 287
Office of the Secretary of the Air Force (OSAF), 69, 123. See also Secretariat
Okinawa, 13–15, 29
Operation Homecoming, 224
Operational Test and Evaluation (OT&E). See Test and Evaluation
Orbcomm, 295, 310
Orbimage, 295, 303
Orbital Sciences Corporation (OSC), 294, 297–98, 303, 310
OrbView, 303
Organization of Petroleum Exporting Countries (OPEC), 231
Osan Air Base, 138
overland radar technology (ORT), 134
over-the-horizon backscatter (OTH-B), 58

Pacific Air Forces (PACAF), 216–17
Pacific Command (PACOM), 216
Packard, David, 64, 73, 82, 98, 148, 186
Palmdale, Calif., 132
Parent-Teacher Association (PTA), 27
Paris, 3, 8, 45, 47–49, 61, 127–28, 222–24, 261–62, 264–65, 274, 284
Paris Air Show, 127–28
Paris, Chris, 8
Pasadena, Calif., 178, 309
Patrick AFB, 242
Patrol Aircraft Technical Service Unit (PATSU), 13
Patuxent Naval Air Test Center, 122

Pave Eagle, 56
Pave Knife, 143
Paveway, 142, 149, 222
Payne, Fred A., 32
Pegasus, 294, 310
Pelton, Joseph N., 278, 281–82, 284–85, 314
Penn State University, 17–20, 27–28
Pentagon, 1, 31–34, 36, 45, 55, 59–60, 63, 65–66, 68–69, 71–73, 75–85, 88, 90, 93, 98, 102–5, 109–11, 113, 117, 128, 133–34, 170–71, 174, 176–77, 180, 187, 213–14, 222, 224, 228, 230–32, 235–36, 239, 244–45, 252–53, 258, 260. *See also* Department of Defense (DOD), Office of the Secretary of Defense (OSD), and individual US armed services
Pentagon Papers, 109
Perkin-Elmer, 176
Perkins, Courtland D., 54
Perle, Richard N., 286
Permissive Action Link (PAL), 146
Perry, William, 20, 133
Phillips, Samuel, 74, 174, 192
Phobos, 291
Pike, Otis, 132
Planetary Society, 291, 300, 309, 312
Planning, Programming, and Budgeting System (PPBS), 33, 37, 60, 75
Ploesti, 88
Plummer, James W., 87–89, 112, 183
Portugal, 118
Potomac Institute for Policy Studies, 303, 313
Pratt and Whitney, 119, 126, 129
precision-guided munitions (PGM), 98, 142
President's Foreign Intelligence Advisory Board (PFIAB), 186
President's Scientific Advisory Committee (PSAC), 37, 174
Price, Melvin, 92
Princeton University, 9, 253
prisoners of war (POW), 139
procurement, 33, 35, 43–44, 47, 78–79, 84, 97–99, 101, 103–4, 110–11, 115–17, 121, 125, 133, 138, 147, 176, 193, 200–1, 204, 206, 236, 260. *See also* acquisition process
Professional Air Traffic Controllers Association (PATCO), 266

Program Objective Memorandum (POM), 37
Project 100,000, 234
Project Focus, 131
Project Transition, 236
Proxmire, William, 87, 115, 146
Puckett, Allen, 49, 106
Pustay, John S., 86

Quesada, Elwood, 25
QuesTech, Inc., 292, 309–10
Quickbird, 303

race relations, 242–43
radar, 8–13, 15, 17–19, 22, 45, 50, 119, 128, 131, 133–35, 138, 146, 233
Radio Research Laboratory, 18
Rafax, 21–22
Ramo, Simon, 106, 192
Ramstein Air Base, 241
RAND Corporation, 21, 84, 107, 109
Rather, Dan, 259
Raymond, Dick, 17–18, 21–22
Raytheon Corp., 20, 29
RCA (Radio Corporation of America), 7, 106, 140, 271
Reagan, Ronald, 105, 266, 283, 311
Reber, James Q., 176
Rechtin, Eberhardt, 78
Reconofax, 21–22, 29
Reed College, 65, 90
Reedy Creek Presbyterian Church, 306
remote sensing, 7, 23, 65, 295, 297, 299, 303, 308–9, 312–13
remotely piloted vehicles (RPV), 97, 137–42, 149. *See also* UAV
Republic of Vietnam. *See* Vietnam
research and development (R&D), 33, 35, 37–39, 45, 47–48, 54, 64, 68, 87, 99–100, 103, 107, 109, 126, 140, 148, 170, 187, 192–93, 195, 204, 206, 259–60, 268, 275, 288, 311
Reserve Officers Training Corps (ROTC), 7, 19, 243, 247
Resor, Stanley, 74
Rice, Donald, 83
Rice, K. P., 38
Richards, Robert, 280
Richardson, Elliott, 81, 85, 95, 244
Richelson, Jeffrey, 198, 207–11
Ride, Sally K., 299, 311
Rizzardi, Kathy, 68

Robb, Mahlon, 24
Roche, James G., 285
Rockefeller, Nelson, 95
Rockwell International, 130
Rockwell, Willard F., 293
Rogers, William, 79
Rolling Thunder, 142–43
Roosevelt, Franklin D., 7
Royal Australian Air Force (RAAF), 114–15
Royal Netherlands Air Force, 126
Rubel, John H., 34
Rumsfeld, Donald, 93, 206, 257
Rush, Kenneth, 79, 94
Rusk, Dean, 59
Russell, Kenneth, 136
Russia, 292. *See also* Soviet Union
Rutan, Burt, 304
Rutgers University, 90
Ryan, John D., 70

Saab-Scania, 126
Sagan, Carl, 300
Sagdeev, Roald, 291–92, 311
Saigon, 57, 128, 214, 216, 228
Saipan, 11–12
San Diego, 15, 25, 312
Sasebo, 15
Satellite Business Systems (SBS), 275
Satellite Television Corporation (STC), 275
Schlesinger, James R., 82–85, 91, 93, 100, 118, 125, 127, 133, 147, 173–74, 185, 217, 228–29, 232, 245, 252, 259, 314
Schriever, Bernard, 41, 62, 98
Schroeder, Patricia, 132
Schultz, George, 114
Schwartz, Robert, 269
Science Park, Pa., 20
Scientific Advisory Board, 47, 53, 65, 149, 205, 211, 215
Scotland, 2, 264, 274–75, 305
Scowcroft, Brent, 95, 190, 258
Seamans, Joe, 190
Seamans, Robert C., 64–67, 70–72, 74, 80–81, 83–87, 94, 97–98, 100–101, 107, 110, 112–15, 117, 119, 121–22, 128, 190, 193–95, 214, 222–23, 225, 227, 244, 246, 254–55
Second World War. *See* World War II
Secretariat (Air Force), 47, 69, 72, 101, 193, 226, 238. *See also* OSAF

secretary of defense (SecDef), 34, 63
secretary of the Air Force (SecAF), 64
secretary of the Navy, 130
secretary of transportation, 257–58, 260, 266, 268
security classifications, 18, 138
Seek Data, 54
Semi-Automated Ground Environment (SAGE), 22, 50–51, 61, 87
Serbia, 112, 143
Shah of Iran, 119, 233
Shanghai, 308
Shedler, Spencer J., 67
short takeoff and landing (STOL), 33, 38–39, 60, 99, 129–30
Shrontz, Frank A., 87
Sigma Pi Sigma, 7
Signal Corps, 16–17, 23
signals intelligence (SIGINT), 21, 137, 172, 183, 208
Singel, Robert D., 177
Singer Corp., 27–29, 31–32, 45
Single Integrated Operational Plan (SIOP), 147
Skylab, 178, 195
Skynet, 200
Slay, Alton D., 101, 220
Smithsonian, 61, 254–55, 259, 279, 304
Social Actions, 242
Society of Satellite Professionals, 281, 284
Solona Beach, Calif., 25
South America, 28
South Vietnam. *See* Vietnam, 254
Southeast Asia, 40, 53, 55–59, 67, 70, 75, 77, 81, 84, 88, 97, 103, 112, 122, 137, 140, 142–43, 213–16, 218, 220–21, 224–25, 228–29, 239, 241, 254
Soviet Academy of Sciences, 290, 309
Soviet Union, 23, 28, 97, 111, 184, 197, 203, 215, 232–33, 288–90, 309
Soyuz Carta, 291
Space Agencies Forum for the International Space Year (SAFISY), 300–1, 312
Space and Missile Systems Organization (SAMSO), 192–94, 200
Space Commerce, 207–9, 211, 277, 284, 293, 297–98, 303, 308–12
Space Detection and Tracking System (SPADATS), 191
Space Future Forum, 291, 298

Space Island Group, 304, 313
Space Launch Complex 6, 195
space launchers. *See* missiles and
    launchers
Space Policy Institute, 300
space shuttle, 187–88, 195, 211, 293, 295
Space Systems Division, 177
Space Transportation System (STS). *See*
    space shuttle
Special Treatment Center, 241, 255
Sperry Corp., 140
Sprey, Pierre, 123–24
sputnik, 291, 311
Sri Lanka, 278–80, 285
Saint George, USS, 9–11, 13–15, 29
Saint Thomas, V.I., 261
Stafford, Thomas, 290
Stanford University, 20
Starbird, Alfred Dodd, 55
State College, Pa., 16, 18–20, 23, 25–26,
    28–29, 32, 45
State College Television Corporation
    (SCTV), 21, 24, 28
Stempler, Jack, 69, 86, 109, 219, 269
Stennis, John, 78, 83
Stewart, James T., 121
Stikker, Dirk U., 46
Strasbourg, 282, 285
Strategic Air Command (SAC), 58, 70–
    71, 115, 132, 137, 144, 147, 180–82,
    196, 202, 216–18, 235
Strategic Arms Limitation Talks/Treaty
    (SALT), 144–45, 172, 189
submarine-launched ballistic missile
    (SLBM), 130
Suez Canal, 181
Sundlun, Bruce G., 292, 310
Sweden, 126, 287
Switzerland, 128
Symington, Stuart, 61, 75, 83
Syncom, 270
system program office (SPO), 104, 119,
    130, 138, 146
Systéme Pour l'Observation de la Terre
    (SPOT), 291

Tacoma, Wash., 9
Tactical Air Command (TAC), 128
Tactical Fighter Experimental. *See* air-
    craft: TFX
Tactical Warfare Programs (TWP), 31,
    33, 36–37, 44

Taiwan, 128, 182, 297
Tarr, Curtis W., 67
Task Force Alpha, 56
Taurus, 294–95
Taylor, Elizabeth, 74
Teets, Peter B., 94
Teledyne Ryan, 137, 140, 149
Telenor Satellite Services, 304
Telsat Canada, 273
Tenerife, 261
test and evaluation (T&E), 101
Tet offensive, 213
Thailand, 40, 54, 57, 142, 215, 217, 221
The Third Man, 283
The United States Logistics Group (TUS-
    LOG), 232
Thompson, David, 294, 311
Thurman, William E., 125
Thurmond, Strom, 84
Timation, 201–2
Time magazine, 250
Tindemans, Leo, 127
titanium, 41–42, 132
Top Gun, 128
Topp Industries, 25–27
Torassay, 2
Toronto, 282, 284
Total Force, 51, 75
Total Package Procurement, 35, 44, 98,
    104, 115–16
Toulouse, 305
Townsend, Guy, 130
Transit, 201
Transportation Dept. *See* Department of
    Transportation
Travis AFB, 242
Triad (nuclear), 130, 145
Trimble Navigation, 295–96, 311
Trimble, Charles, 296
TRW Corp., 106, 176, 189, 198, 200,
    211, 272
Tucker, Gardiner, 174
Tucson, 25, 149
Tulane University, 6, 8
Turkey, 23, 232–33
Turkish Airlines, 261

Udorn, 54, 142
Ulithi Atoll, 13
Unified Command Plan, 216
United Kingdom (UK), 136
United Nations (UN), 278, 280, 300, 312

United Nations Association, 291, 309
United States (as prefix). *See* US
United States, SS, 45
University Corporation for Atmospheric
    Research (UCAR), 295, 297, 311
University of Maryland, 292
University of New Hampshire, 299
unmanned/uninhabited aerial vehicles
    (UAV). *See* remotely piloted vehicles
    (RPV)
US Air Force, 7, 47, 94, 148, 208, 215,
    228, 232
US Air Force Academy, 7, 94
US Association for the International
    Space Year, 299
US Geological Survey, 210, 273
US government. *See* individual agencies
    and departments
US Navy, 9, 14, 121, 138, 201
US Postal Service, 58
US Space Foundation, 295
US Supreme Court, 264
Usher, William R., 72, 86, 233
USSR. *See* Soviet Union

Vance, Cyrus R., 34
Vandenberg AFB, 187, 193
Vela Hotel, 191, 199
Venera, 290
vertical/short takeoff and landing (V/
    STOL), 33, 38–39, 60
Veterans Administration, 6, 241
ViaSat, 304, 314
Vietnam, 22, 29, 36, 38, 40, 54–55, 57–
    60, 62–63, 65, 73, 75, 84, 86, 88, 90,
    97–98, 102, 107, 109, 111, 118, 122,
    128, 137, 139–40, 142–43, 149, 169,
    197, 199, 213–18, 220, 222–30, 237,
    239–40, 249, 253–55, 267
Vietnam War, 22, 29, 40, 54, 58–59, 63,
    65, 73, 90, 97–98, 107, 122, 142–43,
    169, 213–14, 222, 230, 237, 253–54
Vietnamization, 80–81, 214, 222, 224–
    28, 254
Vogt, John W., 221
Von Kármán, Theodore, 47, 61
Vought Aircraft, 43–44

Wakayama Wan, 15
Wallace, Carl, 76
Wallace, George, 63
Ware, Randolph, 295, 297, 310–11, 313

Warner, John, 74–75, 89
Warren, Pa., 9, 16
Warsaw Pact, 215, 230
Washington, D.C., 6, 25, 29, 60–61, 94,
    148–49, 207–8, 210–11, 254–56,
    282–83, 286, 300, 307–8, 312–13
Washington National Airport, 68
Washington Post, 87, 93, 189, 210–11,
    254, 256, 260, 307, 309–10
Watergate, 82, 90, 93–94, 109, 230,
    257, 267
Watson Laboratories, 16
Watson, Hal, 22
Wattendorf, Frank, 47
Weithoner, Charles E., 259
Welch, Larry, 118, 120
Wellesley College, 9
West Germany. *See* Germany
Westinghouse Corp., 119, 135
Westmoreland, William, 121, 213
Wheeler, Earle, 74, 88, 213
Wheelon, Albert "Bud," 106, 171, 279,
    293
Wheelus Air Base, 244
White Sands Missile Range, 23
White, Mike, 68
Whittaker, Phillip N., 67
Whiz Kids, 34
Williams, Robert, 39
Wilson, T., 87, 106, 116
Wolf Trap Center for the Performing
    Arts, 276
Wolf Trap Foundation, 78, 276
Women in the Air Force (WAF), 246
Woodruff, William W., 87, 117
World Communications Year (WCY),
    278–79
World Space Congress, 300, 312
World War I, 3, 74
World War II, 10, 15, 19, 21, 29, 34–35,
    37, 47, 53, 62, 70–71, 74, 77, 107,
    121, 123, 143, 177, 192, 244, 251,
    270
Wright Field/Wright Patterson AFB, 17–
    18, 21–22
WS-117L Advanced Reconnaissance
    System, 178

Xian, 308
X-Prize, 304, 313
Yamato, 13
Yarymovych, Mike, 198, 202

Yom Kippur War, 230–32
York, Robert, 37, 40
Young Presidents' Organization (YPO),
   27–29

Zraket, Charles, 53
Zuckert, Eugene, 41
Zumwalt, Elmo, 89, 94, 248, 251

9 781780 399683